The Preacher King

THE PREACHER KING

Martin Luther King, Jr. and The Word That Moved America

RICHARD LISCHER

New York   Oxford
OXFORD UNIVERSITY PRESS
1995

Oxford University Press

Oxford   New York   Toronto
Delhi   Bombay   Calcutta   Madras   Karachi
Kuala Lumpur   Singapore   Hong Kong   Tokyo
Nairobi   Dar es Salaam   Cape Town
Melbourne   Auckland   Madrid

and associated companies in
Berlin   Ibadan

Published by Oxford University Press, Inc.
200 Madison Avenue, New York, New York 10016

Oxford is a registered trademark of Oxford University Press, Inc.

Library of Congress Cataloging-in-Publication Data

Lischer, Richard.
    The preacher King : Martin Luther King, Jr. and the word that
moved America / Richard Lischer.
        p.   cm.
    Includes bibliographical references and index.
    ISBN 0-19-508779-8
    1. King, Martin Luther, Jr., 1929-1968.   2. Preaching.
3. Sermons, American—Afro-American authors.   4. King, Martin
Luther, Jr., 1929-1968—Oratory.   I. Title.
BV4208.U6L57   1995
251'.0092—dc20   94-30029

We are grateful for permission to reproduce the following:
"Let America Be America Again," in A New Song, 1938.
Copyright © 1938 by Langston Hughes.
Copyright renewed © 1965 by Langston Hughes.

"Lord, I'm Coming Home," "His Eye Is on the Sparrow."
Reprinted from Songs of Zion.

"Never Alone." Arr. copyright © 1960 by Singspiration Music/ASCAP.
All rights reserved. Used by permission of Benson Music Group, Inc.

Various King materials are reprinted by arrangement with
the heirs to the estate of Martin Luther King, Jr.,
c/o Joan Daves Agency as agent for the proprietor.

9 8 7 6 5 4 3 2 1
Printed in the United States of America
on acid-free paper

To
Sarah Kenyon Lischer
with admiration

Preface

I began researching this book after one of my students informed me that the sermons she had read by Dr. King for a class assignment were "pretty dry," by which she meant boring. Although not an expert on King or black preaching, I *had* lived through the Civil Rights Movement of the 1960s, and therefore I remembered what King's voice had meant to the cause of social justice, and I knew that his sermons and speeches as he delivered them were not "dry." Everyone seems to know what I knew, or remembered, then, and many have given testimony to the beauty and power of the spoken word on King's lips. But few have tried to give a rational account of King's prowess as a speaker and preacher of the gospel.

It is now possible to attempt such a task because the Martin Luther King, Jr. Center for Nonviolent Social Change has made the audiotapes and transcripts of King's sermons available to scholars. These, along with materials I have gathered from churches and archives around the country, have provided the basis for a reliable portrait of King the preacher and orator. Perhaps a clearer picture will emerge when the King estate grants further access to his earliest and as yet untranscribed sermons and speeches.

In my account of the preacher King I have followed the audiotapes and transcripts of the sermons, allowing the tapes to "correct" mistakes in transcription. Many of the transcripts lack tapes; from these I have corrected only the typist's most obvious errors. For reasons of concision, I have occasionally paraphrased the transcripts. The essence of King's voice, however, is nowhere obscured.

The style of some parts of the sermons and speeches suggests that they exercised a poetic or musical effect on their audience. Where repetition, alliteration, and other rhetorical clues point to this effect, I have transcribed the audiotapes in poetic form. Where a transcript of the sermon already exists, I have sometimes recast portions of it as poetry.

The reader will notice that the words *Negro*, *black*, and *African American* occur throughout the book, sometimes all in the same paragraph. King used the word *Negro* to describe his predecessors as well as his contemporaries. During his own lifetime, usage turned *Negro* to *black*, and in our own time *African American* is gaining ascendency. I have tried not to violate history in the way I have used these words. *Black* sometimes simply parallels *white*. At other times it evokes the specific usage of the mid to late 1960s. *African American* is both contemporary and historically suitable, but it sounds anachronistic when associated with King and his era. Racial names are always significant, but I continue to hope for the day when they will no longer be definitive.

Durham, N.C.   R. L.
June, 1994

Acknowledgments

During his brief career Martin Luther King, Jr. criss-crossed the nation many times, leaving a trail of sermons and vivid memories in the hearts of those who heard him. Over the past few years it has been my pleasure to follow the trail. The two major repositories for King materials are the Martin Luther King, Jr. Center for Nonviolent Social Change in Atlanta and the Mugar Memorial Library of Boston University. Serious students of King all over the world are indebted to the leadership of the King Center for making available a significant portion of his sermons and speeches. Louise Cook and Diane Ware were especially helpful to me at the King Center, and Howard Gotlieb and Margaret Goostray provided the materials I needed in the Special Collections Section of the Mugar Library at Boston. In Washington, Elinor DesVerney Sinnette, Karen Jefferson, and Esme Bhan guided me through the Ralph J. Bunche Oral History Collection in the Moorland-Spingarn Research Center of Howard University. While I was working in Washington, my colleague Roland Murphy, O.Carm., put me up in his monastery and offered good Carmelite hospitality. Marvin Whiting, Archivist and Curator of Manuscripts in the Birmingham Public Library, provided access to the tapes of the mass-meeting speeches held in Selma and to the Eugene "Bull" Connor Papers containing reports and verbatims of the nightly meetings held in Birmingham.

I am also grateful to the staffs of the Howard Divinity School Tape Recording Collection, the Duke Divinity School Media Center, and the Reigner Recording Library of Union Theological Seminary, Richmond. Dean Lawrence E. Carter of the Martin Luther King, Jr. Memorial Chapel at Morehouse College offered assistance when I turned to him. In the Duke Divinity School Library, I received valuable help from Harriet Leonard and Tom Clark. Relatively late in my research I made contact with Clayborne Carson, the senior editor of the Martin Luther King, Jr. *Papers* project. He and his staff at Stanford University have been most cooperative. I am especially grateful to one of his assistants, Michael Holloran, who guided me to an early King sermon and came to Duke to talk about our common research interests.

Many churches along the trail responded to my queries and supported my work, but none so graciously as Ebenezer Baptist Church in Atlanta. Sarah Reed of Ebenezer provided historical information and audiotapes of the sermons of Martin Luther King, Sr., and coordinated my interviews with members of the congregation. Ebenezer's pastor, Joseph Roberts, also spoke with me about King's relationship to the congregation. I also want to take special note of the generosity of G. Murray Branch, pastor of Dexter Avenue King Memorial Baptist Church in Montgomery, who spent hours with me and coordinated my interviews in Montgomery. Also, thanks to Canon Leonard Freeman of the National Cathedral in Washington, D.C.; Michael Lampen of Grace Cathedral, San Francisco; Mary Eldridge of Central United Methodist Church, Detroit; William Gardner, pastor of Unitarian Church of Germantown, Pennsylvania; and Frank Sims, pastor of Ebenezer Baptist Church, Chicago, all of whom provided tapes or transcripts of King sermons. Henry G. Scott of Cornerstone Baptist Church in Brooklyn made a special effort to send me tapes of Sandy Ray's sermons. Juel Pate Borders graciously sketched the background of her father's ministry at Wheat Street in Atlanta and supplied audiotapes of his sermons. In Philadelphia, Almanina Barbour allowed me to examine her father's sermon outlines and listen to tapes of his sermons. With her insights and stories she offered "hospitality to the stranger" in biblical proportions.

As a part of my project, I wanted to talk to the colleagues who were closest to Martin Luther King, Jr., who not only listened to his sermons and speeches regularly but who also understood his methods of creating them. This list is a most impressive Who's Who of the SCLC: the late Ralph David Abernathy spoke with me at the end of a long day at his church, West Hunter Street Baptist, in Atlanta; the late Bernard Lee, a daily coworker with King who at the time of his death was the mayor's liaison to churches in Washington, D.C., talked to me about King as a person and a preacher; C. T. Vivian of Atlanta, the former SCLC liaison officer who continues to educate Americans about racism, cheerfully and energetically gave hours to this project; Wyatt Tee Walker, the strategist behind the Birmingham Movement, now pastor of Canaan Baptist Church of Christ in Harlem, welcomed me to his church; Jesse Jackson took an hour out of a busy schedule to ride around Durham, North Carolina, with me and talk about King the stylist (thanks to William J. Griffith of Duke University for making that possible); James Bevel reminisced about the mass meetings; J. T. Porter, pastor of Sixth Avenue Baptist Church in Birmingham and a former intern for King and his father, gave freely of his time and memory to characterize the early years of King's ministry; Robert Graetz helped construct a picture of the mass meetings; Gardner C. Taylor welcomed me to his church in Brooklyn and helped me place King into the context of African-American preaching; Prathia Hall Wynn remembered her student days in SNCC; Evans Crawford, Dean of Rankin Chapel at Howard University, reminisced about King and helped me understand the phenomenon of talk-back in a black congregation. My new colleague Sam Proctor, former pastor of Abyssinian Baptist Church in Harlem, filled me in on the formation of a black preacher and on King's Crozer years.

Many others helped along the trail. Gordon Midgette of Atlanta told interesting stories about life around Stockbridge, Georgia, and Martin Luther King, Sr.'s roots in that area (thanks to Fred Craddock for facilitating this interview). The late Kenneth "Snuffy" Smith recalled the circumstances of King's theological education, as did my colleagues Franklin Young and William J. Smith. Other colleagues have been gracious with their help: Jon Michael Spencer, now at the University of North Carolina in Chapel Hill, listened to King tapes with me and patiently tried to explain the musicality of King's voice. William C. Turner was always willing to suggest more books. Ann Hoch helped me with Harry Emerson Fosdick. Stanley Hauerwas talked to me about liberalism. Historian Russell Richey read the first draft of the manuscript. Somehow our friendship survived his critique.

If I was to present a concentrated account of King's preaching, I felt it was necessary to speak with parishioners who heard him Sunday after Sunday. From the ranks of Ebenezer I spoke with Sarah Reed, Lillian Watkins, Minnie Showers, Shirley Showers Barnhart, and Lillian Lewis, who from her office at Atlanta University also helped place King into the context of African-American preaching styles. In Montgomery I interviewed King's former intern J. T. Porter and parishioners R. D. Nesbitt, Ralph Bryson, Zelia Evans, Thelma Rice, and Mary Lucy Williams. They are longtime members of Dexter who shared their memories of a time when Martin Luther King, Jr. was their "Brother Pastor."

The work of several scholars has been invaluable to my thinking about Martin Luther King, Jr. When I began, David Garrow was always accessible and encouraging. He has been a most impressive pioneer in King research. Taylor Branch's biography of King is an inspiring and poetic monument to the King years in America. Henry Mitchell has provided a baseline for all considerations of African-American preaching. James Cone has taught everyone to appreciate King's role as a black theologian. C. Eric Lincoln, more than anyone, has defined the churchly context of King's work. Gayraude Wilmore and Vincent Harding have demonstrated King's continuity with the tradition of black protest in America. These and many, many others have been my teachers along the way.

Some of the material in Chapter 8 has been presented in lecture form under the title, "Martin Luther King: Performing the Scriptures." I gave this lecture one summer at the Free Faculty of Theology in Oslo, at the Institute of Theology at Princeton Theological Seminary, and to the division of theology at Africa University in Mutare, Zimbabwe. In each setting I received not only the warmest of hospitality but also great encouragement in my work.

My research was partially supported by grants from the Association of Theological Schools, Lutheran Brotherhood sabbatical grants program, and the Duke University Research Council. The Dean and administration of Duke Divinity School have been supportive of my efforts from beginning to end. At Duke, Sarah Freedman patiently keyed and rekeyed many versions of this manuscript. Cynthia A. Read, senior editor at Oxford University Press, has offered invaluable advice and support for this book at every stage of its development. To her, as to all who have given so much to this project, I return my deepest thanks.

Contents

10. Bearing "The Gospel of Freedom": The Mass Meeting, **243**

The Preacher King

The Lord gave the word:
great was the company of
the preachers.
    Psalm 68:11
    The Great Bible, 1540

Prologue

"In the quiet recesses of my heart," Martin Luther King, Jr. often said, "I am fundamentally a clergyman, a Baptist preacher." *The Preacher King* may be read as an extended commentary on that confession. Already it seems unlikely that even a nation "under God" could have been so profoundly affected by the minister of a little black church in Montgomery, Alabama. It seems remarkable, in retrospect, that we were willing to listen to his overtly Christian persuasions and that so many of us were moved by them. Yet it is true. At one of its several turning points in the twentieth century, America submitted its laws and customs to the influences of one with the instincts and commitments of a Christian preacher. And he moved "the nation with the soul of a church," as G. K. Chesterton named us, as a preacher moves a congregation.

Nowadays the word *preacher* does not attract much admiration. The word is associated with parochial morality or televised quackery, but in either case the preacher is a rather narrowly defined figure. Martin Luther King was proud of the title, however, because he believed that his religious vocation was essential to the healing of the nation. To him, the preacher symbolized the combination of political and spiritual wisdom that his own church had always required of its leaders. Like the ministers of no other tradition, the African-American preacher harnessed practical necessities to religious power. The black preacher fought for the kingdom of God every day of the week and then celebrated it ecstatically, even poetically, on Sundays. The same one who flexed his muscle in the neighborhood could speak with the tongues of angels in the church. King seized upon this partnership of political acumen and religious eloquence—which as a black man, a southerner, and a Baptist he had inherited from his tradition—and put it to work on America's enduring problem of race.

Like a preacher, he routinely cited the Bible as the authority for his social activities, and cast the Civil Rights Movement in the light of biblical events and characters. King was a creature of contemporary politics, in his element at a press conference or a negotiating session, but he never gave in to the pragmatism of politics. There was always something more, some message from another realm—a spiritual standard that informs and judges this world and ultimately promises to save it from corruption. The language with which he clothed his arguments for a better world was invariably sermonic, which was only fitting, since for fourteen years he preached in his own congregations in Montgomery and Atlanta and in churches around the nation. The substance of these sermons he translated into civil religious addresses and fiery mass-meeting speeches, but it was always *preaching* that he was doing. Even when no text was cited and the deity was not mentioned, the audiences to these speeches considered themselves no less a congregation. King's self-proclaimed mission "to redeem the soul of America" cannot be understood apart from his self-designated identity as a preacher of the gospel.

He succeeded in injecting that gospel into the political debate much in the way the Abolitionists had more than a century before. As no preacher in the twentieth century and no politician since Lincoln, he transposed the Judeo-Christian themes of love, suffering, deliverance, and justice from the sacred shelter of the pulpit into the arena of public policy. How the preacher King accomplished all this is the subject of this book.

The portrait of Martin Luther King, Jr. that will emerge in these pages is fashioned from raw materials that most biographers and critics overlook. The substance of that portrait relies heavily on the unedited audiotapes and transcripts of King's sermons and speeches, including a few recorded by police mobile surveillance units in Birmingham and Selma. In these recorded and transcribed messages his true voice can be heard. Due to a demanding schedule of travel and personal appearances, most of his books and articles were published only with substantial editorial assistance. Even his sermons collected in *Strength to Love* contain many passages borrowed from the printed sermons of other preachers. King and his editors removed all local and personal references from these sermons and polished them up as timeless masterpieces of the pulpit. In their printed form, they are scarcely distinguishable from the liberal commonplaces of the white, mainline pulpit during the Eisenhower era. Anything resembling the African-Baptist gospel in which King was nurtured or the prophetic rage that often seized him was removed in order to lend his utterances universality and to recommend his Movement to as wide a reading audience as possible. In the process, his real preaching and, consequently, something of the real Martin Luther King, Jr. was lost to the public. To the extent it is possible for a book to make a sound, *The Preacher King* will try to restore its hero's voice.

Theologian James Cone and biographer Taylor Branch (along with earlier biographers David Lewis and Stephen Oates) reminded us that King was a product of the black church in America. That assertion was a welcome relief to the many studies of King's *thought*, including King's own brief intellectual biography, "Pilgrimage to Nonviolence," in which he portrayed his moral and intellectual odyssey strictly in terms of academic philosophy and theology. Everyone understands why that self-description was necessary and why it was important for his Movement to accommodate itself to the West's tradition of liberty. But now even the corrective to this overintellectualized profile, the embrace of King's black-church heritage, poses a new and different sort of reductionist danger by dismissing his academic formation as irrelevant. There is, however, a complexity to be captured in King, a tension that lay at the heart of his universal appeal.

This book will focus on that tension. Its interpretive lenses are the African-Baptist tradition that formed him as a preacher and the liberal theological tradition that shaped him as an American religious activist. He was both: a black preacher and a social reformer. He knew the vocabulary and spoke the language of both professions. By analyzing his sermons and speeches, I hope to illumine the brilliance with which he exploited both his inherited and his acquired language. In the process, we should learn much about King's rhetorical, theological, and political agenda for America.

What follows is a critical study. It will report not only what King said but how his total religious performance functioned as a strategy for social and political change. It will celebrate King's personal Christian commitments and, more important, explore how he used the symbols of Christianity to achieve his purposes in the nation. It will be clear that King meant to make a Movement that was Christian, a distinctive purpose that continues to separate him from other prominent civil rights leaders. Discerning the method and ramifications of that purpose is the critical task.

The first part of the book will explore King's formation in the African-Baptist church. That church was a world made up of Atlanta's "Sweet Auburn," Ebenezer Baptist Church, Morehouse College, tutors, mentors, role models, and friends—all who played a part in bringing him through seminary and graduate school to his first plateau: his own pastorate in Montgomery. The first phase of his brief life prepared him to be the public advocate of God's justice for black people in America, which in the African-American tradition meant that he would take a church and preach. From this environment he absorbed key theological strategies for dealing with injustice that he would never relinquish. He learned more from the Negro preacher's methods of sustaining a people and readying it for action than from any of his courses in graduate school; he absorbed more from his own church's identification with the Suffering Servant than from anything he read in Gandhi. What came earliest to him remained longest and enabled him to put a distinctively Christian seal on the struggle for civil rights in the United States.

During this period he also learned to preach—not only to speak but to become an actor for his people and to assume the larger roles of prophet, evangelist, and, last of all, suffering agent of redemption. His first classroom (and stage) was his father's church. At Ebenezer, King was schooled in the authority that God's Word exercises when it is rightly voiced and dutifully heard. When he himself began preaching, he imitated skills he had long admired at Ebenezer and in other African-American congregations. He copied a great variety of techniques, acquired an impressive inventory of "set pieces"—gorgeous thematic formulas—but most important of all, at Ebenezer he learned how to follow the emotional curve of a religious idea as it takes possession of a congregation. *That's* what it was to preach.

He also acquired some basic ideas about the pulpit. These would later guide the strategy for his public utterances. He believed that the preached Word performs a sustaining function for all who are oppressed, and a corrective function for all who know the truth but lead disordered lives. He also believed that the Word of God possesses the power to change hearts of stone. This was not an abstract theology but an empirical experience. He had seen it happen in his father's church.

Who were King's teachers in these matters? The first section of this study will sort through the role models and mentors who taught King and made a difference in his life. We shall hear the actual voices of Benjamin Mays, William Holmes Borders, Sandy Ray, Gardner Taylor, Vernon Johns, Pius Barbour, and, of course, his father, as each makes a distinctive contribution to his apprenticeship in the Word. Some, like Mays and Barbour, mediated to him the African-American slant on the classic liberalism he was imbibing at Crozer Seminary and Boston University. In those schools he was introduced to the talismans of modern thought whose names and theories would decorate his sermons for many years. Perhaps no famous contemporary has amassed so large a troupe of "influences" as King: Hegel, Marx, Thoreau, Freud, Rauschenbusch, Gandhi, Niebuhr, Tillich, and many others—these are the official influences celebrated by King and many of his biographers, and no one can doubt that they played a significant role in his development. During this period he also borrowed without attribution from the published sermons of famous liberal preachers such as Phillips Brooks and others of his own generation. As much as he may have *looked* like his sources in print, however, he never *sounded* like them in church. King never parroted their sermons without adding his own distinctive voice and the unique experience of African Christians in America. Even as he received their themes, he was deconstructing them with the irony and evangelical hope of the black gospel.

This is not to say that King was only pretending to be a liberal or a Boston Personalist or that, as some have suggested, already as a graduate student he was alienated from the philosophical themes he would trade on for the rest of his life. Despite his plagiarism at Boston and the derivative character of his learning in general, the evidence shows him to have fully engaged the Western intellectual tradition. He appears to have embraced it as an alternative to the strong medicine of his own religious tradition, thoroughly absorbed its vocabulary and values, and then come "home" to his own tradition with his horizons considerably widened. For a time, at least, such themes as the infinite worth of human personality or the essential unity of freedom and the human spirit, provided the young graduate student with pegs on which to hang the aspirations of his people.

Early in his ministry he would abandon many of the critical theories about the Bible he learned in religion classes. He reverted to techniques of interpretation that were more ancient than the African-American church, such as allegory and typology, because they allowed his congregations a greater opportunity to identify their struggles with those portrayed in the Bible. The black church not only sought to locate truth *in* the Bible, in order to derive lessons from it, but also extended the Bible into its own worldly experience. King found the ancient methods of interpretation useful in his effort to enroll the Civil Rights Movement in the saga of divine revelation. These techniques he joined to the black church's practice of "performing" the Scripture in its music, its rhythmic pattern of call and response, and a variety of rhetorical adornments—all of which he exported from the church's Sunday worship to political mass meetings around the country. Ironically, the Boston Ph.D. made his mark on the modern world by reviving techniques of interpretation that his professors had dismissed as antiquated.

If his allegory on "The Three Dimensions of a Complete Life" sounded a bit outmoded in affluent white churches, it was his liberal theology that seemed odd in the black church. Most scholars would have published their thoughts on justice and history in learned journals. But the circumstances and choices of King's life were such that the only verbal medium he had at his disposal was the sermon. He hammered out his Christian theology on the anvil of the pulpit. Thus if we want to know what King believed about God or the human condition, the clues are in the sermons that he preached at Ebenezer Baptist Church and in other black congregations. If we want to grasp his hope for America, we have to understand what he believed the Word of God could accomplish in America. If we are still fascinated by his character or personal motivation, the sermons tell us more than the FBI's wiretaps.

A sermon is a cultic performance of a biblical text among people who identify themselves as Christians. King's sermons were that too, and therefore have an self-deprecatory quality about them because he understood that a sermon is meant to be the vehicle of something greater than itself. A public speech serves its own political agenda, but a sermon must follow the Bible's leading into every conceivable corner of life, from gossip in the barbershop to impurity in the bedroom. Because of their very genre, King's tape-recorded and transcribed sermons reveal dimensions of the man and his message not found in his published works. They have a greater degree of intimacy than his other speeches because they presuppose an audience that is also a community of faith. They presume a network of family relations and a context of pastoral care that included baptisms, prayer, admonition, and the normal range of parish activities. In some way, the mundane announcements *after* the sermon—about special offerings, church suppers, coming weddings—give the truest picture. King's sermons at Ebenezer were intimate, not merely because he occasionally let his hair down in them but because at Ebenezer he could utter religious intimacies and ecstasies he could not say elsewhere. There also he could take part in cathartic experiences unavailable in white settings.

Throughout his brief career, King preached, refined, altered, and re-preached a small canon of sermons, fewer than one hundred. Their number is not known because many of the sermons are either lost or unavailable. The exact number of sermons is unimportant because his so-called canon actually consists of an enormous inventory of set pieces, the thematic formulas he patched together in a bewildering number of combinations under a variety of sermon titles. Some of the formulas were of his own devising, many were borrowed. Whatever he used, however, he marked with the stamp of his own genius. His most famous set piece was "I Have a Dream," which he developed from multiple sources, including the prayer of a young SNCC volunteer in Albany, Georgia. In King's repertoire, the piece evolved from a biblically resonant formula, which was most appropriately recited in churches and mass meetings, into one of the most famous civil-religious passages in American oratory. His ear for these pieces and his uncanny ability to punch them into a sermon or speech at precisely the right time was a legacy from the black church and its oral tradition. His originality was an originality of effect, not composition. His ability to create new thoughts and new sermons was curtailed by his public responsibilities, but it was these same pressures that summoned the best from him and made him the most significant American preacher of the modern era.

This book will trace the passage of King's sermonic material into his mass-meeting speeches and civil addresses. When he used his thematic formulas in political settings, they carried religious overtones from the Judeo-Christian tradition and his own "Ebenezer Gospel." As he repeated his formulas from one packed assembly to the next, they exerted the religious effects of faith and discipleship on his audiences. King himself became the sacrament of the Movement in whom his followers could participate by listening and responding to his appeal. Conversions to courage occurred. The Christian doctrine of self-giving love, agape, took to the streets of Birmingham, Selma, and even Chicago. King-led demonstrations became symbolic enactments of Bible stories that his opponents thought safely banished from politics and secular affairs. But on the contrary, to those about to march from Brown's Chapel in Selma, a King sermon announced the Kingdom of God and created a corporate movement toward its realization. The Kingdom had come once again, and, in the urgency of the hour, it seemed its momentum would never be stopped.

One of the more difficult questions this study will address is the impact of King's speaking. Because his sermons and speeches eventually created many audiences, the question of their effect is a complex one. Whatever the effect he produced, it was always the result of a conscious rhetorical strategy related to the composition of his audience. King's rhetorical strategy, as discussed in Chapters 5 and 6, was threefold. Before black and sympathetic white audiences he elevated local conflicts into the titanic battle of universals. He elevated the Movement by clothing it in gorgeous rhetorical apparel and associating it with the most respected themes and thinkers Western civilization has produced. If he was to enlist support for the Movement, he had to show that it was *worth* something to those who had no personal stake in it—those who were *not* earning a pitiful $900 per year or riding at the back of the bus. With something like poetic sorcery, he conjured beautiful sounds that fairly hummed their agreement to Aristotle's advice: a free man should not talk like a slave.

Because of the derivative character of his academic thinking, it is easy to overlook the stunning creativity of King's achievement. I will insist that we do King a disservice to evaluate his originality according to his use of written sources. He practiced the creativity of a preacher and a poet. He had the preacher's (and actor's and politician's) knack of translating every stray piece of information into the dramatic communication of ideas. Although he was a shrewd social strategist, his genius was poetic in nature, for he had the prophet's eye for seeing local injustices in the light of transcendent truths. This is the gift of metaphor, and fretting about sources must not distract us from its appreciation. In King's vision of the world, ordinary southern towns became theaters of divine revelation, and the gospel became a possibility for the renewal of public life. This is what the Kingdom of God will *look like*, he promised a quarter million people at the Lincoln Memorial: like white people and black people from Georgia sitting at table together and acting like kin.

His second rhetorical strategy is related to the first but is more complex and controversial. He practiced identification with his audiences at several levels. Already as a young man, he imaginatively identified with the weariness and rage of his African-American audiences. This, combined with the nobility and eloquence of his style, filled his black audiences with courage and self-respect. King himself modeled a new confidence for the black preacher and a new militancy for the black church in the South.

More complex was his identification of black aspirations and the traditionally white consensus ideals of America. By means of a wealth of literary, biblical, and philosophical allusions, he assured his hearers that history and universal moral law are aligned with the black quest for freedom. He wanted his potentially sympathetic white audiences to recognize the best of their own religious and political values in the mirror of his message. Like a priest, he mediated a covenant with which white moderates and liberals were comfortable. In light of its frightening alternatives, *his* dream was *their* dream too. He reinforced this commonality in many ways—with psychological jargon, popular religious sentiment, the grammar of inclusion, and by a synthesis of biblical and civil-religious rhetoric. For a time, he believed in this synthesis and brilliantly exploited it. At this point in his life, he was not simulating white liberalism, as some critics have asserted, but living out W. E. B. Du Bois's theory of black double-consciousness. He yearned to be fully black and fully American.

After Selma and just before Vietnam, something turned in King. He began to slough off his liberal platitudes and the rhetoric of inclusion. His civil religion was succeeded by its demythologization. By the end of his life, the one true church in which we are *all* brothers and sisters had disappeared, its place taken by the redemptive mission of the bowed but awakening black church. Unmerited suffering and other sacrificial strategies were no longer prominent in his repertoire; he continued steadfastly to practice love as nonviolence, but he also learned to get angry at his opponents and send them to hell.

Those who arrest King's rhetorical development at the identification stage will always be comfortable with him. They will lock arms with their brothers and sisters on his birthday and sway to the strains of *We Shall Overcome*. But if they read the sermons and speeches of his final three years, they will encounter a prophet who was past identifying with the oppressor's values and had gone on to confronting them. Not only his admirers but even his most adamant critics have not fathomed the depths of his militancy. In his latter years he accused America of genocide and compared its conduct of war to the Nazis'. He warned an audience in Montgomery that any country that had treated its natives as America had would not blink at putting blacks in concentration camps. He warned of long hot summers in the ghettos and began calling for radically new ways of distributing wealth. His final rhetorical strategy was the abandonment of strategy and his own surrender to prophetic rage. He reduced the number of authorities in his sermons to one: Almighty God. He began saying, "Thus says the Lord." He soured on liberalism and liberals but never gave up on preaching the Word.

He had long possessed the poet's eye for the beauty of correspondences. With the eye, finally, came the faithful ear. The orator who delighted in the symmetry of words was at the last captivated by his own stigmatic confidence in the Word of God. He believed that the same dynamic of repentance and conversion that he had witnessed at Ebenezer and in little black Baptist churches everywhere was not only *supposed* to work but ultimately *would* work outside the walls of the church. His gospel would either redeem the soul of America or consign it to judgment. In this respect, he was a true evangelist for whom every speaking engagement was a potential altar call. And, in his latter years, he was also a true prophet for whom every conflict invited the wrath of God. Under his leadership, the Civil Rights Movement adopted the black church's joy in the performed word—and also its seriousness with regard to its only law of history: you reap what you sow. With King as its voice, the Civil Rights Movement became a Word of God movement, and the Word, exactly as it is portrayed in the New Testament, became a physical force with its own purposes and momentum.

Perhaps King's greatest spiritual gift was faithfulness to his vocation to preach the Word of God in all circumstances, including personal danger and declining popularity. He adhered to the African-American preaching tradition and, save for the last three years of his life, to the values of political and theological liberalism. He was God's trombone and a doctor of philosophy in one and the same mission. The preacher who once appealed to Billy Graham for advice was the first theological thinker since the Social Gospel movement to forge a synthesis of evangelical and liberal traditions in America, and he did it on a scale that would have been unimaginable to the Social Gospelers. The evangelist in him yearned to "save" our souls, and the liberal reformer in him did so by tapping into the deepest and best of our political and religious reserves. Through a combination of personal gifts and historical circumstances, Martin Luther King, Jr. achieved the very thing that eludes us all today: *He framed a broadly based rationale for the equality, even the kinship, of the races; and he advanced a method for attaining it.* King never produced a social blueprint for America, but, because he was a preacher, he never quit trying to shape a "congregation" of people that would be capable of redeeming the moral and political character of the nation.

I
PREPARATION

1
Surrounded

IN the basement of Dexter Avenue Baptist Church in Montgomery there is a large mural depicting Martin Luther King, Jr.'s ascension into heaven. He is clothed in white raiment and appears to be preaching or giving a benediction as he rises. In the painting he is surrounded by his forebears and teachers, including Frederick Douglass, Harriet Tubman, W. E. B. Du Bois, Vernon Johns, his mother, and the martyrs of the Civil Rights Movement. The mural symbolizes the thicknesses of King's spiritual environment. In the formative period of his life, he was encompassed by a series of concentric circles: first Mother Church, next by the traditions of the black preachers and reformers who came before him, and finally by the circle of mentors and teachers who guided his preparation for ministry. If there were a text beneath the mural it might well be from the Book of Hebrews, "Since we are surrounded by so great a cloud of witnesses . . ."

I

In his "Autobiography of Religious Development" written as a class assignment at Crozer Seminary, twenty-one year-old M. L. King characterized his religious environment as a "universe," a socially constructed world that shaped his identity and outlook on life. The moral and physical center of that universe was the *sanctuary* of Ebenezer Baptist Church in Atlanta where his father presided from the pulpit and the son had been baptized. "The church has always been a second home for me," he wrote. "As far back as I can remember I was in church every Sunday."

In the sanctuary of his father's church, like most Baptist churches, the architecture witnesses to a single line of spiritual authority. At the lowest level in the nave the company of the saved and unsaved, both saints and backsliders, is seated before a humble communion table. Relatives of the ministers, well-to-do supporters of the ministry, deacons, trustees, and the mothers of the church, all sit in their customary places of prominence. Marginal members and visitors cluster toward the rear under the overhang of the balcony. Children and teenagers who have somehow eluded their parents or grandparents congregate noisily in the balcony. Men and women ushers dressed in formal attire monitor the seating arrangements and see to the needs of those who have come for worship.

The power center in the sanctuary is the Georgian-style pulpit, and the worshiper's proximity to it may indicate his or her role or relative stature in the congregation. The floor of the nave slants gently toward the pulpit, which is located in the center of the sanctuary on a stage above the communion table. The stage extends like a runway across the width of the chancel. The pulpit is painted creamy white; its sideboards flare like the prow of a ship. An enormous King James Bible lies open upon it.

Behind the pulpit sits enthroned the son's immediate and earthly source of power: his father and the accumulated wisdom of the Baptist clergy. Above and behind the earthly father on the platform is the church choir seated on risers rank upon rank, symbolizing the angelic hosts of heaven. Above and behind the choir is a heavy, red velvet curtain which when opened reveals the baptismal pool. Still higher, above the curtain, is an ornamental canopy mounted by a white neon cross. Centered above the cross, high on the chancel wall, is the familiar representation in colored glass of Jesus at prayer in Gethsemane. With eyes turned upward he prays to the unseen One from whom and in whom this vast chain of authority subsists.

Whenever Martin Luther King, Jr. ascended the pulpit, he found his mark on the stage—beneath the agony of Gethsemane in the light of a neon cross, flanked by choirs and elders, the Word of God open before him. There he was surrounded by a congregation that had shaped him, sustained him from his childhood, and ordained him to speak on its behalf. He was, as a retired beautician and longtime Ebenezer member put it, "Our M. L." He loved the church and its people, and they loved him like a son. They provided him with more than an audience; they offered him the sanctuary he could find nowhere else.

In the ordered environment of the African Baptist church King discovered his vocation and the strength to carry it out. The architecture of the sanctuary symbolized the cosmos in which King had grown up and

that he never ceased to inhabit. His "world" consisted of two dialecti-
cally opposed realities. The first was a heritage of suffering, which included
forced deportation, enslavement, poverty, segregation, and estrangement
from the dominant culture. The second was the affirmation of God's
purpose for the whole world, but especially for those who bear burdens
imposed by others. God's purpose takes the form of a divinely ruled order
that will ultimately triumph over the chaos of suffering. The sanctuary is
a projection of the community's belief in that order. In it the ultimate
victory is celebrated, as it were, in advance. The events of every Sunday
morning witness to what has been and what will surely be.

The sanctuary dictated the boundaries within which African-American Christians, including Martin Luther King, Jr., sorted out the relationship of suffering and hope. What occurred at Ebenezer and continues to
unfold in innumerable sanctuaries like it represents the African-American church's attempt to build a habitable universe for its people. It offers
a bridge between the struggles of this world and the joys of the world to
come. Such worship does not promise a simple passage from one world
to another and therefore cannot be described as "otherworldly." King's
congregation knew that the future of the Kingdom of God was meant to
be seen and tasted in *this* life. Ecstasy in worship at Ebenezer was never
a forgetting, but an anticipating, even a demanding, of better things now.
Ebenezer's worship (and worship space) not only built a world for Negro
survival but institutionalized a permanent critique of a world in which
survival is all one can hope for.

King fully inhabited this holy space of Ebenezer by internalizing its
sense of order and its spiritual values. The suffering and hope, as well as
countless ritualized anticipations of victory, he made his own in childhood. He did so without intention or decision by simply absorbing the
universe in which he was enveloped. Later, in his public career, what one
scholar calls "the Afro-Baptist sacred cosmos" became portable, like the
Hebrew Ark of the Covenant, to be taken into whatever battle King
engaged.

The rich symbolism of Ebenezer Baptist Church taught young King
many lessons, not the least of which was the provisional quality of all
earthly authority. Jacob's ladder reaches from almighty God through the
ranks of angels and elders to the lowliest backslider. The very existence
of that arrangement acts as a critique of every human law and institution. The enduring reality of God gives hope to those who chafe under
the inauthentic "laws" of the pharaohs of America. The preacher occupies a place in the hierarchy of the divine cosmos as the one who is au-
thorized to proclaim God's lordship over other powers. Because the

preacher has been called directly by God, he also has a privileged perch outside the hierarchy as the one who can "see" how God's purposes are unfolding in the whole world.

II

Throughout his career, King carried the sanctuary with him as a state of mind and soul; he also repeatedly returned to it like a grateful soldier home from the front. One Sunday in the spring of 1963, with the crisis in Birmingham approaching its flash point, King wearily introduced his sermon with a rambling account of all-night negotiating sessions and the bombing of the motel where he had been staying by some "vicious mobsters." At the conclusion of the service, the preacher himself can be heard leading the choir and congregation in singing *Lord, I'm Coming Home*:

> My soul is sick, my heart is sore
> Now I'm coming home;
> My strength renew, my hope restore
> Lord, I'm coming home.

Incidents of this nature, tape-recorded in his home church, could be multiplied many times over: King telling about the terrors of his week, asking for his congregation's prayers, and always, with evident sincerity, thanking God for allowing him to come home one more Sunday. The tapes confirm his self-portrait as "a minister of the gospel, who loves the church [and] who was nurtured in its bosom. . . ."

If the inner world was a sanctuary, his holy-of-holies, its outer boundaries encompassed a wider and more complicated scene. King grew up in a Negro ghetto just east of downtown Atlanta known as "Sweet Auburn." Outside the bosom of the sanctuary, Sweet Auburn was the larger lab where he observed and experienced the church's bittersweet existence in the world.

For young King, Sweet Auburn illustrated the irony of black life in America, for the organic and self-sustaining features of its community, qualities every American village likes to claim for itself, depended upon its being and remaining a ghetto. Here the races had lived in uneasy coexistence until the riot of 1906 split the city and Atlanta copied other cities' segregation laws that established white and Negro neighborhoods. By the time Martin Jr., or, as his birth certificate read, Michael Luther King, Jr., came into the world, Auburn Avenue's twelve-block stretch of shops, churches, and substantial homes had become the most important residential and business center in black Atlanta. Like Beale Street in Memphis, U Street in Washington, or Lenox Avenue in Harlem, Sweet Auburn was much more than a tarnished reflection of middle-class white society. It had a style all its own. Although its residents were only too aware that their segregated homes and shops, though refined and successful by Negro standards, were far from "sweet," Auburn Avenue was a world that belonged to its own people and, as such, it was a haven from the frontal indignities inflicted by white society. More important, it was a laboratory in which youngsters like Michael Jr. could observe models of black success within the larger world of segregation.

In his autobiographical statement young King attributed his secure sense of self to the view of God mediated to him by his environment:

> It is quite easy for me to think of a God of love mainly because I grew up in a family where love was central and where lovely relationships were ever present. It is quite easy for me to think of the universe as basically friendly because of my uplifting heredity and environmental circumstances. It is quite easy for me to lean more toward optimism than pessimism about human nature mainly because of my childhood experiences.

That he discovered "lovely relationships" in a segregated church in a segregated neighborhood might have been reminder enough of the injustice that surrounded and besieged the island of Sweet Auburn. His account does not mention his father's terrible temper or his own impulsive suicide attempts at age twelve. Yet his "Autobiography," laboriously written in a childish scrawl, does not deny the pain faced by a black youngster growing up in America. In the document he tells the story of his personal discovery of what W. E. B. Du Bois had called "the Veil," when, because of his color, he first perceived himself to be, as Du Bois put it, a "problem" to white society around him. Without cause or warning, the parents of his white friend forbade their child to continue playing with six-year-old Mike King. The friendship was over. It is a simple story with literary parallels in Marcus Garvey, James Weldon Johnson, James Farmer, and the autobiographies of many African Americans. "For the lucky," says Farmer of this recognition, "it is sudden, like a bolt of lightning, striking one to his knees. For others, a gradual dying, a liver of meanness working its way to the heart."

At the center of King's universe lay his place of refuge, Ebenezer Baptist Church. As a Negro church, its very existence reminded King and all its members that the same Veil that divided the social and economic realm divided the spiritual as well. Although all Baptists were brothers and sisters in Christ, and all could sing lustily,

> Hold up the Baptist finger
> Hold up the Baptist hand,
> When I get in the Heavens,
> Going a-join the Baptist Band,

no member of Ebenezer would have dared to attend the white First Baptist Church of Atlanta or any other white Baptist congregation. Everyone at both ends of Peachtree Street understood that. In one of his sermons King would later reflect on the irony of the white Baptist congregation that sent thousands of dollars to Africa to missionize the natives but fired its pastor for allowing a black man to sing in the choir.

Against this backdrop of alienation, the congregational life of Ebenezer in which, as King later put it, "everybody [was] the same standing before a common master and Savior," came to model an alternative to the racist and classist society around it. The "Beloved Community" he would later discover in the writings of the philosophical idealists preexisted in the earthly community at the crossroads of Sweet Auburn. It was not that Ebenezer eschewed the patterns of secular society; in many ways it replicated them but in a sanctified form. As a result of the exclusion of Negroes from political life in America, the church became the *polis* for its members. Its benevolent societies planted the seeds of large Negro insurance companies and burial associations. Its network of administrative boards and auxiliaries copied and often surpassed in efficiency the ward organizations in large cities. The congregation's fund-raising mechanisms, such as cooperative ventures with insurance men and a variety of intramural competitions, solidified its infrastructure and put Ebenezer on a sound footing during the Great Depression.

In this church everybody could be "somebody." Those of humble occupation could claim the dignity and positions of leadership that were denied them in white society. In the black church a barber or a redcap could become an elder; a seamstress might lead the Woman's Missionary Union. King would later poke fun at black bourgeois congregations that bragged, "We have *so* many doctors. . . ." Of Ebenezer's four thousand members in the late 1930s and 1940s, most were skilled, semiskilled, or unskilled workers, many of the personal-service type. Some were domestics, shoemakers, hairdressers, postal workers, public school teachers; far fewer were doctors, lawyers, professors, or businessmen. (Even the relatively privileged Negro physician in those days was barred from

membership in county and state medical societies, as well as from intern-
ships, residencies, and fellowships in the municipal hospitals.) A few top-
echelon people, like C. A. Scott, editor of black Atlanta's *Daily World*,
and Jesse Blayton, president of Citizens Bank, both of whom would later
obstruct King's efforts to desegregate their city, rounded out the little
world at the corner of Jackson Street and Auburn Avenue.

The democratic spirit extended to the congregation's worship and
polity. Within the social hierarchy of Ebenezer, elements of storefront
Pentecostalism coexisted with the upper levels of black society in Atlanta.
Sociologists St. Clair Drake and Horace Cayton's classic study of "Bronze-
ville" in *Black Metropolis* identified congregations like Ebenezer as "mixed-
type" churches, "in which the tone is set by people with middle-class
aspirations, but in which some concession is made to the 'shouters.'" At
Ebenezer they sang the metered hymns with lugubrious dignity but saved
their excitement for the gospel songs and preaching. Ebenezer's complex system of organization also fit Drake and Cayton's description of
the typical large urban congregation whose range of activities is as broad
as the social makeup of its members. Where Ebenezer clashed with the
sociologists' profile was in the education and personality of its pastor,
who in the large urban church is usually a college-trained member of the
middle class. King's father was a shrewd leader who eventually earned a
college degree, but his style of leadership broke the mold of the black
bourgeoisie. Despite his congregation's influential membership, Daddy
King saw to it that Ebenezer never lost its mass-identity as a talk-back,
whooping, gospel-singing, workingman's church.

Nowhere was the pneumatic, New Testament quality of the church
more evident than in what Du Bois called "the Frenzy" that descends upon
black worship. Like his African forebears, King Sr. held that the great God
binds himself to special communities and discloses his presence through
acts of power in the assembly. The Spirit who once dwelt in the rivers of
Africa was available by immersion in the pool behind the mysterious
curtain at Ebenezer and in the preaching of charismatic leaders. Even
King's more educated son used the language of physical possession to
describe the act of preaching: "The Word of God is *upon me*," he once
cried, "like a fire shut up in my bones and when God gets *upon me*, I've
got to say it!"

At Ebenezer, the divine presence took the form of a dialogic pattern
of call and response between preacher and congregation. A descendent
of the slaves' "ring" or "shout," this practice created a communications
field of interacting sounds and rhythms in the sanctuary. As dialogue gave
way to what Du Bois described as a "Pythian madness," the natural sepa-

ration of leader and audience was broken down. The Word became the achievement of the group.

A version of the ring was still being practiced by ministers in the Stockbridge, Georgia, area where King Sr. was baptized and ordained. As late as the 1950s, ministers and churches met once a year on Emancipation Day for a festival of preaching in which the ministers would whoop and turn in circles while the audience shouted and kept time. King Jr. would later disdain the preaching of "whooping, loud, emotional" sermons and churches whose members have "more religion in the hands and feet than in their hearts and souls," but he was a product of this communal approach to celebration and in his own preaching and oratory used it to stunning effect.

At Ebenezer King absorbed a synthesis of traditions regarding the Word. There the rafters rang with the power of *nommo*, the creative spoken word, only three generations removed from Africa. Of course the power did not belong to language in general but exclusively to the *deus loquens*, the talking God who was revealed in the Christian Bible. At Ebenezer young King learned that when the preacher assumes his proper place in the hierarchy above the people and beneath the cross—and says what God wants him to say—the entire organism hums with celestial power. The people had better pay attention.

If Ebenezer was democratic in character, it was a democracy orchestrated by a powerful leader. From a young age, King understood himself to be the successor of charismatic leaders who were called to govern the church and advance the race. In 1894 King's maternal grandfather, A. D. Williams, became the second pastor of the seventeen people who made up Ebenezer Baptist Church. He guided the small congregation at a number of locations on and around Auburn Avenue, including a storefront on Edgewood Avenue, until after eight years of worship in the basement of the present building, the great doors of the new sanctuary were opened, and the people entered with joy.

The son of a slave exhorter, Williams was a rough-hewn man, and the church he built reflects his simplicity and functional approach to religion. The building is a red-brick imitation of a fine church, resting on a battleship-gray foundation of concrete block. It lacks the commanding bulk of Big Bethel A.M.E. up the street and cannot match the more expensive stone masonry of Wheat Street Baptist a block or so to the west. Ebenezer's two amputated towers underscore its absence of majesty. Above the entrance, the chevron and cross with EBENEZER in neon confirms its ordinariness. Suspended over the sidewalk like a theater marquee or chiropractor's shingle, the neon EBENEZER invites the down-and-outer or the sojourner to come into the church and to discover how "the Lord has helped us."

More important than the building was the congregation's tradition of advocacy on behalf of Negroes in Atlanta. Under Williams's leadership the church took an active role in local politics. Having many times watched the Ku Klux Klan parade down Auburn Avenue, he became a charter member and later president of the Atlanta chapter of the NAACP. When a city bond issue failed to provide for the building of a public school for Negroes, Williams led a successful revolt to defeat the bond issue. As a result Atlanta got its first black high school, Booker T. Washington High School, which his grandson would later attend. When an inflammatory local newspaper blamed the bond issue failure on "dirty and ignorant" Negroes, Williams led a boycott that eventually closed the paper down. He took Sweet Auburn as his parish and was often seen sitting on his chair on the sidewalk in front of the church keeping an eye on the comings and goings on the avenue. In 1929, the year of Martin Jr.'s birth and the forty-third anniversary of the congregation, Pastor Williams announced that henceforth Ebenezer would dedicate itself to the support of Negroes and to "every righteous and social movement."

When Williams died in 1931, his son-in-law and heir apparent, Michael Luther King, Sr., assumed leadership of the congregation. The only Ebenezer Martin Jr. ever knew was the church fiercely governed by his father. The elder King was a perfect instance of W. E. B. Du Bois's assertion that "the Preacher is the most unique personality developed by the Negro on American soil. A leader, a politician, an orator, a 'boss,' an intriguer, an idealist—all these he is, and ever, too, the centre of a group of men, now twenty, now a thousand in number." Michael King had come out of Floyd Chapel in the village of Stockbridge, some thirty miles southeast of Atlanta. He left behind a good mother, an abusive, drunken father, and the back-breaking life of a sharecropper and currier of mules. In the big city, he confronted and overcame every obstacle he faced with the same explosive mixture of enthusiasm, ambition, hard work, and bluster. A near-illiterate, he was determined to break into Morehouse College and to become a refined pastor. At age twenty-one he was a fifth-grader in a remedial school in Atlanta. For several years he served a tiny church in Atlanta called Traveler's Rest where he labored with a reading ability "barely beyond a rank beginner" and preached to deacons who "didn't know the alphabet." A member of the lowest caste of rural Negroes (his schoolmates had taunted him unmercifully for smelling like a mule), he set his cap to marry the princess of Ebenezer, the boss's daughter, Alberta Williams. A poor man with dung on his boots, he dreamed of living in a brick house with a fine porch and shutters on the windows.

When King took over Ebenezer in the fall of 1931, the Depression had seized the city, and the great church doors were literally padlocked by a local bank. His first job was to save the church. Historian Taylor Branch gives a colorful account of King's bold financial measures, which included the centralization of the treasuries of the various auxiliaries and the publication of each member's contributions—amounts plus names. King also established birth-month clubs whose members engaged in friendly fund-raising competitions to the benefit of the whole congregation.

King's shrewdest financial maneuver he learned from the successful insurance companies on Auburn Avenue. Because their clients were not able to afford large policies, insurance salesmen collected small premiums on a monthly or even weekly basis. King recruited the insurance people as members who in effect became collection agents for the church. The salesmen acquired a new field of customers, and the church benefited in many ways from a new system that not only increased contributions but also created a network for communication and home care.

If Mike King had his critics, they were soon silenced by the church's dramatic financial recovery as well as the irrefutable passion with which he willed one thing: the advancement of Ebenezer. Whether by Machiavellian business instincts or blunt-edged buffoonery, Mike King got his way. When he addressed the members of the congregation, he treated them as a single person, a child, to be admonished and cajoled: "I want to tell you this morning, Ebenezer, you can do it." Within a year Ebenezer had done it. King had set it on a trajectory that would take the congregation from a membership of four hundred in 1931 to a high of four thousand in 1940. And Ebenezer responded by making the former currier of mules the highest paid Negro preacher in Atlanta.

Like his father-in-law before him, King linked the salvation of the church to the economic and social health of the Negro in Atlanta. In this respect, Ebenezer fit the pattern of the great Negro urban churches of the early twentieth century whose pastors, such as Reverdy Ransom of Chicago and Adam Clayton Powell, Sr. of Harlem, had been influenced by the Social Gospel movement. Rather than elaborating a theology of ministry, which King Sr. was not inclined or equipped to do, he chose to flex his political muscle in Atlanta. He once led a voters registration march down Auburn Avenue to City Hall, and later in his ministry fought for equal pay for Negro schoolteachers and the hiring of Negro policemen. As a leading representative of the black community, he came to have some influence in the white machine that ran Atlanta. But his protests, like virtually all those of his era, took place on a field whose bases and boundaries, it was understood by all, were fixed.

III

In 1934 Ebenezer gave its successful pastor a summerlong tour of Europe, Africa, and the Holy Land. The journey climaxed in the Baptist World Alliance meeting in Berlin, in the land of Luther, whose defiant "Here I stand" sparked the Protestant Reformation. When King returned from Germany, he changed his own name and that of his son from Michael to Martin and had his son rebaptized Martin Luther King, Jr.

There is a special place in the hierarchy of the black Baptist church for the "preacher's kid." In his autobiography, *Lay Bare the Heart*, James Farmer recalls, "Every part of the PK's life is colored by the fact that he is a preacher's child. He breathes it in the air, suckles it from his mother's breast. Relentless eyes and ears and tongues demand that he be somehow larger than childhood." The PK plays a dual role, and the trick is to maintain both roles in healthy tension with one another. The preacher's child is torn between being as good as the congregation expects and as aggressive as the father's position demands. If goodness predominates, the child turns out to be a goody-goody and distrusted by his peers. If aggressiveness wins out, the child becomes the proverbial parsonage-bred hell-raiser. The poles of goodness and power create a magnetic field from which it is difficult for the child to escape. Farmer concludes, "Naturally, rebellion is the most common response to the unnatural life-style of the PK. . . ."

The son of a preacher may turn to his mother for comfort in his struggle for goodness but only to his father as a source of potency. Farmer's profile of the preacher and his wife does some justice to the household in which Martin grew up: "The black preacher, especially in the South, is king in a private kingdom. Whether learned or ignorant, he is both oracle and soothsayer, showman and pontiff, [and] father image to all. . . ." Of the preacher's wife, he continues,

> [S]he is mother, cook, housekeeper, and unobtrusive helpmate—laughing at his wit no matter how stale, nodding dutifully at his sermons, and clinging to his arm when he decides that she should be there. She cannot be aggressive or the women of the church will resent her. She mustn't be jealous lest she deprive some pillar of the church of his presence. Above all, she must take care not to detract from his image of power, or the hypnotic spell will be broken, the mystique shattered.

In his fifteen-page, handwritten autobiography, Martin Jr. repeatedly characterizes his father as a "real father." Of Alberta King he adds, almost as an afterthought, "Our mother has also been behind the scene setting forth those motherly cares, the lack of which leaves a missing link in life." Thus from an early age the preacher's son learned that women are for background support. What they have to offer they give from the sidelines out of their infinite capacity to comfort men against the rigors of their "real" work in the marketplace. Women's nurture is important, but ultimately it must be set aside in favor of stronger lessons from the source of authority in the family.

From his earliest years King bore well the peculiar burdens and privileges of being a PK on thriving Auburn Avenue. His "Mother Dear" taught him hymns and songs like *I Want to be More Like Jesus*, which he performed at the drop of a hat or the passing of a plate. The minutes of the National Baptist Convention meeting at Mount Vernon Baptist Church in Newman, Georgia, contain the following notation: "Master M. L. King, Jr., age five, accompanied by his mother, sang for the Convention and was given a rising vote of thanks."

However nurtured by the love of his mother, the youngster more readily identified with the clerical interests of his father. A letter from eleven-year-old Martin to his father indicates how quickly the PK assumed his father's role: "I am being a good boy while you are away. We had good church services all day today. No one joined the church but Rev. Edward brought too [sic] good sermon. Mother Holleys funral was today and Rev. Sim and Rev. C. S. Jackson preached the funral."

Little Mike quietly observed and absorbed the power that came from being Big Mike's son. He also learned not to violate the goodness that was expected of him. He hated conflict for the feelings of guilt it produced in him and gradually developed a smoothness and cleverness of tongue that enabled him to avoid confrontation with anyone, above all, with his father. Later in life when conflict, ironically, became his bread and butter, he skillfully isolated only the worst elements of racism in American society to attack frontally.

In his autobiographical statement King acknowledges that he, the son, was never so good or spiritually motivated as his place in the hierarchy might have suggested. He confesses that even his baptism at the hands of a visiting evangelist was motivated not so much by conversion as a desire to keep up with his older sister. He recalls, "Conversion for me was never an abrupt something. I have never experienced the so-called 'crisis moment.' Religion has just been something that I grew up in. Conversion for me has been the gradual intaking of the noblest ideals set forth in my family."

When the inevitable rebellion came, it took the form of occasional scuffles at the pool hall, lounging at the marble-topped fountain of the Yates & Milton Drug Store, or jitterbugging (toward which his father was most disapproving in the 1940s but more lenient in the 1950s). Martin Jr.'s theological precociousness led to a more serious rebellion. As a youngster he grew progressively annoyed with the "fundamentalist" bent of his Sunday school teachers until, at the age of thirteen, he shocked his class by "denying the bodily resurrection of Jesus." From that age doubts came "unrelentingly."

But instead of culminating in a grand rebellion against his father and the church, his religious struggle led him to a thoughtful and unemotional embrace of "the noble moral and ethical ideal" by which he had been raised. When he finally turned from rebellion to goodness, it was not, young King remembered, because his father had forced him into it:

> [B]ut my admiration for him was the great moving factor; he set forth a noble example that I didn't mine [sic] following. Today [1950] I differ a great deal with my father theologically, but that admiration for a real father still remains.

By the time Martin Jr. was a senior in high school he was resolved to enter the ministry. But he lacked one thing, the only formal prerequisite for entrance into the Baptist ministry: the soul-searing call. Moses and all the prophets had experienced it. Without exception the slave preachers heard (and visualized) it as a real voice in the night of sin or despair. What young King had was a gradual appropriation of his father's power and a corresponding awakening of his own gifts to lead people and to move them with his words. The latter discovery he made one summer when he served as the informal devotional leader for a group of chums at a summer work camp. At Morehouse College he came under the influence of men like George Kelsey and Benjamin E. Mays, who modeled for him and a generation of Negro students a ministry based on reason and learning rather than emotion. Thus by the standards of the black church his call was a relatively tame event: "My call to the ministry was not a miraculous or supernatural something, on the contrary it was an inner urge calling me to serve humanity." For the searing touch of God King would have to wait until later in his ministry.

Nevertheless, King's account of his call satisfied his father. The elder King arranged for his son's trial sermon at Ebenezer. When the church

basement couldn't hold the crowd, the event was moved to the main
sanctuary. Older members recall that the youngster spoke little of Jesus
but delivered his fancy words with great authority. No one knew or cared
that the eighteen-year-old preacher had borrowed his first sermon from
Harry Emerson Fosdick's "Life Is What You Make It." Young Martin's
reabsorption into the life of the black Baptist church was made complete
on February 25, 1948, when the son knelt before his father in the sanctuary and, surrounded by a great company of dark-suited Baptist preachers, was ordained to the public ministry of the gospel.

IV

During the course of his formation King came to the realization that he
was not the first to preach the gospel under oppressive conditions. He
discovered that he was surrounded by teachers and examples. When he
considered his vocation, he tended to view it through the lens of the men
in his family: "This is my being and my heritage for I am also the son of
a Baptist preacher, the grandson of a Baptist preacher and the great-grandson of a Baptist preacher." His great-grandfather was a slave exhorter
named Willis Williams from Greene County, Georgia. King's preacherly
lineage represented not only his own bloodline but also the long line of
witnesses whom homiletician Henry Mitchell calls the Black Fathers. The
witnesses also included pioneering women preachers such as Jarena Lee,
Zilpha Elaw, and Julia Foote, each of whom left an autobiographial account of her ministry. King's predecessors were the pastors and evange-
lists who from the late eighteenth century onward served churches or
denominations that were primarily under black control.

There were actually two lines of progenitors trailing behind King,
the Sustainers and the Reformers. The longer line was the Sustainers, who,
like Great-grandfather Williams, ministered to the spiritual needs of enslaved and segregated people but never attempted to revolutionize the
conditions under which they lived. The second and much shorter line
was the Reformers, who, like his grandfather and to a lesser extent his
father, were willing to raise hell for the freedom of the race. These were
the so-called race men who knew how to translate congregational and
charismatic power into political clout. King would be true to both traditions, those represented by his slave-exhorting great-grandfather and his
activist grandfather: he would use Christian symbols as medicine for the
wounded *and* as weapons in the fight against racism.

Ministry in Sweet Auburn, as throughout a segregated nation, had proved to be a *strategy* of survival and advancement. How were Negro Christians to push up to the table in this society without violating their religious principles—or getting themselves killed? What resources did Christianity offer in the struggle and what sacrifices and risks did it demand? More important than the names of King's individual heroes who had wrestled with these questions were the collective strategies he inherited from them. Before he read academic theology, he had already internalized a set of strategies—pastoral, theological, and prophetic—for ministry in a hostile land.

From the Sustainers he had inherited an exceedingly complex pastoral strategy designed to stimulate hope while deferring its reward. The Sustainers were not theologians, but if they had been, they would have justified their strategy with the doctrine and images of Christian eschatology. Under the rubric of the new creation, the Sustainers imagined a New Jerusalem of harmony and peace. It was not the apocalyptic wrenching of heaven and earth that they foresaw but the placid aftermath of Armageddon. Before the time was ripe for a holy war, the slave preachers and later the free Negro preachers projected a heavenly vision against the dark, low ceilings of slavery and segregation. That vision glowed with gold and pearl and the beauty of absolute holiness. There would be plenty to eat and drink, and the slaves' quarters would be transformed into mansions that outshone the finest plantations. Gayraude Wilmore believes that that vision promised a reckoning for those who were not yet able to act and thereby averted the disaster of premature revolution. Otherworldly preaching, he contends, was an "interim strategy" designed to provide comfort and assurance of final victory. The long history of tyranny throughout the world seems to support Wilmore's sage observation: "Oppressors have never been able to relax in the presence of this kind of otherworldliness."

Are Wilmore, James Cone, and other theologians correct in their reading of the otherworldly preachers? Was it a real strategy, and did it make the oppressors nervous? The more conventional approach has interpreted otherworldly religion as the opium of the people, the original "high" that permitted (and continues to permit) a temporary escape from social and economic oppression and from the responsibility to do something about it. For most of his life King's criticism of his own tradition— with its focus on the "mansions in the sky"—lay in the conventional camp. It was no accident that one of his earliest sermons, "The Three Dimensions of a Complete Life," was a psychological reinterpretation of the symbolic picture of heaven in the Apocalypse. For the most part, he reiterated the conclusions of studies like those of his mentor Benjamin Mays and Joseph Nicholson, who in the 1930s demonstrated, with some chagrin, that otherworldliness continued to dominate the Negro church's preaching. In the 609 urban congregations they visited, three-quarters of the stenographically reported sermons dealt with "the other-worldly aspect of religion and life." They esti-mated the percentage to be higher in the country churches. Coretta Scott King remembers the rural Alabama preachers of her childhood much in the same way:

> Seldom if ever did the preachers of that period deal directly with the plight of their people. Occasionally, when some black person had been beaten or otherwise badly treated by whites, there would be a reference to it from the pulpit. The preacher's role was to keep hope alive in nearly hopeless situations. "God loves us all, and people will reap what they sow," they would tell us. . . . They never preached what we would now call "the social gospel," neither did they discuss from their pulpits Negro rights or the issues of segregation. It was too dangerous.

Instead, the preacher could only paint a picture of life on the other side of the Jordan, a practice that continues to this day. When the preacher begins rhetorically to "cross over," the congregation passes over with him. Together, in the context of praise and ecstasy, they both arrive at a place that may nourish a day or an entire week in a zone where all suffering has ceased and joy reigns.

Those who dismiss such preaching as an escape valve may have a legitimate criticism of emotionally self-indulgent worship. But in times of persecution or unrelieved want, the symbol assumes a mysterious power to actualize its own content. The slave's experience of the "Lord's victory" in the timelessness of worship will hasten the coming of that day in history. The ritualized cry for deliverance provides the form if not the substance of things to come. This was as true at Ebenezer as it was in the poor rural congregations of Georgia. When a simple Macon County coun-try preacher uses the metaphor of the *dream* to imagine the future, the dream ignites his congregation's hopes for a better life no less than when Martin Luther King, Jr. moved his national congregation with the same symbol.

King explicitly rejected otherworldly preaching but admired the Sustainers' strategy of affirming the worth of the oppressed. This they did implicitly on the basis of the doctrine of creation. Their logic was simple: if God made you, you are his children and precious in his sight. As early as 1950, the seminarian King was using his newfound psycho-logical jargon to express his admiration for the slave preacher who "would come to his triumphant climax saying: 'You—you are not niggers. You—you are not slaves. You are God's children.' This established for them a true ground of personal dignity. The awareness of being a child of God tends to stabilize the ego and bring new courage."

King also appreciated the skill with which the Sustainers, who often preached before admiring white or mixed congregations, were able to give the biblical story a double voice, one for the oppressed and another for the oppressor. Although the slave and old-time Negro preachers could not make a frontal assault on oppression, they knew the difference between the tranquilizing and sustaining functions of a sermon. They turned to the great stories of the Bible and playfully "voiced" them in such a way that the stories themselves made the points that they could not express. Folk preachers such as Lemuel Haynes, Uncle Jack, Henry Evans, and C. R. Walker used vernacular dramatizations of the Bible to portray the conflicts between the powerful of this world and God's powerless people.

One of the great pulpit storytellers of the nineteenth century and an excellent example of the sustaining function of the Negro clergy was John Jasper of Richmond, who exemplified the "old-time Negro preacher" described by James Weldon Johnson as "an orator, and in good measure an actor." In the following passage from a sermon on the plagues in Egypt, Jasper embellishes the narrative in uproarious fashion and tacks on a moral at the end:

I tell you, my brethren, this scheme did the business for Pharaoh. He come from ridin' one day and when he get in the palace the whole hall is full of frogs. They is scampering and hopping round till they fairly cover the ground and Pharaoh put his big foot and squashed 'em on the marble floor. He run into his parlor trying to get away from them. They was all around; on the fine chairs, on the lounges, in the piano. It shocked the king till he get sick. Just then the dinner bell ring, and in he go to get his dinner. Ha, ha, ha! It's frogs, frogs, frogs all around! When he sat down he felt the frogs squirmin' in the chair; the frogs on the plates, squattin' up on the meat, playing over the bread, and when he pick up his glass to drink the water the little frogs is swimmin' in the tumbler. When he tried to stick up a pickle his fork stuck in a frog; he felt them runnin' down his back. The queen she cried, and 'most fainted and told Pharaoh that she would quit the palace before sundown if he didn't do something to clear them frogs out'n the house. She say she know what is the matter; 'twas the God of them low-down Hebrews, and she wanted him to get 'em out of the country. Pharaoh say he would, but he was an awful liar; just as they tell me that most of the politicians is.

Jasper's remark about politicians illustrates a danger faced by early Christian preachers, the Sustainers, and later by King and his associates. Of the Sustainers, Eugene Genovese writes, "They had to speak a language defiant enough to hold the high-spirited among their flock but neither so inflammatory as to rouse them to battles they could not win nor so ominous as to rouse the ire of ruling powers." They were obliged to speak with two voices simultaneously, which is one of several verbal skills traditionally associated with "signifying." The historical circumstances of black preachers made it necessary for them to please one audience and sustain (or alert) another, sometimes in one and the same speech. Signifying is a black thing because in America white preachers or politicians were rarely under the cultural pressure to pull it off in order to survive. The Sustainers were caught in the tension between audience and prophetic purpose; the result was often a double-edged sermon. They preached a message of loyalty for the benefit of the many whites who overheard them and, by means of a paralanguage of tone, gesture, rhythm, and vernacular, addressed a more nuanced call for resistance to those with ears to hear.

V

Like the Sustainer, the Reformer remained within the symbols and moral teachings of Christianity but trained them, as it were, *against* Christianity and the caste system it had underwritten. The Reformers did not follow Nat Turner and the insurrectionists in plotting armed attacks against slavery, but they were no less strategic in their opposition. Instead of exploding the religious cosmos in which a real if tenuous spiritual relationship between the races was maintained, they exploited it. Africans in America had received and embraced a religion of love from people who wanted them out of the way. The Sustainers had turned the Christian message like a gem against the light until the religion of pacification yielded resources for survival. The Reformers made a further turn. Under their ministrations, the religion of survival yielded up the resources for liberation.

If King had been asked to place himself within the African-American preaching tradition, he would have undoubtedly identified with the Reformers—with Richard Allen, Henry Highland Garnet, Harriet Tubman, Alexander Crummell, Edward Blyden, the fiery emigrationist Henry McNeal Turner, and the nonpreachers David Walker and Frederick Douglass. As a young man he professed admiration for the apocalypticists Nat Turner and Denmark Vesey, who planned and participated in violent insurrections—which King's first biographer reported in a volume pointedly entitled *Crusader Without Violence*.

The extent of King's intellectual engagement with the major figures of his own tradition is difficult to determine. The editor of his *Papers* notes that Morehouse College's records are incomplete and that, as a fifteen-year-old college freshman, King was not all that serious a student anyway. He took ten courses with sociologist Walter Chivers, but his greatest exposure to the ideal of reformist Christianity doubtless occurred at Ebenezer, where famous big-city preachers regularly made guest appearances in the pulpit. They were living proponents of the linkage between religion and racial justice. As with the Reformers, the ideal they represented was that of the master preacher who demands a full accounting of Christianity and who skillfully uses Christian resources to liberate the race.

Through them King acquired further strategies for a ministry of liberation. The most profoundly theological of the strategies came to him from Richard Allen, the first ordained preacher of African descent in America and the founder of the African Methodist Episcopal Church. Allen was not his *source* in a literary or academic sense; indeed, King never mentions his name. But most of King's arguments for freedom as well as his theologically freighted strategy of social change preexisted in Allen and his associates. In doctrine Allen remained a loyal Methodist and a Sustainer of his people, encouraging them to loyalty and affection for their masters. In his church polity, Allen pursued a more aggressive line. He was a bold Reformer who shrewdly manipulated his white Methodist antagonists and attacked slaveholding Christians on the basis of the Christians' Bible.

There are few arguments or sermon texts in King that were not anticipated by Allen or the other Reformers. Long before King defended the native intelligence and essential humanity of his people, Richard Allen had made the same passionate argument: the very masters who divide our families, withhold education from us, and otherwise degrade us are the same critics who stigmatize us as immoral, stupid, and inhuman. Allen wrote, "We can tell you from a degree of experience that a black man, although reduced to the most abject state human nature is capable of— short of real madness—can think, reflect, and feel injuries, although it may not be with the same degree of keen resentment and revenge that you, who have been and are our great oppressors, would manifest if reduced to the pitiable condition of a slave." Long before King identified the God of the Exodus with the suffering and segregated of his race, Allen and many others had used the same story to prove, in Allen's words, "that God himself was the first pleader of the cause of the slaves." Long before King proclaimed the liberal credo, "the Fatherhood of God and the brotherhood of man," Allen had founded a denomination whose motto was "God our Father, Christ our Redeemer, Man our Brother." Long before King encouraged his followers to win the enemy through love and the example of noble suffering, Allen had suggested the same tactic.

It is at this last point where the links between Allen and King appear almost genetic. After King became famous, he would make much of unmerited suffering as a means of spiritual redemption. Although King himself later traced its intellectual lineage to concepts with which he had become familiar in graduate school, the doctrine of unmerited suffering was nowhere more clearly taught than by Allen and one of his associates, a preacher named Daniel Coker. In his *A Dialogue between a Virginian and an African Minister*, written in 1810, Coker creates two symbolic figures, a slaveholding, white Virginian and a distinguished minister of African descent. The African argues that slaves should submit to unjust suffering, as Jesus did, in order to win over the cruel master. Quoting an obedient slave, "I am tied and stretched on the ground, as my blessed master was, and suffer the owner of my body to cut my flesh . . . and bear it without murmuring." The slave's soul is purified through obedience, and the master's heart is changed by the slave's sacrificial witness. By the end of the pamphlet, the Virginian is clearly won over and announces his intention to manumit his fifty-five slaves.

Coker's points were the spiritual trademarks of King's most famous campaigns. Only, instead of the bleeding slave, it was the segregated Negro who would now assume the cruciform position. Evil must always be resisted, but any suffering, however undeserved, that facilitates the good of the whole must be embraced. The acceptance of unmerited suffering identifies the victim with the purposes of God. The victim pulls all the moral levers. Early in his career King may have quoted Gandhi, but he was adhering to Allen, Coker, and one strand of his tradition's theology of redemption when he wrote, "Do to us what you will and we will still love you. . . . But we will soon wear you down by our capacity to suffer. And in winning our freedom we will so appeal to your heart and conscience that we will win you in the process."

Allen and Coker's theology was not inconsistent with the pastoral strategies of the slave preachers. Both were carefully nuanced applications of Christian doctrine to situations of social injustice. Both were designed to hold black and white together. But the final strategy King inherited was more frankly prophetic. Instead of manipulating the nuances of Christian doctrine, the prophetic tradition issued a call to arms and leveled a blast at its opponents. Its only criterion of speech was the justice of almighty God; its only models were the Hebrew prophets. Late in his career, when King began to give full vent to his rage, it was the holy rage of the prophetic tradition, and not the resentment or vengefulness of his nationalist critics, to which he lent his voice. His prophetic forebears, as it were, taught him *how* to get mad, what themes to press, and what language to use.

The dynamic of sin → suffering → redemption → reconciliation moved powerfully beneath the surface of both the sustaining and reformist traditions of black protest. Even those like the provocateur David Walker who reviled the church were driven by the same logic, with one major addition: between the elements of suffering and redemption Walker planted the firecracker of revolt. He anticipated Marx's truth that the step from misery to liberation is neither natural nor automatic but requires an act of will.

Walker said little of the slaves' duty to love their masters, but on the basis of God's deliverance of Israel he developed a theology of God's justice. The only true religion is adherence to the demonstrated character of God. Walker turned his fiercest anger against the hypocrisy of the white church preachers who "form societies against Free Masonry and Intemperance, and write against Sabbath breaking, Sabbath mails [and] Infidelity," but who justify the "bloody and murderous head" from which springs the great American evil. "But oh Americans! Americans!! I warn you in the name of the Lord (whether you will hear, or forbear) to repent and reform, or you are ruined!!!!!!"

Although not a preacher himself, Walker's influence on the nineteenth-century clergy was enormous. Black Presbyterian pastor Henry Highland Garnet revered Walker, intensified his criticism of white Christianity, and echoed his call for insurrection. Garnet's most famous speech was given at the National Convention of Colored Citizens in Buffalo in 1843. In it he rehearsed the familiar saga of slavery in America and shocked the assembly with a demand for direct action against the slave-holding "heartless tyrants." "You cannot be more oppressed than you have been—you cannot suffer greater cruelties than you have already. Rather die freemen than live to be slaves." One can hear echoes of this rage in King's perorations. Whenever he hammered away at the white church's preoccupation with private morality above social justice, he was echoing Walker, Garnet, and the prophetic tradition. Whenever he warned America of impending doom, as he had planned to do in his final unpreached sermon, "Why America May Go to Hell," he was expressing an inherited rage, one that did not come to him straight off the pages of the Old Testament but was mediated to him by generations of angry prophets.

Garnet's abolitionism was typical of northern black preachers like

J. W. C. Pennington, Samuel Ringgold Ward, C. B. Ray, and Daniel Payne. They were motivated by their faith in America and in the God who directs the destinies of nations. Southern black preachers, who were on the whole less educated than their northern counterparts, took a more evangelical line and tended to rely on the personal conversion of their racist oppressors. Both traditions preached a God whose universal love was larger than racial or regional differences. King would echo the universalist hope, eloquently expressed by Garnet before the U.S. House of Representatives in 1865, that the nation would heed "the voices of universal human nature." But he would also continue the methods of the revivalist, the methods of his father, and press the nation for repentance and conversion.

Like Garnet, Allen, Coker, and many others before him, King would devote much of his energy to negotiating the demands of two audiences, black and white. Allen and Coker bent over backward to move and mollify their white audiences. David Walker's *Appeal*, though ostensibly addressed to "the Coloured Citizens of the World," was a jeremiad meant to be overheard by whites in America. Even Garnet's radical platform was located in a historically white denomination, the Presbyterian Church, and Garnet himself died an officer of the U.S. government. More skillfully than any, King would employ their theological and prophetic strategies, along with the Sustainer's pastoral care of the downtrodden, in order to cultivate the support of whites while advancing the cause of blacks.

For King, as for many of the preachers who preceded him, the strategies were circumscribed by the claims of the Christian gospel. As faithful rebels, they found themselves in a double bind of loyalty to the race and fidelity to what was often a conflicted understanding of the Christian mission. Nothing in King's education adequately prepared him for resolving the internal debate between the beauty of suffering and the power of justice that he found at the heart of his own tradition. For that resolution, he would wait upon events.

King's formation in the African-Baptist church cannot be separated from the far-reaching social shifts that, to the eye of the historian, appear to have "prepared" the way for his historic success: black migration and urbanization, the broadening of horizons occasioned by World War II, the rise of political liberalism, the coming of television. But deeper and more telling than the body of information at his disposal or the sum of social and personal "influences" wrought upon him was the history of indignities borne by the race and encoded in each of its members. Of such is knowledge, not information, and it is collective. King bore in his body all the marks of this collective history of weariness and protest. When he went into the streets with his campaigns, his actions were a continuation of his father's and grandfather's early efforts and of all others who had gone before. When he rose to preach, thousand of long-suffering Sustainers and fiery Reformers rose with him. As a Sustainer he comforted his people with the gospel of their inestimable worth; as a Reformer he astonished them with the word of deliverance. His preaching was an electrifying moment, but only a moment, in which he merged his voice with the tradition's continual cry.

2

Apprenticed to the Word

KING rarely spoke publicly about the Sustainers and Reformers who preceded him or the influential mentors who
formed him as a preacher. Perhaps he understood that American culture
demands utter originality of its leaders or feared that if he acknowledged
his debt to the black tradition, he would lose his credibility with white
audiences. When he did reflect on his own rhetorical gifts, he never identified his real teachers, never praised the black Fathers or spoke of Africa,
Auburn Avenue, or his Daddy's awesome power, but instead repeated
the conventional principles of oratory as if he had merely plowed in the
furrows of Cicero and Quintilian. The only on-the-record statement King
ever made regarding his own preaching and oratory occurred in Atlanta
a week after John Kennedy's assassination. King was sick and bedridden
with the flu but nevertheless granted a long interview to a young graduate
student in speech named Donald H. Smith. The interview exemplifies
King's generosity as well as his reticence: he opened his home to an un-
strategic academician but gave him a superficial account of his training that
bypassed his apprenticeship in the African-American preaching tradition.

In the course of the interview, which was interrupted several times
by the telephone and his baby son Dexter ("Dexter does not have the
virtue of restraint"), King acknowledged that the spoken word was essential to the success of the Movement and that the style of any speech
may be altered according to the makeup of the audience—maxims common to Aristotle, Quintilian, and any introductory text in public speaking. He also alluded to the venerable three Ps: proving an appeal to the
intellect, painting an appeal to the imagination, and persuading by an
appeal to the heart.

When Smith asked him how he was affected by the verbal responses of the audience, King disingenuously professed to understand the phenomenon ("Baptist churches are highly emotional") but to be "outright annoyed" by it—this in the year of "I Have a Dream" and his electrifying mass-meeting speeches in Birmingham. King could not have been indifferent to the power generated by the response of his audiences. Nor could he have been unmindful of the mentor-preachers who surrounded him in his youth, when he answered Smith's question about the origin of his prowess: "It's something I have just grown up with."

I

The child of the African-American congregation grows up in an atmosphere of signals and effects that hums with the authority of the performed word. The fledgling preacher's first teacher is, in fact, that atmosphere, which, according to one prominent preacher, the youngster absorbs by "osmosis." The first moment in the decision to preach is recognition of the power of the spoken word. Traditionally, this power has been celebrated in preliterate or transitional cultures in which verbal performance exercises a potency not yet tempered by knowledge of writing or printing. Plato was afraid of the orators because he knew that a properly constructed sequence of words has the inherent power of compelling assent. Aristotle did the first scientific study of speaking and concluded that the greatest thing by far is to be a master of metaphor, for such a master makes connections that others cannot make and thereby captures language for his own political ends. Among the Yoruba of Nigeria, intricacies of verbal expression are understood as evidence of genius if not divinity.

Africans in America brought with them their delight in *nommo*, the spoken word whose rhythms and prolific powers reflected those of nature itself. Their joy was made a cruel necessity by the antiliteracy laws the South applied to slaves. The delight in verbal performance, which is today very much in evidence as a way of establishing "yo rep" (-utation) among African-American youths in the form of folk sayings, "lying," jokes, proverbs, "signifying," "toasting," rapping, mimicry, and many others, is traceable or at least attributable to the verbally gifted slave exhorters and Negro preachers who held their illiterate hearers in rapt amazement, and whose mastery of the spoken word enabled them to dissemble with the Man. In a touching passage of his autobiography, Malcolm X traces the origins of his own verbal firepower to his acquaintanceship with a fellow prisoner in the Charleston Prison, a man named Bimbi, who, Malcolm

remembers, "was the first man I had ever seen command total respect
. . . with his words."

Two of King's closest associates, C. T. Vivian and Fred Shuttlesworth,
both captivating speakers in their own right, remember King as a man
who understood the power of oratory. Vivian: "They said Martin used to
practice in front of a mirror as a kid. Martin actually believed, from what
I gathered, . . . that if you had the right set of words you could move
the world." In a 1968 interview Shuttlesworth sounds as if he has read
his Aristotle as he explains King's leadership:

Q: In other words, why did people follow him?
S: That's elementary. The man with the marbles is the one who domi-
    nates the game.
Q: What marbles did he have?
S: The ability to articulate what people want. . . .

In the canon of King family legends is the story of little Martin who
one Sunday in church looked up to his "Mother Dear" and said, "When
I grow up I'm going to get me some big words." The first words he learned
were of the hymns his mother taught him. When he sang them, the people
wept and "rocked with joy." Like his father before him, Martin memorized passages from the King James Version of the Bible and practiced
them until he had incorporated its Elizabethan cadences into his own
pattern of speech. With his young church friends he studied other preachers in the neighborhood, not out of piety as much as fascination with
their technique and the power they wielded over their congregations. By
the time he was in high school, he was distinguishing himself in oratory
contests, the most sophisticated and competitive of which was the Youth
Section of the National Baptist Convention's preaching competitions.
Many of the candidates were PKS like Martin, eager young products of
the predominantly oral culture of the Negro church. An old Baptist hand
remembers one such contest with awe: "Let me tell you, when you hear
those young people speak, it's like hearing seasoned people do it. Martin
won that thing."

When he enrolled at Morehouse College, his period of osmosis was
over. Now he would enter into rhetorical training with no less purpose
or dedication than that of a bright boy in ancient Greece. Although rhetorical training among the Greeks was reserved for the privileged, its
preoccupation with "style" has truer resonances to the churchly culture
of Sweet Auburn or any black ghetto than to an affluent white system of
education. The Greek view—and its African-American counterpart—is
summarized by Richard Lanham in this way: For the ancients, training in

rhetoric was for the young. The Greek boy wholly dedicated himself to the word, how to write it, remember it, and pronounce it. He looked *at* words for the sensual enjoyment they afforded as manipulable objects and sounds before he looked *through* them for their underlying meaning. The student could perform a Homeric scene (or the young preacher render a sermon from the black Fathers) long before he understood the philosophical or theological vision from which it was hewn. In place of original thought, which is always somewhat embarrassing in a teenager, the student (or young preacher) cultivated the ability to organize and exercise the cleverness of others. In doing this, he practiced a full range of verbal and histrionic adornments that he would eventually incorporate into his own style. With every shift of mood or occcasion, the classically trained speaker could call up thematic formulas, or commonplaces, and deftly insert them into any speech for the purpose of any argument. These units of wisdom were stored by means of elaborate memory schemes that allowed their recall at a moment's notice. Ultimately, the student was required to accommodate his personality to the demands of the rhetorical situation, so that the public manifestation of personality might always match the public occasion. The greatest sin was to slip out of character in a public performance. For the gifted young preacher no less than the student of rhetoric, the effect of such a regimen was the supplantation of true spontaneity in public life by its premeditated affectation. The preacher enjoyed freedom from the role only when he was away from the eye of the congregation or in the unguarded company of his fellow preachers.

II

In the church's first textbook on preaching, *On Christian Doctrine*, Augustine insisted that the interpretation of the Bible is work for men and that rhetoric is pleasure for boys. Young King also took his subjects in that order. The sixteen-year-old's first rhetoric teacher was Gladstone Louis Chandler, who had taught his Daddy before him and who acted the role of rhetorical drill sergeant to the young recruits who came to Morehouse. Chandler had been a Morehouse institution since 1931. The rigor of his speech course was a tradition the recruits both dreaded and enjoyed. He taught out of the old Bryant and Wallace speech text, *Fundamentals of Public Speaking*, but in his hands the rules of motivation, audience analysis, animation of voice, and gesture became a holy crusade for the civilizing of the raw material before him. He taught by his own sophisticated example and by incessant drill. Clarity, unity, coherence, and emphasis—these were his "word gods." He prefaced his vocabulary drills with, "Gentlemen, we are going to establish a 'GLC Word Bank.'" When asked, "How are you," Chandler's boys would reply, "Cogitating with the cosmic universe, I surmise that my physical equilibrium is organically quiescent." Chandler longed for the good old days when "Morehouse men" were shaped on the anvil of declamation and a weekly debate. He was in charge of the Webb Oratorical Contest, which was of some interest to black Atlanta. King's second-place finish may well have been related to his pronunciation, which Chandler judged to fall short of "national" standards of purity. The old rhetorician undoubtedly had something to do with King's abiding love of words as well as his weakness for flowery and bombastic expression. Years later, when King was a prisoner in the Fulton County Jail and given permission to request books, he reminded his wife to send his constant companion, *Increasing Your Word Power*.

By the time King arrived at Morehouse, the weekly debate had become a forum for the school's dynamic president, Benjamin E. Mays. If Chandler's goal for the boys was civility, May's was that and a more educated ministry. In 1890 Booker T. Washington had proclaimed three-fourths of the black Baptist ministers and two-thirds of the black Methodists morally and intellectually unfit to preach. Forty-three years later Mays himself in a pioneering sociological study, *The Negro's Church*, confirmed Washington's judgment and noted that 80 percent of the Negro clergy were not college graduates. As late as 1963 the situation had worsened, and Mays estimated that in a nation of fifty-five thousand black churches, no more than two hundred blacks were enrolled in seminaries. At Morehouse, Mays tackled the problem by using the Tuesday assemblies as a forum for his self-improvement propaganda, in which he methodically sketched a picture of—as he puts it in one such session—"the kind of men I think Morehouse College should produce." They should be "literate men" who are able to "think logically and profoundly," "intellectually tough . . . and rugged." By becoming "masters in their chosen field," they will "overcome the accidents of birth and the handicaps imposed upon us by a segregated society." Mays insists that a Morehouse Man should wear good clothes, drive a good car, have money in the bank, and own some stocks and bonds. But most of all he should be known by his works. If the Morehouse Man is to be written up in *Jet* magazine, Mays continues, let it be for his latest book rather than his latest car or his ability in bridge (an audible groan is heard from the bridge players). He then turns to the future clergy in the audience and advises, "Be known by the sermons you preach," not by the opulence of your latest Cadillac or Imperial. (King would echo this: "How many Negro preachers do we find who are really more concerned about the size of the wheel base on their automobile . . . than they are about the size of their service to humanity?") Mays encourages the students to take a "sane, prophetic view of religion" characterized by "poise in the midst of chaos" and the unshakeable conviction that in the end good will triumph. He warns the students not to set themselves against the accumulated experience of the ages. He encourages them to maintain their respect for the dignity of human personality, for of such is the essence of the Christian religion.

King had a keen ear for Mays's more eloquent phrases and quotations and wove them into his own sermons and speeches throughout his career. On the interrelatedness of all people Mays said, "[T]he destiny of each individual wherever he resides on the earth is tied up with the destiny of all men that inhabit the globe." Mays probably borrowed the sentiment and some of the phrasing from Harry Emerson Fosdick's assertion, "We are intermeshed in an inescapable mutuality," which eventually appears in many King sermons and addresses in the following formula: "We are caught in an inescapable network of mutuality, tied in a single garment of destiny. Whatever affects one directly affects all indirectly." On this theme all three conclude with the obligatory quotation from Donne, "No man is an island." Any number of Mays's favorite poems turn up in King's sermons and speeches. For example:

> Fleecy locks and dark complexion
> Cannot forfeit nature's claim.
> Skin may differ but affection
> Dwells in black and white the same.
> Were I so tall as to reach the pole
> Or to grasp the ocean at a span,
> I must be measured by my soul,
> The mind is the standard of the man.

It was from Mays that King first heard the challenge "Clearly, then, it isn't how long one lives that is important, but how well he lives, what he contributes to mankind and how noble the goals toward which he strives. Longevity is good . . . but longevity is not all-important." King paraphrased this sentiment many times in his career, perhaps most poignantly in his speech in Memphis the night before his death.

Mays's (and later King's) sermons give the impression of enormous research and erudition when in fact, like most of the liberal sermons of their era, they are creative assemblages of borrowed materials. King often quotes Mays's favorite quotations and draws on his teacher's supply of homiletical clichés. King and Mays were immersed in a sea of popular tradition. There was no question of whether they would use the tradition—since preaching is the transmission of traditional wisdom—but for young King the question was how extensively he would borrow from his teacher and other mentors until he developed his own style.

More important than the occasional borrowed phrase was the perspective on ministry and life that King also learned from Mays. King spent most of his time at Morehouse attempting to catch up on an inferior high school education. What he acquired from Mays and his religion teacher George D. Kelsey was not technical proficiency in theology or philosophy but an expanded worldview, and from Mays, a role model for an educated ministry. Mays was a liberal who believed that human largesse of spirit would eventually overcome ignorance and prejudice and usher in a new era of understanding. History and human nature are like two magnificent stallions yoked together on a long and arduous journey toward inevitable success.

From Mays, King acquired the habit of sprinkling his speeches with the names of heroic examples of human progress: Moses, Socrates, Galileo, Lincoln, Albert Schweitzer, Churchill, and a host of others. Jesus' name was on the list because he too, in a special way, symbolizes the aspirations of all the heroes. "Jesus was powerful," said Mays, "because he identified himself with the people" and taught by word and example the sacredness of human personality. In Mays King possessed the student's greatest treasure: a teacher who believes in truth, refuses to qualify it, and crabbedly insists on it as the perspective from which all else must be viewed. Mays was a moralist with a simple message: eschew smallness of spirit, avoid cramped and orthodox thinking, and, most important, don't get caught on the wrong side of history.

Whether Morehouse College measured up to Mays's aspirations for it is less important than the effect that the Morehouse mystique exercised on young men like Martin Jr. who had emerged from the secure world of the black bourgeoisie still haunted by suspicions of their own inferiority. If the black church had taught them "You are God's children" and therefore spiritually equal to whites, it was the mission of Morehouse and men like Chandler and Mays to drum it into these grandsons of slaves that they were the intellectual inferiors of no one.

Mays was doubly important to young Martin because he offered a viable alternative to his father's style of preaching. Martin could honor his father's ambition for him by entering the ministry, but he would exercise that office in another idiom. Mays was an intellectual whose speaking style, though full-voiced and dramatic, did not compromise his fundamentally reasoned interpretation of life and religion. It would later become a point of pride to King that he had never gotten up before an audience and pretended to be ignorant. This respect for intellect he learned from Mays, not his father.

Of the father, King's biographer David Lewis writes: "The King family belonged to what is known as the school of hard preaching, of which cult of personality, an occasional pinch of exploitation, and sulfurous evangelism are indispensable ingredients." The senior King was a loose cannon in the church who translated his natural volatility into a style of preaching that left his audiences sometimes overpowered, often offended, always entertained, and occasionally moved to deeper spiritual awareness.

King Sr. could do with his voice what technicians can do with a public address system: he would alternate between a soothing, mellifluous tone with only a hint of roughness and a piercing shriek that assaulted the entire sensory apparatus. The sounds he made could be heard but also *felt* inside the head and just beneath the sternum. To hear the lullaby and the shout in one speech was to experience the loving daddy and the angry father who throughout his adult relationship with his son attempted to control him by alternating between bathos and domination. The shrieks came without warning and often without relation to the content of the speech. In King Sr.'s use, the technique of alternation was a way of keeping the congregation in an unsettled and agitated state and of bullying those who might hold an opinion contrary to the preacher's. King Sr. was a whooper, which means that in moments of high intensity he gave certain words and phrases both a rhythmic and a musical value. He was not a "moaner," one who establishes a more sustained and predictable pattern of intonation. But the music in his preaching emerged from the percussiveness of his overpowering shout.

For the content of his sermons, King Sr. relied less on biblical texts than he did on his own opinions. It was his habit throughout his ministry to say what he pleased when he felt like it. If he saw a good-looking woman in a pretty hat, that observation found its way into his sermon. ("They can't kill ya for looking.") Usually his ad libs were related to his own opinions and preoccupations. On one occasion when he announced that a regularly attending member, who used to sit "right over there," had died of a stroke, he cried, "My God, it could'a been me!" Although he doted on Martin Jr.'s presence in the pulpit, his son's preaching placed an intolerable burden on him, and more than once he interrupted Martin's benediction by shouting with exaggerated pathos, "Aren't you gonna let me have anything to say in my own pulpit?" In his younger days, King Sr. would sometimes dash Billy Sunday style from one side of the platform to the other or "walk the benches," which entailed climbing over the pew-railing while maniacally shrieking at a congregation whose amusement was rapidly giving way to terror. He was mentally agile enough to use any person, including an assisting minister or an unsuspecting visitor, as an object lesson or prop. On one occasion he wound up preaching from the lap of a startled young assistant. One of his former student ministers, who had the distinction of serving as an intern to King Sr. in Atlanta and King Jr. in Montgomery, described the father as a "journeyman" with a "natural gut style" and an inimitable flair for drama. His son, on the other hand, was a "straight stand-up" preacher—"no antics."

However much young King admired his "real father" and understood his need to engage in such antics ("to develop the church"), he never approved of undignified tactics and regularly criticized younger preachers who employed them:

> I was talking to a young man not long ago, and he was—after the sermon I was talking with him—and he was just jumping all over the pulpit and jumping out and spitting all over everything and screaming with his tune, and moaning and groaning, and I said, "Now I just can't understand you, young man; you're getting an education. I can understand that there were certain cultural patterns in the past that caused many of the older ministers to do this, but I don't understand why you feel you have to do this." And he said, "Well, you know, I got to get Aunt Jane." And he has reduced the gospel to showmanship, reduced the gospel to playing in order to get Aunt Jane. I say to you this morning, the gospel isn't to be played with.

What did King learn from his father about preaching? Before he learned a particular technique from his father, who was rough and unpredictable in ways foreign to his son, he inherited from him as well as his grandfather and great- grandfather a composite impression of the pulpit's authority in the community. Although he reacted against his father's lack of discipline in the pulpit, he knew there was something to learn from him. Just as the elder King had gotten the job done one way or another in the community, the old man dominated the pulpit too and used it as an instrument of power. When he wasn't bullying his child, Ebenezer, into submission, he was romancing her with a seldom-appreciated lyricism. Like so many experienced preachers, Daddy King retained a great number of formulaic pieces by which he could transform an ordinary harangue into a memorable event. The formulas were simply *there* in a collective limbo waiting to be chosen and resuscitated in the perfor-

mance of the sermon. Like his father, young King would learn to tend
the tradition and manipulate it wisely. In one of King Sr.'s sermons he
refers to an older preacher who used to sing a song about the Lord:

> It moves me everytime I think about it . . .
> "I lo-o-o-ve the Lord. . . ."

He sings the word "love" in a surprisingly youthful voice. He then moves
into a traditional set piece on providence with which he brings the sermon to its climax:

> I made up my mi-ind
> That I'm goin' on wid Jesus . . .
> He is bread for a starving man . . .
> My Father, Jesus, has . . . a thousand hills
> And cattle on every one of 'em.
> If you be worrying about a quart o' milk,
> You must not be my Father's child. . . .

He concludes by singing portions of a formula that is found everywhere
in the black church:

> I know him to be a shield.
> I know him to be a rock.
> I know he is a-b-l-e [chanting]
> To raise us up when we are down.
> He walks in the midst of all we do.
> In fact, this is my Father's world.

While King Jr. may have eschewed his father's excesses, his filial piety
never wavered. Long after he had eclipsed his father, he remembered the
source of his own power. Even when he had effectively become his father's
claim to fame, he honored his father as the greater preacher. When his
Dad would complain that M. L. had become a better preacher than he,
Martin Jr. would only put his arm around his neck and say, "Awww, Dad."
A 1954 letter from father to son reflects a mentoring relationship that
was never fully broken. The young graduate student was making a name
for himself as a preacher. His father uses the occasion of one of Martin's
recent successes to warn him of the temptations that beset those who
can move others with the Word:

> Alexander called me yesterday just to tell me about how you swept them
> at Friendship [Baptist Church] Sunday. Every way I turn people are con-
> gratulating me for you. You see, young man, you are becoming popu-
> lar. As I told you, you must be much in prayer. Persons like yourself are
> the ones the devil turns all of his forces aloose to destroy.

III

Between the mind of Mays and the unpolished force of his father, King discovered three mediating influences who, like Mays, appreciated a good theological argument and, like King Sr., sat astride enormous urban congregations. They were William Holmes Borders, Sandy Ray, and Gardner C. Taylor.

Six years after Daddy King became the senior pastor of Ebenezer, William Holmes Borders came to Wheat Street Baptist, a large and influential congregation located just one block to the west of Ebenezer on Auburn Avenue (formerly Wheat Street). Borders quickly established himself as a force to be reckoned with on Auburn Avenue. Unlike King Sr., whose Morehouse degree was eked out over many years, Borders had distinguished himself academically both at Morehouse, where he later served on the faculty, and at Garrett Evangelical Seminary in Evanston, Illinois. Just as King Sr. had saved his congregation during the Depression, so Borders rescued Wheat Street from ruin by building a larger and more impressive sanctuary than Ebenezer's and by establishing himself as the premier black preacher in Atlanta. He would later build the first church-sponsored housing project in the United States as a part of the Wheat Street Complex, which included apartments for the elderly, a shopping center, and a supermarket. With his interest in better housing and the rehabilitation of vagrants whom he invited into his posh sanctuary, Borders was said to have "changed the ideal of worship on Auburn Avenue." It was black-church karma that the two vainglorious preachers should be enemies. The King Sr.-Borders feud was a bitter one that would persist to old age, long after their wives had grown to be friends. The son of dirt-poor parents, Borders had come too far to settle for anything less than being the "highest, biggest, and best" of Atlanta's black clergy.

Much of what Borders had to teach about preaching, the younger King had already learned from his father. Both were dynamos of energy whose next word or dramatic move could not be predicted. Both dominated an audience like no other preachers in Atlanta. Yet when Martin's interest in the ministry was beginning to waken, he and his friend Walter McCall would slip away from Ebenezer on Sunday morning to sit in the balcony of Wheat Street to study Borders. Young King knew that something was happening up the street that was not happening at Ebenezer. A handsome young intellectual was outdrawing his Daddy by nearly 4 to 1, and seminarians from around the city were coming to apprentice themselves to this powerful man. Young Martin began to feel that his own church "couldn't be the whole answer" to his many religious and vocational questions.

What young King heard in Borders was an educated man who had learned to speak the language of his people. Borders did his exegetical homework and prepared an exposition of the text in manuscript form that he delivered in a reedy baritone not nearly as powerful as Daddy King's. Then he would literally step back and let it out, and the stories would begin. He told stories with the vividness and abandon of a child. In Borders's pulpit the great Old Testament stories came to life, their setting translated from ancient Palestine to black Atlanta, their heroes the familiar denizens of Auburn Avenue. David was a "country boy," said Borders in a late sermon, who whipped a big man with a "Cassius Clay tongue." When the boy slung his smooth stone at Goliath, everybody in the church could hear Borders's imitation of a stone whistling through the air and splatting into the giant's forehead. Soon the manuscript would be forgotten, and Borders would be crouching on the stage to fetch five smooth stones, lurking in wait behind the organ, or lying full length on the sanctuary floor like a dead giant. As Borders lost himself in the event of preaching, he transported his audience to another realm. After the smoke had cleared in such a sermon, Borders never failed to provide a moral. In his sermon on David and Goliath, for example, he concluded, "God is in everything you do to give you success." It was not the moral that filled the Wheat Street Church Sunday after Sunday.

In 1943 on the national radio broadcast, "Wings Over Jordan," Borders preached his most famous sermon, "I Am Somebody." The sermon evoked hundreds of congratulatory telegrams from around the country. The "praise poem to the Negro race" epitomized everything the slave preachers had promised their downtrodden congregations. Its refrain became a staple in the repertoires of the soul singer James Brown, Martin Luther King, Jr., who coined the noun "Somebodyness," Jesse Jackson, and many other civil rights orators. In 1982 at the age of seventy-seven, Borders presided over "the very best" performance of his poem at Wheat Street Church. He served as "Master Tutor" for each three-hour rehearsal and at the conclusion of each "taught, lectured and preached on the greatness of the Black man."

The second mediating influence on young King's formation was an old family friend, Sandy Ray, pastor of Brooklyn's Cornerstone Baptist Church. Like Borders, Ray was a contemporary of King Sr. and yet another Morehouse man who had gone on to distinguish himself in service to the church and community. After leaving school, Ray had pastored churches in Ohio, where he wielded enough influence to get himself elected to the state legislature. He came to Cornerstone in 1944 and there he built a succession of fine buildings and a reputation for pastoral and social leadership. When the Montgomery Bus Boycott settled into the long winter of 1955-1956, "Uncle Sandy" and another New York preacher, Tom Kilgore of Friendship Church, coordinated the city's black church support of the boycott. Over a five-year period they raised more than a million dollars. Ray and Kilgore also served as bagmen for the effort, literally toting duffle bags of small bills directly from the offering plates of New York City churches to Montgomery, Alabama.

Sandy Ray was known for the humility of his preaching style. He was not an intellectual like Mays or a performer of the caliber of King Sr. or Borders. Ray's "folk ear" allowed him to hear the Bible in the idiom of his people. His illustrations were homely stories of a mule that got stuck in the mud or a dying deacon who didn't want to pray or a man whose fence fell down for want of sound supports. When he preached, Ray struck each word like a man breaking rocks with a sledge. By relentless pounding of a single, simple theme, his sermons achieved a power for which the contemporary hearer, perhaps expecting ecstasy, is unprepared. One can still hear Ray on tape, exhorting his people to live as Christians, the sounds of fire engines, sirens, and other Brooklyn calamities in the background. Whenever Martin Luther King, Jr. used homely illustrations—and he did from time to time in sermons that were not published or broadcast, stories of a masseur at the Butler Street YMCA or a barber on Auburn Avenue—one hears most clearly the humility and rooted sense of place of young Martin's "Uncle Sandy."

What Borders (and King Sr.) lacked in self-discipline and Ray in eloquence was abundantly present in the consummate black preacher of King's and our own day, Gardner C. Taylor. Like Ray, Taylor was an old family friend with a power base in the North, the huge Brooklyn parish Concord Baptist Church. Taylor's high baritone possessed all the power of Daddy King's but without the shriek. His emotional range exceeded that of Borders but without the histrionics. Both Taylor and King Jr. achieved profound emotional contact through natural timbre; both could soar in a disciplined tremolo, both could use their resonators to toy with sounds, and both could "gravel" like Louis Armstrong. But Taylor's high is purer, his low more richly resonant, and the mastery of his vocal instrument more complete than King's. Taylor can elicit applause from an audience by reading the text. On one occasion as he read some of the proper names in Luke 3 (Tiberius, Ituraea, Trachonitis), members of the congregation began responding, "My Lord, My Lord!"

Taylor and King Jr. were trained in the same rhetorical tricks—the ponderous beginning, the vowels distended three times their normal length, the affected stutter—and both were adroit at manipulating inherited commonplaces. Taylor's allusions, however, are more organic to his sermons, his metaphors more original and intellectually satisfying, and his powers of biblical reportrayal far more vivid than young King's. Like Benjamin Mays, Taylor infuses his sermons with principles drawn from the liberal view of human nature and history. But unlike Mays (and King Jr.), Taylor holds to an explicitly evangelical doctrine of salvation centered in the substitutionary atonement of Christ. When King reverted to evangelical formulas, one senses that they were just that for him, traditional formulas rather than, as in Taylor's view, assertions of an objective fact.

Taylor's influence on King was pervasive but difficult to document. Benjamin Mays's phrases turn up in King's sermons, but by most accounts Taylor was his model of the ideal preacher. His was the grand style to which King and many preachers of his generation aspired but never quite attained. If such style can be acquired only through drill, as the Greek rhetoricians had taught, Taylor nevertheless managed to pull it off without a hint of practiced art. If, as Aristotle said, the great speaker must appear to be a good and competent man, Taylor seemed to have forsaken appearances for the Good itself. Taylor was one of the few who could generate passion while retaining his composure. To King he was an example of John Chrysostom's observation: he is a rare preacher who can move the masses without losing his soul.

IV

When King entered Crozer Seminary in the autumn of 1948, he began in earnest his study of theology and preaching. Auburn Avenue, Morehouse, and, most of all, Ebenezer had provided him a rich preparation for the ministry; for many young Baptist preachers the apprenticeship young King had served through his nineteenth birthday would have sufficed for a lifetime. But King had only begun.

As a young Negro in an overwhelmingly white school, King's education would follow two predictable tracks. On the first, he would become acquainted with the tradition and vocabulary of Western theology. He would internalize the perennial antinomies of grace and nature, providence and fate, being and nonbeing, and liberty and bondage, that since Augustine have provided the scaffolding of theological discourse. His main focus would not be social philosophy, as he later intimated that it had been, but systematic theology with an enormous concentration of courses on the person and work of Jesus Christ. He would submit comprehensive papers on theological subjects written with the studied unoriginality that is frequently expected of novices in any professional discipline. Young King would learn his lessons well, so well that he would intellectually engage the claims of the Western Christian tradition and graduate first in his class, the only student to be granted honors in his comprehensive examinations.

The second track of King's education is familiar to African Americans who study at white liberal seminaries like Crozer. Black students quickly discover that they are not being prepared for ministry in the world of the black church. Liberal Protestantism stresses the importance of democracy in church organization and warns against centralization of power in the hands of the pastor. It favors the priesthood of all believers over various forms of hierarchicalism in the church and therefore fails to appreciate the strong, charismatic preacher who is the cornerstone of the African-American congregation. Liberal Protestantism's embrace of higher criticism in the early twentieth century transformed Christianity into a set of universal principles—brotherhood, disinterested love, the sacredness of personality, and the like. What was lost to liberalism in this move was the *story* and its celebration in worship that had sustained the Negro church for generations. These discrepancies between the principles of liberalism and the needs of the black church led King and the other Negro students to seek another option. In the dingy, blue-collar city of Chester, Pennsylvania, the second track of their education was apprenticeship to the most learned and influential Negro preacher in town, J. Pius Barbour.

Of the first track much is known. King's move to Crozer and later to Boston University, both self-consciously liberal institutions, appears to have signaled a repudiation of his own churchly heritage. The first-year seminarian did as a matter of fact let slip his embarrassment at the "fundamentalism" he had learned at Ebenezer, but this unripe theological critique in no way justifies the standard account of King's theological development that altogether omits his rootage in the black church, his heritage from the Sustainers and Reformers, and his apprenticeship to those who taught him to preach the Word. If it is important that Martin Luther King, Jr. be displayed primarily as a thinker, a shaper of ideas, then the interpreter will be drawn like a mouse to cheese to the ideas, intellectual figures, and graduate school professors that shaped him. King's academic career invites such an investigation; he even encouraged it by publishing a brief account of his intellectual development entitled "Pilgrimage to Nonviolence," in which he traces his steps (and the steps of so many of his generation) from Hegel and Marx to Rauschenbusch's social gospel, through the obligatory encounter with Reinhold Niebuhr's Christian realism—inspired all the way by Gandhi, Jesus, and Thoreau—

to his final conviction that people are meant to be free and to live in harmony with one another.

Given the nature of the schools King attended, the itinerary and its conclusion were predictable. Indeed, it was King's ability to articulate the terms of the mainstream intellectual agenda of the West and to place himself squarely in its midst that afforded him instant credibility with the white liberal establishment in America. The liberals understood the chain of ideas that culminated in King's concerns; it was their thing too. They too had pondered Freud, dallied with Marx, converted to Niebuhr. They too had avoided positivism and scientism by reverting to the infinite value of "personality." Like the great nineteenth-century liberals Adolf von Harnack and Albrecht Ritschl, they too considered the religion of Jesus to be an illustration of a more universal law of ethics. Liberalism's chain of ideas and influences was whole and unseverable. It would not admit of the kind of interferences that King had experienced on Auburn Avenue or in the blazing sanctuary of Ebenezer.

In seminary and graduate school King internalized the vocabulary and values of theological liberalism; he did not become a liberal but embraced a new language with which to rationalize his more original religious instincts. No matter how many times he repeated the liberal platitudes about the laws of human nature, morality, and history, King could not *be* a liberal because liberalism's Enlightenment vision of the harmony of humanity, nature, and God skips a step that is essential to the development of black identity. It has little experience of the evil and suffering borne by enslaved and segregated people in America. Liberalism is ignorant—even innocent—of matters African-American children understand before their seventh birthday.

King had grown up in a viciously segregated world whose evils, he later came to believe, can be transformed only through conflict and suffering. He represented a race that had collectively bypassed the Enlightenment and that consequently knew nothing of the ideal of individual autonomy but a great deal about the freedom of a people delivered at the Red Sea and redeemed by the blood of Jesus. Like Gandhi, King was undoubtedly supported and encouraged by Western ethical thinkers from Plato to Thoreau, but, again like Gandhi, he was not decisively shaped by any of them. "The more we study Mahatma Gandhi's own life and teaching," writes Charles Andrews, "the more certain it becomes that Hindu Religion has been the greatest of all influences in shaping his ideas and actions." Substitute "black-church gospel" for Hinduism, and the same observation holds true for Martin Luther King, Jr.

Despite their different worldviews, liberalism and the black gospel tradition share a number of key assumptions about God and the world. At Crozer Seminary, a Baptist school, King began to make some connections. There he was introduced to the greatest American Baptist of them all, Walter Rauschenbusch, who fifty years earlier had merged the heat of evangelicalism and the rationality of socialism into the Social Gospel movement. King is usually compared to Rauschenbusch on the basis of the latter's this-worldly understanding of the Kingdom of God ("the millennial hope is the social hope") and his representation of the Prophets as moral reformers. King did in fact accept both characterizations from Rauschenbusch with no questions asked. But even more germane to the Civil Rights Movement was the view of Jesus that Rauschenbusch passed on to King.

Liberalism was more interested in the human characteristics of Jesus than the traditional dogmas of Christ. Popular nineteenth-century preachers like Henry Ward Beecher and Phillips Brooks promoted the social relevance of Christianity by brilliantly bringing to life the personality of Jesus and showing how his love ethic could be applied to personal and social relationships. In liberal thought Jesus accomplishes God's will by exercising the moral influence of his person on society. In *Christianity and the Social Crisis*, Rauschenbusch wrote:

> Jesus, like all the prophets and like all his spiritually minded country-men, lived in the hope of a great transformation of the national, social, and religious life about him. He shared the substance of that hope with his people, but by his profounder insight and his loftier faith he elevated and transformed the common hope. He rejected all violent means and thereby transferred the inevitable conflict from the field of battle to the antagonism of mind against mind, and of heart against lack of heart. By his moral influence and superior consciousness of God—not his divinity as such—Jesus becomes society's teacher. Over the centuries he assimilates to himself like-minded people who infiltrate the culture and organically join themselves to it in order to change it.

In the early years of his civil rights career, it was the moral-influence theory of change that lay at the bottom of King's famous attempt to shame the conscience of the oppressor. But King's knowledge of Rauschenbusch was tempered by his experience of another Jesus, not the moral example of liberalism but Richard Allen's and Daniel Coker's Suffering Servant whose death is a redemptive event in which the believer participates by obedience. In the beginning, King spoke of the necessity of accepting suffering as a tactic for shaming the opposition, but as he was drawn into the vortex of the Movement, the moral-influence theory reverted to something more real and terrible, something that Gandhi or the Social Gospelers did not divine, namely, the necessity of conforming one's own suffering to the twisted agony of the crucified Christ.

What Rauschenbusch and the black tradition had in common was a willingness to *use* Jesus to get things done. During the Protestant Reformation Luther had encouraged Christians to "use" Christ in order to partake of his spiritual benefits. It remained for later generations of Christians, whether slave preachers or Social Gospelers, to translate the religion of Jesus into survival tactics on the plantation or social principles in a newly urbanized and industrialized nation. It is no mystery why Rauschenbusch rang a bell with young King. Had not his father and other progressives like William Holmes Borders translated Jesus into jobs and other political improvements on Auburn Avenue? Had they not also heard the demand of the Kingdom in the cries of the poor? For discouraged blacks and hopeful white liberals like Rauschenbusch, Jesus was an example to follow and, especially for blacks, a refuge and very present help in trouble.

The liberalism King learned from his favorite teacher at Crozer, George Washington Davis, which Davis called "evangelical liberalism," stressed the unity of all truth, including that of revealed religion and personal experience. At Crozer King learned that universal principles, acceptable to *all* people of goodwill, underlay the doctrines of the church. Religion cannot be removed from the so-called secular world because the same life animates them both. "[R]eligion has been real to me and closely knitted to life," the seminarian wrote in a paper. "In fact the two cannot be separated; religion for me is life." To which his professor could only add, "EXCELLENT."

All other, particularist approaches to revelation young King tended to lump with "fundamentalism." Hence Crozer's (and King's) distaste for the Swiss theologian Karl Barth's theology of the Word of God. Barth's utter reliance on the Word and his dark pessimism with regard to all human effort contradicted not only everything King was learning about liberal theology but also everything he had banked from Ebenezer. Barth was too imperious in his rejection of progressive notions to suit Crozer and too remote from race improvement to help the black church. The young seminarian dismissed Barth's critique of human morality as "preposterous."

For liberals the stumbling block of Christianity had long been its particularism, its embarrassing insistence on the supremacy of one God, one church, and one revelation. Nineteenth-century thinkers strained out the offensive particularities and presented a religious Esperanto whose basic vocabulary and tenets were compatible with the beliefs and prac-

tices of reasonable people everywhere. "But when these literal interpretations are removed," King later argued in his doctoral dissertation, "Christian doctrines are found to have a symbolic value that is indispensable
for living religion."

Liberal theology and black Christianity agreed on the unity of truth
but disagreed on where to find it. Liberalism honored such Christian
values as love and personality for their alleged conformity to the laws of
the universe, which the famous preacher Harry Emerson Fosdick cheerfully characterized as "friendly, purposeful, personal, and good." The
African-American church found its version of the unity of truth elsewhere:
not in universal principles or propositions but in the entirety of its life as
the people of God. It celebrated God's deliverance in the sanctuary and
tried to keep it alive on Auburn Avenue.

Despite their different expressions of faith, both traditions were eager
to relate God to the secular world. Liberalism was more interested in transforming the whole society into the Kingdom of God than in preserving
the church as a distinctive community of witnesses. It therefore welcomed
scientific advances (King peppered his early sermons with pop psychology) and interpreted the emergent social sciences as further signs of
humankind's progress toward self-knowledge.

The black church also rejected the distinction between the sacred
and the secular, though the roots of its experience of unity lay not in
Enlightenment Europe but in Africa and in the daily necessity of using
God to advance the race. Liberalism's insistence on the unity of knowledge led to the gradual secularization of previously sacred enclaves of
belief and action. Although Fosdick's famous sermon was entitled "The
Sacred and the Secular Are Inseparable," his work was more an ode to
modernity than a witness to sacred truth. The black church would ratify
Fosdick's thesis about the wholeness of human life, but it did not celebrate the unity of experience by extinguishing every burning bush. The
black church instictively resisted the acids of modernity, for its preachers knew that one does not move oppressed people to courage or hope
by depriving them of the symbols that have sustained them. While mod-
ernist theology was busy secularizing the world, clearing it of mystery
the way forested land is stripped for paving, the black preachers were
even busier sacralizing the godforsaken places in the lives of their people.
King's sermonic rhetoric would later transmute shabby little towns like
Selma, Alabama, and Philadelphia, Mississippi, into holy ground fit for
the revelation of God. The boldness with which he hurled God into the
teeth of raging social problems was less a stratagem than an instinct
granted by the black church's long battle for survival.

Finally, liberalism and the black gospel tradition were optimistic about the human future but on entirely different grounds. Liberals viewed history as a continuum that could not be thrown askew by divine interventions because history was itself governed by divinely sanctioned laws. The primary source of these laws, however, was not the particular message of any religion but a more general revelation manifest not in burning bushes or inspired prophets but in the noblest ideals of thinking people everywhere (the West). The so-called laws of history, by which liberals really meant to indicate their moral prescriptions for the future, were nothing but the tentacles of human nature. What you see in the very best of human nature is what you will get, for the future is a predictable extension of the present. King had been raised on hope, though in his seminary years he referred to it as "optimism" and attributed it not to the black church's profound trust in the providence of God but to his own psychological makeup. In one of his student term papers he wrote, "Also my liberal leaning may root back to the great imprit [sic] that many liberal theologians have left upon me and to my ever present desire to be optimistic about human nature."

The black experience in America left no room for optimism about human nature, but the Negro church did coin its own "laws" of history, the most popular of which was Saint Paul's "God is not mocked, for whatsoever a man soweth, that shall he also reap." In its worship the African-American church anticipates the final vindication of this law, which is nothing less than the vindication of God, by means of music and celebration. The black church's estimate of human nature, change, and the future is not gradualist like the liberal belief in progress but catastrophic or, to use a theological word, eschatological. The church does not imagine its own future by extrapolating from human ideals but by relying on God's intervention. King knew both views and unconsciously superimposed one upon the other until toward the end of his career the liberal optimism was blown away, exposing once again the bedrock of black eschatology.

The liberal creed was codified in the philosophy of personalism, most notably in the Boston Personalism of its founder, Borden Parker Bowne, and his successors, Edgar Sheffield Brightman and his disciple L. Harold DeWolf. King was introduced to personalism by his favorite teacher at Crozer, George Washington Davis. He later elected to do his Ph.D. in theology at Boston University (after acceptance by Edinburgh and rejection by Yale) in order to study under Brightman, whose *Nature and Values* and *Moral Laws* had become Bible to King. When Brightman died during King's first year at Boston, he continued his studies under DeWolf, whose work provided a convenient if not overly imaginative synthesis of liberal theology.

Boston University in the late 1940s and the 1950s was a center of liberal theological thought. The oldest and most liberal Methodist theological school, Boston opened its doors to Negro graduate students in impressive numbers, among them future leaders and scholars such as Sam Proctor, Major J. Jones, C. Eric Lincoln, and Evans Crawford. Brightman had a reputation for being a fussbudget whose knowledge was indexed on the hundreds of three-by-five-inch cards he used in his lectures. DeWolf was a more engaging lecturer and open to dialogue with the students. In his course Philosophy of Personalism, DeWolf and King occasionally became embroiled in long dialogues while the rest of the class looked on. King was the leader of the Negro students and one of the brightest of all the graduate students in religion. He is remembered by one of his classmates as a self-assured and smooth young man, though far from flamboyant, who immersed himself in his academic program and, like a typical graduate student, was never happier than when earnestly arguing ideas. In the circle of graduate students that met regularly in his apartment on Massachusetts Avenue, the Negroes endlessly debated the pros and cons of doing race-related dissertations. In the fifteen years from 1942 to 1957 only five Boston students completed doctoral dissertations on race-related topics. King was not among them.

Harold DeWolf made a large and lasting impression on King. In his systematic theology course, King laboriously copied into his notebook DeWolf's definitions of religion: "The quintessence of religion is devotion to supreme ideals. Also devotion to a personal deity." Unlike the earlier liberalism of the Social Gospel, which maintained an evangelical approach to God and Christ, Boston Personalism held that God is the ideal personality. Under the heading "What we have learned in this course" King wrote of Jesus: "The revelation of God in Christ is not dissimilar to the revelation of God in other men but in Christ the revelation of God reaches its peak."

Martin Luther King, Jr. was a fourth-generation Boston Personalist in the line of the liberal "fathers" Bowne, Brightman, and DeWolf. The line actually reaches back to the great German liberals of the nineteenth century, Albrecht Ritschl, Adolf von Harnack, and Friedrich Schleiermacher, who, respectively, rendered Jesus a moralist, an idealist, and a romantic, intuitive genius. Behind these lay the immense authority of Kant, whose theory of phenomena decisively segregated knowledge of the supernatural from scientific inquiry. In America, the personalists answered the challenges of Darwinian evolutionism and modern society by locating moral value exclusively in human personality. The essential human being or the "fundamentum," they insisted, was not subject to the laws and vicissitudes of nature. The human being occupied a privileged realm of spirit that scientific naturalism could not touch.

Personalism also salvaged a special place for Jesus, about whom King had written in his class notes, "Although we in the liberal camp have difficulty with the pre-existent idea and supernatural generation idea, we must somehow come to some view of the divinity of Jesus." According to the Christology King was taught at Boston, Jesus does not incarnate God in the orthodox Christian sense but represents the best thinking about God the world has known to date. King's notes continue, "It was the warmness of his devotion to God and the intimatecy [sic] of his trust in God that accounted for his divinity. (There is a release of inner resources of power when one comes into association with Jesus. In this sense Jesus saves.)" Ironically, the personalists reacted to the claims of modern science by codifying their beliefs in a pseudoscientific series of "Laws," such as Brightman's Law of Altruism, the Law of the Ideal of Personality ("[I]f no person is benefited, then there is no moral value"), and many others, traces of which remained in King's sermons like radioactive waste long after he had given up the whole liberal project.

Whether King consciously reflected on the distance between the personalism he learned at Boston and the practice of Christianity as he had known it in the black church, we don't know. Formally, at least, personalism said all the right things about morality and the sovereignty of the individual and therefore appeared to offer just the intellectual foundation for Negro progress that King had been seeking. The personalists argued that humanity is essentially free to make ethical choices that transcend the constraints of the situation. This is possible only because at the core of each person is the inextinguishable flame of spirit. One hundred years before King put on his doctoral hood, his slave forebears believed the very same thing about the "little me" who lives in each of us untouched by the oppressive powers around it. The graduate student intuitively made the connection between the values of personalism and the black church's proclamation of "somebodyness" when he wrote in his notebook, "We cannot [doubt] the fact that Jesus considered personality the supreme thing in the universe. Jesus considered everybody somebody." The mature preacher King obliquely echoed the "little me" theme when he referred to the real but immaterial quality of his own personality. In a 1959 sermon he said, "You look at me and you think you see Martin Luther King. You do not see Martin Luther King; you see my body, but, you must understand, my body can't think, my body can't reason.

You don't see the 'me' that makes me *me*. You can never see my personality." The philosophy of personalism guaranteed the reality of a human
essence unvanquished by suffering and oppression, but it also unwittingly
sponsored a new gnosticism: not even the worst sins of the flesh could
stain the essential spirit.

That personalism was devoid of the soul of religion, which is mystery, and could nurture no common life or activate no shared memory of
suffering among its adherents did not trouble King in those days. The
academician in him might well have preferred the philosophical projection that made God an ideal personality to the slave preacher's anthropomorphic projection that made him "de Big Massa." Having grown up
with images of a God whose nostrils flare in anger against his enemies,
young King experienced the God of personalism as a liberating relief from
the religion of his father.

The thinker most credited with ending the reign of liberalism in the
United States was Reinhold Niebuhr, whose *Moral Man and Immoral
Society* had burst upon the scene sixteen years before King entered Crozer
Seminary. No student of the "doctrine of man" in that period could ascend to the liberal heaven or sink to the Barthian hell without passing
through the purgatory of Niebuhr's Christian realism. Niebuhr taught his
generation to question the optimistic estimate of human nature without
giving up the struggle for justice or giving in to cynicism or despair. Like
the liberals, he spoke of personality, but because of the sinfulness in-
grained in all social and political systems, he effectively restricted the perfection of persons to a dimension beyond society. His critique of Gandhian
types of moral perfection revealed the subtle forms of coercion inherent
in even the purest exercises of nonviolent resistance. Having begun with
Rauschenbusch, for whom the Kingdom was a quest for the Holy Grail,
King was "confused" by Niebuhr's embrace of mixed motives and morally tainted methods for bringing about if not the Kingdom at least a more
just society.

Niebuhr's social realism is clearly illustrated in one of the most prophetic passages of the twentieth century, with which he concluded *Moral
Man and Immoral Society*. Writing in 1932, Niebuhr predicted the emancipation of the Negro race through nonviolent direct action. He reminded
his readers (young King among them) that the oppressor race would not
relinquish its power from a heightened moral sense or as a result of violent revolution, the latter posing grave dangers to an outnumbered mi-
nority. He continued, "The technique of non-violence will not eliminate
all these perils. But it will reduce them. It will, if persisted in with the
same patience and discipline attained by Mr. Gandhi and his followers,

achieve a degree of justice which neither pure moral suasion nor violence could gain."

King was chastened by his encounter with Niebuhr but he never converted. He criticized Niebuhr's system for scoffing at the very thing he himself would die trying to accomplish, namely, the injection of Christian love into the social and political process. The graduate student's critique was simple: Niebuhr's brand of Christian realism made no provision for spiritual development on any but an individual level. Niebuhr was too ready to consign the social order to the hell of power politics and piecemeal engineering. Young King was loath to relinguish a vision of the whole, what the Social Gospelers had called "the beloved community" on earth. King's eventual break with liberalism had less to do with the influence of Reinhold Niebuhr than it did with his disgust at liberalism's paternalism in the Civil Rights Movement and its failure of nerve on Vietnam. By the end of his life King was disillusioned not only with theological and political liberalism but with liberals who betrayed him in his hour of greatest need.

Profound changes in the graduate student's thinking cannot be attributed to Niebuhr despite the mature King's need to make it appear that Niebuhr had once made a decisive difference. Such was the dominance of Niebuhr: one was virtually obligated to retroject Niebuhr into one's intellectual formation and stake out a position in relation to his, which is precisely what King did in his brief 1958 sketch, "Pilgrimage to Nonviolence," in which he credits Niebuhr for dampening his "superficial optimism concerning human nature." King's comments about Niebuhr were a part of his larger strategy to magnify the sociopolitical side of his training at the expense of the more parochial concerns of theology.

As it turned out, Niebuhr's most enduring influence on King was more rhetorical than theological. Niebuhrian realism provided the rhetorical counterbalance to the assertion of every great hope. At Boston King had studied the philosopher Hegel in great detail, but the only piece of Hegel he ever displayed in public was his thesis-antithesis-synthesis scheme known to most beginners. The Hegelian triad not only was congenial to King's political goal to clarify both sides of an issue in order to reach a resolution but also activated his personal need to avoid frontal conflict and to please others if at all possible. In Niebuhr King found the dark side of every issue, the antithesis that warned, "Beware of human treachery," which lent moral complexity to his natural bent toward optimism.

The weight of Niebuhr (and Hegel) even contributed to the shape of King's sermons, which was not merely the three-points-and-a-poem

favored by so many preachers but the statement of a hope or goal, coun-
tered by the realities of personal or social sinfulness, completed by a
resolution that reaffirms the goal without denying the imperfection of
those who must attain it. Niebuhr's was a rhetorical contribution because
in King the opposing terms in the conflict usually amounted to little more
than superficial abstractions—straw positions. In a term paper he wrote,
"I have attempted to synthesize the best in liberal theology with the best
in neo-orthodox theology and come to an understanding of man." King
would apply the same scheme to every conceivable issue, including individualism versus collectivism = Kingdom of God; acquiescence versus
violence = nonviolent resistance; pessimism versus optimism = humanism. The ultimate equation, of course, was black versus white = not conflagration but, because of shared values, peaceable community. It doesn't
matter that the equations do not hold up under analysis or that they fail
to persuade historians or philosophers. They were never meant to be
analyzed. They are rhetorical devices that enabled King to speak and act
in situations of conflict. As the voice of the Movement, he was always
seeking a rationale for both confronting his opponents and accommo-
dating people of goodwill. For that, Niebuhr and Hegel held him in good
stead.

King's Crozer papers reveal a highly derivative style of thinking and
a pattern of citation that often transgresses the boundary between mere
unoriginality and outright plagiarism. Students at Crozer, like those of
many seminaries, were expected to master and summarize the positions
of important theologians, often with a minimum of critical reflection. King
mastered not only the theological options before him but also the conventional format for presenting them. His papers reproduce sentences
or groups of sentences that any professor would or should know did not
originate in the mind of a seminary student. He habitually cites important sources, lists them in a bibliography, but omits quotation marks and
otherwise underrepresents his reliance upon them. The same pattern persisted at Boston, where, according to the editors of his *Papers*, he extensively and skillfully plagiarized portions of his doctoral dissertation.

Why did King do it? The reasons may never be known, but it is important to differentiate between the carelessness of his Crozer papers and
the obviously more intentional plagiarism of his doctoral dissertation.
His work at Crozer is not untypical of the student who knows what the
professor wants and will regurgitate it for a grade. By the time he arrived
in Boston, however, he had logged seven years of higher education. He
knew the rules of footnoting and the seriousness of his infractions.

David Garrow suggests that young King was a southern-church black man who was essentially out of his element in an academic environment. He was "first and foremost a young dandy whose efforts to play the role of a worldly, sophisticated young philosopher were in good part a way of coping with an intellectual setting that was radically different from his own heritage. . . ." He reminds us that when Coretta Scott first met her future husband she was unimpressed by his "intellectual jive." The problem with this portrait is that it does not square with King's superior academic record at Crozer, including his top examination scores. Nor does it comport with the general intellectual requirements of a Ph.D. program that King satisfied. His friends at Boston do not remember him as one who was out of his element or "going through the motions" (as Garrow puts it) of white academic theology while harboring deep-seated alienation from his environment. They remember the leading Negro student who carefully selected his major professors, was immersed in his studies, and who had the intellectual capacity to carry on prolonged theological debates with his professor while the rest of the class respectfully watched.

Least compelling is the theory that in his plagiarism King was simply adhering to the standards of African-American communication, especially African-American preaching. It is one thing to assert that in the black tradition language is a shared commodity, which is true enough in the church, but it is quite another to translate that generalization into a rationale for academic falsification. Garrow writes,

> Any argument that King simply carried over from one context into a second the learning style he had acquired in the first, without appreciating or understanding that what he was doing was both academically inappropriate and ethically improper, is so unrespectful of both King's impressive intelligence and the top-notch undergraduate training to which he was exposed at Atlanta's Morehouse College as to be highly implausible.

Black scholars such as David Lewis also find incredible any "explanation" of King's citation habits that views them as a simple extension of African-American patterns of communication.

The dilemma cannot be resolved, but for the purposes of understanding King's preaching two points must be made. The first is that despite carelessness and lapses in academic honesty, King's immersion in academic theology was real and significant for his development as a preacher. Originality is a scarce commodity in a seminarian or a graduate student. Its absence does not invalidate King's exposure to many theological issues or cancel out the work he invested in many examinations at both

Crozer and Boston. By whatever methods it was acquired, King's theological education provided the vocabulary and conceptual framework of his sermons at Ebenezer and his larger message to the nation. He never wholly abandoned his personalist vocabulary, the Hegelian clichés, or his penchant for casting social conflict in academic theological categories. Although the originality of King's scholarship may be dismissed, the influence of his theological education upon his mode of expression should not be underestimated.

The second point is closely related to this observation. He approached all intellectual learning as raw material for the rhetoric of his sermons. Scholars have remarked upon the superficiality of his learning, some intimating that already as a student he harbored a premeditated design for self-falsification. In fact, what he harbored was the preacher's habit of reading all knowledge and experience in terms of its fecundity for moving people with language. The philosopher's body housed the instincts and the intellect of a rhetor. King could and did explain complex theological options to the satisfaction of his professors and examiners. But his greater gift was what an early biographer called the dramatization of ideas, the translation of abstract theological options, many of which were as dead as the "greats" who had uttered them, into the life-or-death urgency of a sermon. This is not to say that his rhetorical instincts *caused* him to sit loose to the rules of scholarship but only that whatever knowledge he acquired—regardless of the depth of its rootage—he *used* for rhetorical and homiletical purposes.

V

At the same time that King was learning the vocabulary of liberalism at Crozer, he was receiving this first classroom instruction in homiletics. He took a whopping nine courses in homiletics, most of them with Robert Keighton, a journeyman instructor who, judged by his later comments, does not appear to have appreciated the gifts and potential of his talented pupil. Keighton's approach to preaching was heavy with nineteenth-century poetry, saturated with sentiment, and driven by the liberal agenda for preaching articulated by Henry Ward Beecher fifty years earlier: that "of moving men from a lower to a higher life . . . of inspiring them toward a nobler manhood." On King's handwritten bibliography for his homiletics course, he placed a star beside Halford Luccock's popular *In the Minister's Workshop*. The aim of many of the texts, including Luccock's, was to locate preaching within a larger and nobler cultural enterprise.

The manuals of the 1930s and '40s were saturated with quotations from the cream of nineteenth-century Western belles lettres and seasoned with anecdotes from the lives of heroic Scottish preachers. The texts make for embarrassing reading today not only because of their distance from the church's earliest preaching but because in their tone of high-minded nobility they are oblivious to the social ferment and material misery that existed at the time of their writing.

Liberal preaching stood at the same distance from the Bible as liberal theology. It therefore favored topical preaching above textual, for the topic usually represented a universal truth abstracted from the thicket of the Bible's narratives and primitive eschatologies. According to Luccock, the sermon should begin with a "felt difficulty," which the preacher then would define, classify, and solve. Instead of following the contours of the biblical text, the topical sermon identified and organized an important idea. Keighton devoted an inordinate amount of time to sermon forms, such as those catalogued by Andrew Blackwood, William Sangster, Luccock, and the other leading teachers of the era. In Keighton's class the students learned techniques like "faceting," by which the preacher cuts the gem of an idea according to a preconceived pattern. Luccock and his generation refined the medieval art of dividing topics (almost universally into threes) and produced such patterns as the Roman Candle sermon (one subject with several predicates); the Chase or Question sermon ("Is it this or this, no, but this"); the Hegelian thesis-antithesis-synthesis sermon (to which King would become devoted); and the Classification sermon, which divides any topic into three parts. Most of King's sermons bear the marks of these or some other flagrant design scheme. For example, in his published sermon on the Good Samaritan, he first identifies the "theme" of that story as "Altruism" and proceeds to divide it into "Universal Altruism," "Dangerous Altruism," and "Excessive Altruism." In another of his published sermons, "The Answer to a Perplexing Question," the concept of which he borrowed from a sermon by Phillips Brooks, King follows the Chase form and Hegelian form. "How can evil be cast out?" he asks. He suggests two inadequate answers, namely, reliance on human power and do-nothing dependence upon God, before providing his own solution, which is a synthesis of the two rejected alternatives.

However dutifully he played the outline game, King never relied on a pattern of reasoning for the power of his preaching. Because of its intellectualization of the Bible, white homiletics stressed the rhetoric of *form* with which to develop religious ideas. The white liberal sermon's closest parallel was a well-organized, written essay that was spoken by the preacher. Black preaching, however, employed a different rhetoric, that of *sound—*

tonality, timbre, rhythm—which produced a musical drama rather than a spoken essay and invited participation rather than intellectual appreciation. King was schooled in both rhetorics, but in the actual delivery of his sermons he chose to heed the poet's warning: "I gave up fire for form till I was cold." The outline was a convenient if crude ladle for burning coals. It was a rationalistic setup for the power that would inevitably burst the form.

Even before he came to Crozer, King had been initiated into the books of sermons and sermon illustrations, such as the *Best Sermons* series, which he and his pals had been sharing like cigarettes on the sly. The annual editions of *Best Sermons* were the popular expressions of the religion and morality of the era. They also reflected the churches' comfortable relationship with American culture and politics (the 1955 edition was dedicated to Dwight D. Eisenhower). King would later dip into the sermon volumes of a prolific Methodist minister in Florida, Wallace Hamilton, and the Yale raconteur Luccock, as well as the black mystic and personalist, Howard Thurman, whose *Jesus and the Disinherited* King is said to have carried with him to the day he died. At Crozer he began to understand the liberal theology of Harry Emerson Fosdick, George Buttrick, and other pulpit greats whose sermons he had been preaching since he was eighteen. In addition to Fosdick and Buttrick, King devoured the nineteenth-century "prince of the pulpit," Phillips Brooks, from whom he borrowed the concept of his perennial "The Three Dimensions of a Complete Life," which he preached as his trial sermon at Dexter Avenue Baptist Church in Montgomery and every year thereafter for the rest of his life. Coretta Scott King remembers that she was glad to hear Martin preach "Three Dimensions" at London's Saint Paul's Cathedral on his way to the Nobel ceremony because he had been preaching it since his student days, and it was her favorite.

King did not learn to preach from Robert Keighton. The many forms he learned provided the scaffolding for the gifts he had acquired elsewhere. He appears to have used his preaching classes as cushions against the harsher demands of his academic courses. More important, Keighton's class provided a laboratory in which he could practice the expression of his newfound liberal theology. There he learned how to package conventional liberal sentiment in conventional rhetorical structures, all of which provided the young preacher with an intellectually viable antithesis to the thesis, which was the Frenzy of his own tradition. In his student sermons at Ebenezer in the late 1940s and early '50s, he worked out a synthesis between the rhetoric of form and the rhetoric of sound that would hold him in good stead throughout his life.

VI

At Crozer Seminary the second, black track of King's education in theology and homiletics was offered at "Barbour University," the name the Negro students gave to the parsonage of the most learned and influential black-church preacher in Chester, J. Pius Barbour. Barbour and his wife were family friends of the senior Kings, their friendship hailing back to their Morehouse and Spelman days in Atlanta. Now Barbour was in a position to help prepare Martin Jr. for the ministry and to keep an eye on him for Daddy King.

Barbour was a round-faced, full-orbed raconteur, whose rimless glasses were held monocle-style at the tip of his nose and whose Roman collar and five-thousand-volume library set him apart from the ordinary run of Baptist preachers in Chester. He was a lover of learning, a disputant in any matter relating to divinity and the human situation, and, above all, a teacher of anyone who was open to instruction. The first Negro graduate of Crozer, Barbour had gone on to complete a master's degree in theology in a joint program at Crozer and the University of Pennsylvania. His dissertation examined theories of religious knowledge in Ritschl, Troeltsch, Durkheim, James, Feuerbach, Sabatier, Freud, and Jung. It was most likely from Barbour that King learned that the roof will not fall in, even in a black church in a blue-collar town, if the preacher discusses economics, alludes to Freud, and otherwise explores the big ideas and influential thinkers in Western history. At the same time King was learning socialist theory in the classroom he was apprenticing under a pastor who was preaching it in his pulpit and applying it to the economic problems of his congregation. King's mature preference for socialism is as traceable to Calvary Baptist Church as it is to any of the academic institutions he attended.

Barbour was a "God man" who preached Jesus but also a "race man" who argued socialism as the means of reversing white economic exploitation of Negroes. His sermon notes read like a legal brief in which the preacher has sifted the philosophical, political, and economic questions in order to pinpoint the place at which the God issue has its greatest impact on the race. He believed that his people needed a leader who could master secular issues and at the same time draw from the well of Negro spirituality. At a Detroit rally in the late 1940s he called for a national Negro leader who was not a sports figure. While acknowledging widespread Negro distrust of the traditional Negro preacher, he nevertheless proposed the church as the natural context from which that leader would emerge. This is so, he said, because the church is the only Negro institution that is financially independent of whites. "Negroes pay my salary," he said, though he did not often admit how little.

Barbour was devoted to the church's ministry. He enjoyed people and would visit with parishioners for hours or hang around Greasy Phillips's diner in town to make conversation. But in forty-one years of ministry at Calvary he never made more than $6,000 a year, and this led him to acquire several paid political appointments in Chester. In Pius Barbour, young King had an opportunity to watch the black preacher operate in his historic role as an intermediary between the black and white communities. He communicated black problems to the city fathers, and they relayed their responses, sometimes in the form of jobs or paroles, through him to the black community. Like the slave preacher a century earlier, the black urban preacher was an effective go-between only so long as he was trusted by both sides. From Barbour, King learned the importance of trustworthiness in the work of communication between the races. At a national level he would play the same historic role and, through prophetic utterance and action, transcend it.

Theologically, Barbour was himself caught between the liberalism and Niebuhrian realism of his era. Many of his sermon titles, such as "Rising Above Circumstances" or "How Can You Get Strength When You Feel Yourself Going Down?" reflect popular liberal concerns. Although he criticized emotionalism in Negro religion, he frequently—and paradoxically—expressed liberalism's confidence in experience, as in this 1949 sermon: "My authority in religion is my experience with God." Like the personalists, Barbour insisted that the universe is upheld by moral laws and that Jesus represents the "moral ideal" for all generations, but these views were tempered by his realistic assessment of the immorality of racism and economic oppression. Like Niebuhr, he rejected nonviolence as a social philosophy and would later chide King for his "poor little ole' me" tactics that would lead to exercises in false nobility and needless suffering for Negroes.

King would later copy Barbour's uncanny ability to combine theological erudition with old-time religion. In a 1951 sermon Barbour preached a sophisticated analysis of the ways of knowing God, from the rational to the experiential and mystical, concluding with a series of "steps toward certainty" that included study, prayer, and meditation. A few Sundays later he took off his philosopher's gloves and lambasted the corruption of the black church—"Preachers with so many women . . . they have harems," "missionary sisters with each other's husbands," ushers visiting "juke joints"—and he perorated, "You have sinned. The Standard of Jesus is Absolute Moral Perfection. Absolutely clean. No dancing; no card playing; no movies; no romance. . . . Nothing but Absolutely Moral Cleanliness and Devotion to God." Perhaps it was this austere Jesus of "Absolute Moral Perfection" about whom Barbour once poignantly confessed (alone with a tape recorder), "Somehow I couldn't warm up to Christ."

Stylistically as well as theologically, there was much to learn from Pius Barbour. King not only acquired the art of philosophical generalization from Barbour (and Mays) but he also admired Barbour's skillful deployment of the thesis, antithesis, and synthesis in the organization of a sermon. In a 1948 sermon on the relation of science and religion, Barbour announces that he will "make a synthesis," just as John the Evangelist had done when he portrayed Jesus as the synthesis of the Greek logos and the Jewish personal God. King also followed Barbour's pattern of literary quotation and allusion. For example, in a November 1950 sermon, Barbour ticks off "Aeschylus, Sophocles, Euripides," names King would later repeat for their delicious sound value whenever he alluded to Greek culture. In the same sermon Barbour quotes the "Sound and Fury" soliloquy from Macbeth, adds a Schopenhauer quotation, and concludes with Ecclesiastes' "Vanity of Vanities" speech—in the exact order King would follow in later sermons. It was also from Barbour that King borrowed his jibe at "the paralysis of analysis" and other one-liners. Under Barbour's influence King's long-standing fascination with language continued at Crozer. The black preacher never quits playing with words because their beauty is integral to the message itself. On the inside cover of one of his class notebooks, when he should have been taking notes on Augustine or Aquinas, young King was doodling pretty sentences for his own delight: "We are experiencing cold and whistling winds of despair in a world sparked by turbulence."

Gardner Taylor may have been King's idol, but Pius Barbour was his teacher, his rhetor in the ancient sense, who subjected King and other Negro seminarians to his own system of homiletical drill. The seminarians often joked about his blustery self-importance and various affectations, some of which may have had to do with his exclusion from the white theological faculty of Crozer, and they sensed that the frustrated "professor" was, as one of them remembers, "living his life through us," but every Sunday they were glad to be in his presence. For King, who was not obliged to hold a weekend job, the drill would begin on Saturday when he came to the parsonage to eat at the Barbour table and to practice his sermon before a mirror in the parlor. He is remembered to have worked harder on pronunciation and memorization than the others. (Only the white students joked that he practiced reading the telephone book.) All the Negro students would return on Sunday for church, another sumptuous meal, and then, in the overstuffed comfort of the Barbour parlor, the "Doc" would conduct an intensive course in black homiletics.

Of all the preachers who passed through the Barbour parlor over the years, Martin was the most promising, save perhaps for Sam Proctor, who would eventually succeed Adam Clayton Powell at Abyssinian Baptist Church in Harlem. Barbour would painstakingly lead the group through each movement of that morning's sermon, pausing over transitions, phrasing, and imagery. He encouraged them to be logical in their delineation of ideas but imaginative and evangelical in their elaboration. In the matter of the climax, which is all-important in the black sermon, Barbour advised the young men to "tell what God's power has done in your life" and then in the inevitable moment of rhetorical abandonment to "use your God-given assets in the peroration." In his own sermon notes he reminded himself to do this by writing in the margins, "Paint!" or "Preach!"

Barbour had his own system for evaluating the students' sermons as well as his own. He reviewed their sermons under three headings: content, delivery, and audience reaction, the last category evoking the most spirited discussions. The science of "audience reaction" reinforced Aristotle's pragmatics of rhetoric: there is no truth unless an audience counts it true. Barbour pressed them to understand why a particular comment or metaphor in the sermon provoked a highly vocal response (or no response at all). What do these people of God want and need? What is true for them? Occasionally he would go so far as to telephone one of the deacons for his critique so that the seminarians could hear it for themselves. The entire congregation understood itself to be a laboratory for these bright young men, and occasionally a member would complain good-naturedly, "Oh, he's got those students preaching on us."

Barbour was a stern master rhetor for the seminarians but hardest of all on himself. After service every Sunday he wrote three grades on his own sermon outline and made additional evaluative remarks. The grades and comments are the touching chronicle of a laborer who was worthy of his hire, whose meticulous attention to the matter and craft of preaching is broadly reflective of the grand tradition of black preaching. "One joined," he wrote. "D." "No wit B- One joined." One day he wrote, "Content C, Delivery B, Audience Reaction A" and added with humility, "?"

Of Barbour, King once said, "He made the gospel live for me." A significant dimension of that gospel was Barbour's confidence in the inner resources God had given to the black church in America. He encouraged his students to identify those resources of intellect and spirit in themselves and to preach out of the strength that was within them. To that end he required each young preacher to name which of the Bible's animals they wished to be. He himself chose the eagle.

On the first track of his theological education King learned a new language for his inherited convictions about the dignity of persons and the glory of God. In his preaching classes he learned how to channel some of the old fire from Ebenezer and Auburn Avenue. As he began to notice the white students hanging around outside the preaching labs on the days he was assigned to preach, it dawned on him that he had been given the ability to speak the gospel in universally acceptable terms and, like some of the slave preachers, the gift of moving white audiences. Of all that he learned in seminary and graduate school, his most important gain was knowledge of the self and his own capacity to bend the will of others with the Word. On the second track of his education at "Barbour University" he not only learned how that Word works in the laboratory of the congregation but reclaimed a deeper sense of the viability and power of the old convictions themselves.

3

Dexter Avenue and "The Daybreak of Freedom"

THE old city of Montgomery, Alabama, is laid out like a Lionel Train model village. Its squares are bordered by two-storied antebellum residences, some of which have been converted to boutiques and dentists' offices, by Queen Anne doll houses that were exercises in nostalgia when they were built, and by granite and marble government buildings. The chief building is the Alabama State Capitol. Its great white dome is topped by three flags: the Stars and Stripes, the Alabama state flag, and the Confederate flag. A statue of Jefferson Davis stands sentry at the entrance. The trees along the squares are mostly magnolia and cypress; one hotel maintains a few palm trees. Montgomery lies on the Alabama River at the edge of a swamp that menacingly threatens to engulf the entire Lionel village in Spanish moss.

The renovations done to the old city are so perfect and explicit that they fail to convince: Montgomery is not a "New South" city and never was. In the 1950s the city's seventy thousand whites and fifty thousand blacks were locked in an uneasy social and economic embrace. Four years before the young Reverend Martin Luther King and his wife arrived, the median annual income for Montgomery's Negroes was $970. Sixty-three percent of the Negro women were domestics, and 48 percent of the men were laborers or domestics. Despite the enforced intimacy of the races, a rigid caste system, buttressed by dozens of local statutes, forbade blacks and whites to acknowledge the life they in fact held in common. A local statute went so far as to bar whites and blacks from playing cards, dice, checkers, or dominoes together. Restrooms and drinking fountains were clearly marked. By law, a white person and a Negro could not share a taxi. The segregation of restaurants and public transportation was carried out with a routine cruelty that left the black citizens of Montgomery, like those of most southern cities, humiliated and burning with resentment.

In 1952 a white woman in Montgomery accused a Negro teenager of raping her. His name was Jeremiah Reeves, and he was a drummer in the Negro high-school band. A white court found him guilty and, after five years of legal appeals and protests, he was executed. The Jeremiah Reeves story reminded Montgomery Negroes of the more famous Emmett Till case. Two Mississippi white men were acquitted of mutilating and murdering the black teenager, Emmett Till, for allegedly flirting with a white woman. The Reeves episode represented the predictable course of white justice in and around Montgomery. Whites were never brought to justice for sexual assault or violence against Negroes, but Negroes accused of crimes against whites were always prosecuted and sometimes lynched.

The entire city, from its magnolia-lined squares to its massive public buildings, churches, and monuments, was built on a carefully organized system of injustice. The system in turn rested on a bottomless marsh of hatreds and fears created by one of America's most original ideas: race. Like other cities in the South, Montgomery knew no religious commonality or human universality that could cancel or even mitigate considerations of race. When the bus movement proposed to do just that, the city found itself without moral or religious resources from which to act. Of Montgomery at middecade James Baldwin wrote, "I think that I have never been in a town so aimlessly hostile, so baffled and demoralized."

I

King arrived in Montgomery in 1954 to assume the pastorate of the most distinguished Negro church in the city, Dexter Avenue Baptist Church. Dexter was built during Reconstruction on the site of one of the city's four slave pens. As a black church, it therefore occupies an incongruously central location in the old city of Montgomery. It sits three-quarters of the way up Goat Hill from the main business district, a scant block below the front portico of the Capitol, where in 1861 it was proclaimed of Jefferson Davis, "The man and the hour are met" and where, less than one hundred years later, George Wallace challenged the forces of integration, "I draw the line in the dust. . . ."

Dexter is a dowdy, reddish-brick building with a peeling frame cu-
pola topped by a copper roof and a weathervane. It dwells in the midst
of power and intrigue, surrounded by the offices and agencies that should
have protected it, but it is out of harmony with its environment. It is not
a high-steepled monument that complements the Public Safety Building
next door or the massive Judicial Department Building across the street,
both of which agencies, when given the opportunity, tried to crush Dexter
and its pastor. During King's pastorate the Dexter congregation emerged
as a symbol of the true condition of the church in the world: its vulnerability to its powerful neighbors constituted its first witness against them.

Ebenezer had taught King that the basic unit of Christianity in the
world is the congregation. Although he had absorbed the universal principles of liberalism, when the time came for him to embark upon a career
he turned again to the congregation as the only vehicle of redemption he
knew. Perhaps he understood that Christianity was never meant to work
in the lecture hall or at the level of abstract principles but, rather, among
a community that is joined by race, family, neighborhood, and economics, but whose truest identity transcends all of these. Africans in America
learned quickly what the liberals and individualists never grasped: that there
is no strength in solitude. The power of Jesus is in the church. The congregation is the laboratory for the love commanded by God and the instrument of his justice. The black preacher knows that if it isn't happening
there, it isn't happening.

Martin Luther King, Jr. approached Dexter Avenue Baptist Church
as the first test of all that he had learned from the church and his mentors. Even before the Boycott of 1955–1956, Dexter had proved to be every
bit the challenge he was looking for. He was following a strong and controversial minister, Vernon Johns, who already in King's day had become
a legend in the Negro church. Johns was legendary for his erudition (it
was said he had *Thanatopsis* and the Book of Romans by heart) and his
eccentricity. The erudition he had acquired at Oberlin College in Ohio
and from his voracious study of the classics and nineteenth-century
poetry. Johns could be brilliant when parrying epigrammatic sayings or
occasionally bombastic when expositing the meaning of life: "Debris piles
on the faces of queens and kings, and seashells are left stranded on moun-
tain tops. Our health, our wealth, our friends, our ascendency go whirling away in the current of the years." It was his militant eccentricity,
however, that qualified Johns as an African-American original. As an educated man himself, he had the credentials to poke fun at the snobbery
and intellectual pretensions of his "class" church members, many of whom
were teachers in the local black state college. He not only insisted on

singing spirituals in the worship service, which was strictly prohibited in the staid atmosphere of Dexter, but wondered out loud if his members were ashamed of their heritage. When Dr. H. Councill Trenholm, president of Alabama State College, entered the church one Sunday as though he were a visiting head of state, Johns said from the pulpit, "I want to pause here in the service until Dr. Trenholm can get himself seated here on his semi-annual visit to the church." Trenholm never came back during Johns's tenure. The gambit for which Johns is best remembered by Dexter members was his practice of selling farm produce outside the church on Dexter Avenue—onions, potatoes, cabbage, watermelons—this to the embarrassment of his own people and the amusement of the neighboring white Methodists. Johns resigned in a huff no fewer than five times. The fifth time, the deacons and congregation firmly accepted.

Johns's eccentricity was his way of acting out his economic and social convictions about Negroes in America. He was a militant both in his provocation of his own people and in his defiance of the white oppressors in Montgomery. To his own congregation he never tired of preaching the distinctiveness of being black and the importance of hard work and economic self-reliance. Sitting on wagons of fresh vegetables and other farm produce was his way of preaching the gospel of black agricultural capitalism. One of his favorite expressions was "If every Negro in the U.S.A. dropped dead today, it would not affect significantly any important business operation." Toward the white power structure of his city he turned an angry prophetic face. After a lynching he once posted as his sermon topic on the bulletin board outside the church, "It's Safe to Murder Negroes in Alabama." Another Sunday morning topic: "When the Rapist Is White." His strategy of attacking the passivity of his own members and the racism of white society quickly got him into trouble with both groups.

King's biographer, Taylor Branch, has done much to retrieve Vernon Johns as a prototype of the militant black minister and an influential character in the Civil Rights Movement. There is no question that his brilliance and eccentricity were legendary in the network of important black churches and colleges on the East Coast. Wyatt Tee Walker, former King associate and now pastor of a large Baptist church in Harlem, remembers hearing many of the notable black preachers in his college chapel: H. Lawrence McNeil, Gardner Taylor, Thomas Kilgore, Benjamin Mays, J. Raymond Henderson, William Holmes Borders, and Vernon Johns, whom he "worshipped." Of Johns he adds, "I still preach his sermons." The influence of Johns's preaching on his young successor is more difficult to trace. King undoubtedly had some of Johns's material by the grapevine, the informal channels through which the good illustration or catchy phrase is transmitted from pulpit to pulpit. For a few pieces Johns and King probably share a common source, for example, Fosdick's allusion to the scientist Haeckel's question, "Is the Universe Friendly?" Because so few of the great black preachers' sermons were in printed form (Johns was the first Negro to publish a sermon in *Best Sermons*), younger preachers like King picked up what they could by ear, by attending conferences, visiting the important pulpits, and by culling the gems from dormitory bull sessions and other informal occasions of preacher talk.

Although Johns was undeniably militant, he never disciplined his militancy into a force for social change in Montgomery. The true forerunners of the Montgomery protest were the few who organized for direct action against the American apartheid. In the 1940s the newly formed Congress of Racial Equality was experimenting with nonviolence as a technique of occupying restaurants. James Farmer led sit-ins in the Jack Spratt Coffee House in Chicago as early as 1942. Two years before the Montgomery Boycott, the Reverend Theodore Jemison of Baton Rouge organized a brief but successful boycott of that city's buses. In Montgomery, those who laid the groundwork for the protest were Rosa Parks, a seamstress who before her arrest had attended a conference at the politically radical Highlander Folk School in Tennessee; E. D. Nixon, an activist in the Brotherhood of Sleeping Car Porters who was acknowledged to be the most militant Negro in Montgomery; and Jo Ann Robinson, an English professor at Alabama State College and the president of the Women's Political Council. Robinson and the council had been protesting city bus policies for years. Vernon Johns, too, stirred in these waters, but his knack for antagonizing his own members reduced his effectiveness as an agent for change in Montgomery.

The Dexter congregation was a "deacons' church" whose educated and independent-minded members could make life miserable for even the most legendary of pastoral leaders. Unlike Ebenezer Church in Atlanta, which had been governed by only three pastors in its sixty-eight-year history, Dexter had gone through nineteen pastors in seventy-seven years. The graduate student King arrived with less imposing credentials than many of his predecessors, among whom were numbered scholars, religious authors, and the president of a Negro college. Young King was installed on Reformation Day, October 31, 1954, in a service presided over by a layperson, the powerful Deacon T. H. Randall. When the new pastor gave the customary benediction at the end of the service, the congregation breathed a collective sigh of relief. At last it had found a polite young man who would attend to his religious duties without antagonizing his parishioners and the rest of the city.

Initially, King did not disappoint his congregation. He asserted his authority as the God-called leader of the congregation, but he did so well within the bounds of black Baptist expectations. The new pastor announced,

> When a minister is called to the pastorate of a church, the main presupposition is that he is vested with a degree of authority. The source of this authority is twofold. First of all, his authority originates with God. Inherent in the call itself is the presupposition that God directed that such a call be made. This fact makes it crystal clear that the pastor's authority is not merely humanly conferred, but divinely sanctioned. . . . Implied in the call is the unconditioned willingness of the people to accept the pastor's leadership. This means that the leadership never ascends from the pew to the pulpit, but it invariably descends from the pulpit to the pew.

Although he adds that the congregation need not "genuflect" before the pastor as if he possessed "some infallible or superhuman attributes," the message is clear. Power descends from the pulpit, as it did from his Daddy's and grandfather's pulpit at Ebenezer. With a few sure strokes, King established his authority to speak for God. To how many more he would speak and to what effect neither the preacher nor his flock imagined.

During his first year King confirmed his theological authority with a dazzling flurry of administrative and pastoral activities. He reorganized the congregation along the lines of Ebenezer into twelve birth clubs. The friendly competition among the clubs fostered increased giving and service in the congregation. The fellowship within the clubs cemented many friendships among the members. The clubs sold turkeys, held bazaars, stocked a new church library, and even sponsored a summer musical. King was particularly proud of his new Social Political Action Committee whose first project had been the recruitment of members for the NAACP. In a move that even Vernon Johns had not dared to make, the new pastor made registration to vote a prerequisite for membership in the congregation.

In the Dexter undercroft two old photographs tell the story of the halcyon days of that first year when young King was simply "Brother Pastor" to his people. In one he stands hip-deep in the baptismal pool in his special waders, about to perform a baptism by immersion. In the other, he stands smiling on the church steps, the honored authority figure in the obligatory group photo of the summer church school children. The pastor filled his days with the care and cultivation of his flock. He baptized and taught, visited the hospitals, sought new members, and admonished the backsliders. He was a systematic pastor with an eye for details. He called on the entire membership of the congregation in rough alphabetical order. He attended between five and ten parish meetings per week. Like many ministers, he was preoccupied with church attendance. When preaching in other churches, he was known to call home on Sunday night to get the figures. By all accounts, King was a successful shepherd of his flock, who, beyond his penchant for organization, appears to have been moved by a genuine and uncomplicated liking for the people he served. He specialized in what the classical tradition would have called "pastoral conversation." There was time to talk to everyone and to do so in a way that was edifying and encouraging. A former parishioner remembers, "He was always looking for an opportunity to commend people."

The commendations were traditionally given in the annual newsletter of the church. The letter contained a report from the previous year and the pastor's "Recommendations" for the coming year. King's first letter contained no fewer than thirty-four detailed recommendations, from building renovation (carpeting, "electric cold water fountain," painting) to a cultural committee to give encouragement to promising artists. He also established a unified church treasury that stripped the auxiliary organizations of their cash pots and centralized them in the hands of the church treasurer. Should the congregation adopt these reforms, the pastor added in a now-familiar rhetorical flourish, "success will be as inevitable as the rising sun." At the end of his first year, King could report the implementation of most of his recommendations plus the enrollment of thirty new members. The financial report revealed M. L. King, Jr. to be the single highest contributor at $335.25, which, along with Coretta's offerings, represented a tithe of their $4,410 annual income. The young pastor reported 87 pastoral visits, 49 sick visits, 12 baptisms, 5 marriages, and 5 funerals. To his academically oriented congregation he added 26 books read, 102 periodicals read, and one doctoral dissertation completed!

Despite his many accomplishments, King's pastorate was affected by tensions between him and the deacons of the church. His early need to be away for research and writing on his dissertation produced some irritation among the leadership of the congregation. When the deacons announced they would deduct from his salary the cost of hiring his pulpit replacements, he absolutely refused. During his first year Deacon Randall literally kept a "little black book" on the pastor, filling it with the complaints of members and observed shortcomings. As King's involvement in the Boycott deepened, he was forced to be away from his congregation more frequently. His annual reports in the Dexter *Echo* became annual apologies. In 1957 he confessed, "I am not doing anything well" and complains of "this almost impossible schedule under which I am forced to live." By 1959 he could report only eight new members. That year there was no message and no recommendations.

King's relationship to Dexter Church was marked by ambiguity. His parishioners loved and respected him deeply, yet the leaders were not willing to relinquish their traditional proprietorship of the church and its pastor. The congregation's support for its pastor is touchingly demonstrated by a collection of letters and verses from the members of his birth club, the January club, when he was facing trial in Montgomery. Dated March 19, 1956, the messages consist of Bible verses and inspirational poems; some, however, filled with foreboding:

> Know that love has chosen you,
> the hard-beset, the sorely tried,
> to live his difficult purposes
> Though hatreds rally,
> Know that love has chosen you
> And will not pamper you nor spare. . . .

When he returned from New York after his near-fatal stabbing, a delegation of deacons received him as a head of state at the Montgomery airport, and one of its number, Deacon Ralph Bryson, read the congregation's official welcome to its pastor.

King drew enormous strength from his parishioners who encouraged him, prayed for him, and guarded his parsonage after it was bombed. But in the end, the demands upon King became more than Dexter could bear. Despite the fame of its pastor, the church was not thriving. In response to King's absenteeism and his delegation of his duties to others, the power of the deacons reasserted itself, and the pastor found himself "under fire." King was encouraged either to cut back on his outside commitments or to leave Dexter. When his responsibilities in the Movement led him back to Atlanta and his father's church, he left a congregation both saddened and relieved by his departure. Many in the congregation would have agreed with his first biographer's optimistic if foreshortened assessment of their pastor's temporary career as a civil rights activist: "In time the preoccupation with racial integration will presumably ease and the Rev. Dr. King will be allowed to address himself to some of the other great problems of life."

II

In his leadership of worship at Dexter, King maintained the liberal and dignified traditions of the congregation while adding a few of his own special touches. Unlike most Baptist ministers (including his father), he practiced "open baptism," which meant that he did not insist on the rebaptism of converts from other denominations. When he did baptize, the ceremony was held at the beginning of the service. King and the candidate donned special garments and wading boots; the pulpit was slid to the side, revealing the pool underneath, and both pastor and candidate entered the pool for the traditional triune immersion.

For the Lord's Supper, which was celebrated on the first Sunday evening of every month, King incorporated a touch of drama to the liturgy. With the deacons around him in dark suits and white gloves (for the holiness of the ordinance), King would assume the role of Jesus as he narrated the account of the Last Supper. He would solemnly intone, "And He said unto them, 'One of you shall betray me,'" at which point the organ would cease its mood-setting background and the congregation would be plunged into a moment of awed silence before the pastor continued his recitation.

King's style of liturgical leadership was well suited to the staid formality of his congregation. During his first year in Montgomery, King presided in a church that was usually one-half to two-thirds filled. The word around black Montgomery on the young pastor was that he was a good but not great preacher. It wouldn't have mattered how good he was, for attendance was based on Dexter's reputation as an educated "class" church and not on the emotional appeal of the preaching or the music. Dexter did not have a gospel choir, and it rarely celebrated its heritage by singing a spiritual. Like many class churches, Dexter had assimilated the standardized features of the white Methodist ritual. Its liturgy included a *Gloria Patri* and organ preludes by Handel, Mendelssohn, and Bach. Dexter members found verbal response and talk-back inappropriate and annoying. They occasionally referred to "Mother _____," who was the "last person who shouted" in the worship service. King's former intern, J. T. Porter, remembers the atmosphere in the sanctuary: "silence, perfect, total silence." The deadness of the place had galled Vernon Johns. He accused the congregation of systematically suppressing its African heritage. One Sunday he blew up in the pulpit: "Why do you sit there like bronze Buddhas?"

King came to his first congregation with a cache of sermon manuscripts he had developed in his college and graduate school days. During his first year he worked very hard at producing and memorizing new manuscripts, which he pointedly left on his chair when he rose to enter the pulpit. He also brought with him a repertoire of poetic verses and longer set pieces already committed to memory and distributed throughout his body of sermons. Parishioners still remember the contrast between the assaultive rhetoric of Vernon Johns and the uplifting and encouraging messages of his successor. J. T. Porter goes so far as to characterize King's first year of preaching as "very positive [thinking], [Robert] Schuller type thing," by which Porter means to indicate a style of sermon that affirms the goodness and potential of human nature at the expense of confrontation with radical evil.

During his five years at Dexter, King established his canon of sermons. As his national responsibilities increased, his schedule did not permit him to prepare new messages every week. Modifications and developments in his thinking as well as changes in current events he simply integrated into the old familiar sermons, which he repeated again and again. In terms of form, outline, and thematic set pieces, the vessels remained remarkably constant. What gradually changed was the elaboration of the sermon heads, which took him in new directions, and the character of the climax, which even in the most identical sermons was never duplicable. Many of the titles he had employed as a graduate student and a young pastor he continued to use throughout his career. Others he modified in order to camouflage the essential sameness of the sermon's content. Reflecting on the sermons and speeches of King's maturity, J. T. Porter, who was also with him later in the Birmingham Movement, remembers, "Everything else was a spin-off of what I heard the first year."

His trial sermon at Dexter was "The Three Dimensions of a Complete Life," which he had been preaching during his student days at Boston. In an allusion to the then current craze for 3-D movies, he changed the title to "3-D in Religion." The sermon was a great success ("That did it," said a deacon), and King was elected pastor by acclamation. "3-D in Religion" articulated the broad themes of liberalism with which its educated audience would have been familiar. Just as the length, breadth, and height of the New Jerusalem are equal, so the "city of ideal humanity is not an unbalanced entity but it is complete on all sides." The complete life is the well-balanced life. The *length* of the New Jerusalem stands for the individual's cultivation of his or her greatest powers. Love of self is the sine qua non of the healthy life. The *breadth* of life is concern for the neighbor. The story of the Good Samaritan teaches the importance of "dangerous altruism" over against those who are in need. Love of neighbor entails the ability to project the "I" into the "Thou"—an allusion to the philosophy of Martin Buber. Finally, the *height* of the heavenly city represents humanity's need for God. As H. G. Wells argued, religion creates a life of balance and completeness. Therefore, King concludes, seek God and in him discover the well-rounded life.

"3-D in Religion" was also well received because it eschewed the confrontational smite 'em style of Vernon Johns. It contained no prophetic denunciations, no mention of race, no specific remedies, no challenges, and therefore no emotional peroration. It did not allude to the radical demands of the prophets or the radical grace of Jesus of Nazareth. Its young preacher had once wondered "whether religion, with its emotionalism in Negro churches, could be intellectually respectable as well as emotionally satisfying." He went on to confess, "I revolt against the emotionalism of Negro religion, the shouting and the stamping. I don't understand it and it embarrasses me." In short, "3-D in Religion" represented the balanced considerations of a sensible young scholar.

But there was another side of the candidate that was hidden from Dexter's eyes. We now have documentary evidence to confirm what King's old friends have been informally reporting for years: his trial sermon and indeed his staid Dexter Avenue style did not represent the "original" M. L. A few weeks after his trial sermon at Dexter and eighteen months before his oratorical breakthrough at the beginning of the Bus Boycott, King was invited to preach at Second Baptist Church of Detroit. There the young Ph.D. candidate reached back beyond Boston to Ebenezer and delivered an African-American gospel sermon entitled "Rediscovering Lost Values."

The audiotape of this sermon reveals elements of King's preaching style that he retained for the remainder of his life. Although prepared by a Ph.D. candidate, the sermon's exegesis naively relies on allegory and moralism. The story of Mary and Joseph's search for the boy Jesus, whom they find teaching in the Temple, yields *lessons* in "recovering lost values," the most important of which is love. More to the point, already in February 1954 King was preaching against "hate," as though he were already embroiled in the bus crisis and urging his people not to return the hate of their persecutors! It's *wrong* to hate, he says:

> It always has been wrong and it always will be wrong! [*Amen*] It's wrong in America, it's wrong in Germany, it's wrong in Russia, it's wrong in China! [*Lord help him*] It was wrong in two-thousand B.C. and it's wrong in nineteen-fifty-four A.D.! It always has been wrong, [*That's right*] and it always will be wrong! [*That's right*]

King continues to use the word *wrong* (twenty times in ninety seconds), but unlike his later sermons in which repetition lent power to his message, in this sermon it begins to fizzle out, prompting a member of the congregation to shout, "Go ahead," that is, move on.

The sermon also reveals King's lifelong habit of mixing learned allusions to the subjects of "reality," evolutionary "process," and the "moral universe" with memorized quotations from poets such as William Cullen Bryant and James Russell Lowell. How long King had had their verses by heart is impossible to say, but he was quoting the same poems fourteen years later in his last Sunday sermon at the National Cathedral in Washington.

The sermon also features an emotional dialogue with the congregation culminating in an old-fashioned gospel climax: "I'm not going to put my ultimate faith in the little gods that can be destroyed in an atomic age, [*Yes*] but the God who has been our help in ages past, [*Come on*] and our hope for years to come, [*All right*] and our shelter in the time of storm, [*Oh yes*] and our eternal home." In the climax the preacher proclaims the greatness of God. He does so by joining his voice to the chorus of those who have sung "Our God, Our Help in Ages Past." This is *our* song. It is *we* who have been sheltered in the time of storm. Moreover, the black preacher personalizes this gospel by allowing his faith to represent the faith of all ("I'm putting *my* ultimate faith . . .") The call-and-response pattern comes to the young graduate student from Africa by way of the brush arbors of the South, the clapboard meetin' houses, and Ebenezer Baptist Church. The congregation's participation makes the climax a joint celebration of God's mercy to this race and this congregation.

Clearly, the preacher is trying his wings to see what will and will not work on this audience. With what power will God be present this morning? How well is the preacher cooperating with the Holy Spirit (and the expectations of the people)? One can almost see the shadowy figure of Pius Barbour at the rear of the church taking in the whole performance and giving his apprentice a grade for content, delivery, and audience response.

During the summer and fall of 1955 Pastor King reverted to a more philosophical style of preaching. He delivered well-rounded statements on the meaning of life, such as "Discerning the Signs of History," "The Death of Evil upon the Seashore," and "The One-Sided Approach of the Good Samaritan." During the first year he rarely attacked the problem of racism in Montgomery, though he did encourage and finally require NAACP membership and voter registration. When the bus crisis broke in December of that year, he suddenly found a focus and a climax for his sermons. The abstractions give way to the demands of the struggle. *The* sign of history *par excellence* is liberation. The evil that must die upon the seashore is segregation. The Good Samaritan now teaches not merely love but a *dangerous* love between the races. Everything had changed.

His 1957 sermon "The Birth of a New Nation" marks a watershed in the development of his preaching. Although it contains references to liberal principles, the sermon's power comes from the skillful way in which the preacher joins the Montgomery situation to an African nation's quest for independence and merges both into the more primal longings of the American slave. King's journey to the independence celebration in Ghana in 1957 afforded him an opportunity to teach his congregation about colonialism and to apply the lessons of Ghana to the United States. In "The Birth of a New Nation" he combines a history of colonialism in Africa with a rambling account of his personal experiences at the celebration in Accra. When the new flag of Ghana went up, King remembers, "Before I knew, knew it, I started weeping. I was crying for joy . . . and I could hear that old Negro spiritual once more crying out,

> Free at last, free at last
> Great God almighty, I'm free at last!

In the sermon King narrates at length the story of Kwame Nkrumah, the first prime minister of Ghana. What impresses him most about Nkrumah is the moral authority he had earned by his imprisonment at the hands of the British. When Nkrumah addressed the nation on the eve of independence, King remembers, he wore his prison garments. King (and his opponents) would never forget the symbolic importance of going to jail.

"The Birth of a New Nation" is a significant example of the change that took place in King's preaching after the Boycott. The emergence of liberation movements in Ghana and other African nations as well as in the American South confirmed what King had always believed intellectually about freedom. What Brightman theorized in his Law of Autonomy King witnessed in the streets of Accra and Montgomery: "There seems to be a throbbing desire, there seems to be an internal desire for freedom within the soul of every man." More than that, the birth of a new nation corresponds to Israel's deliverance at the Red Sea. Both philosophy and biblical revelation have found their confirmation in the stirrings of freedom around the world. Ghana reminds us that "the oppressor never voluntarily gives freedom to the oppressed." If Nkrumah had believed that, the Gold Coast would still be a colony in the British Empire, and if Negroes in Montgomery believe that, they will remain in the back of the bus. The Law of Freedom does not imply inevitability, for "freedom only comes through persistent revolt, through persistent agitation." In this sermon the realism of Niebuhr has triumphed over the idealism of Brightman.

The spirit of Vernon Johns has forever vanquished the power of positive thinking. God has already delivered Israel and Ghana by breaking the backbone of their oppressors. Now "we find ourselves breaking aloose from an evil Egypt, trying to move through the wilderness towards the promised land of cultural integration." Will not our God also deliver the African peoples of America?

III

After the Boycott had commenced, King's Sunday morning sermons found a new purpose and vitality. The specificity of *race*, which he had assiduously avoided in his graduate education, now sharpened the point of his biblical interpretation and preaching. No one sermon captured the transformation that was taking place in him, but his first major rhetorical triumph, the address to the massed protesters at the Holt Street Baptist Church in Montgomery, left him changed utterly.

At the age of twenty-six, King had unexpectedly been elected president of the hastily formed Montgomery Improvement Association and had been chosen to give the Negro community's response to the arrest of Mrs. Rosa Parks. With little time to prepare his speech, King found himself driving toward the run-down section of Montgomery where the church was located. He was now well removed from the ambience of Dexter with its attractive neighborhood and dignified reserve. The barnlike Holt Church was filled with common people who were waiting for something to happen. Loudspeakers had been set up outside to accommodate the overflow. One-tenth of the Negro population of Montgomery was there, and those in the church and those on the lawn and in the street were singing *Onward Christian Soldiers*. King's performance would influence and perhaps determine the community's response to its leaders' call for a boycott.

King was poised to respond to this moment of urgency with all the intellectual and spiritual resources his history had to offer. In retrospect, the power of his address is heightened by the inscrutability of its origins. What the folk in Nazareth said of Jesus the preacher might have been asked of King: "Where did this man get all this?" Many of the images and phrases of the Holt Street address sound like permanent installations in the orator's vocabulary, as if the twenty-six-year-old had been saying these things for many years. Some of the phrasing may have been lifted from his sermons already preached, but most of the speech belonged to the moment. In his later accounts of the Holt Street address, King dwelt on his lack of preparation. In doing so, he was unconsciously borrowing the persona of the recently converted slave preacher for whom conversion (or in King's case *election*) almost always entailed prophecy. In account after account of the novice's first sermon, the slave preacher confesses to a total lack of preparation so as to give even greater glory to the Holy Spirit who opens "locked jaws." The now-historic success of the Holt Street address may be attributed to two factors: the substance of the speech itself and the joy with which the audience was prepared to receive it. King made the speech. History and the Holy Spirit prepared the joy.

In substance, the speech exhorts the community to support the Bus Boycott and to do so within the bounds of the law. Its appeal has a three-fold basis: in the guarantees of the nation's democratic traditions, in the integrity of the oppressed people themselves, and in the teachings of Jesus. King begins by addressing the crowd as "American citizens" who have "love for democracy." He insists throughout the speech that all the Negroes want is the rights they already possess under the Constitution, for "democracy transformed from thin paper to thick action is the greatest form of government on earth." To those who are afraid of where this protest might lead, he promises that no one will "defy the Constitution of this nation."

King's appeal to the integrity of the Negro people was more complex. In this his first public address, he implicitly honors a principle of communication that he never forsook, namely, the importance of character in the persuasion of an audience. "Mrs. Rosa Parks is a fine person. And since it had to happen, I'm happy it happened to a person like Mrs. Parks, for nobody can doubt the boundless outreach of her integrity. Nobody can doubt the height of her character, nobody can doubt the depth of her Christian commitment and devotion to the teaching of Jesus." He also appeals to the integrity of the entire Negro audience, which, like that of Mrs. Parks, is related to its faith in Jesus. "I want it to be known throughout Montgomery and throughout this nation that we are Christian people. . . . We believe in the teachings of Jesus." King's purpose in appealing to the integrity and Christianity of his audience is to assure his listeners that he will not ask them to do anything immoral or illegal: "There will be no crosses burned at any bus stops in Montgomery. There will be no white persons pulled out of their homes and taken out on some distant road and murdered. . . . We are not . . . advocating violence. We have overcome that." "*Repeat that! Repeat that!*" the crowd chorused back. In this first speech, his appeal to nonviolence is based on no theory other than the simple assumption that Christians will do no harm to those who have harmed them.

King's appeal to the teachings of Jesus combined Christian doctrine with the rhetorical idiom of the traditional Negro preacher. After telling the story of a good woman who is insulted after a long day's work, King skillfully assumes a persona and seizes a theme as old and as potent as the slave preacher's:

> You know my friends there comes a time when
>   people get *tired*
> Of being trampled over by the iron feet
>   of oppression. [Thundering applause]
> There comes a time my friends when people get
>   *tired*
> Of being flung across the abyss of humiliation
>   where they experience the bleakness of
>   nagging despair. [*Keep talking!*]
> There comes a time when people get *tired*
> Of being pushed out of the glittering sunlight
>   of life's July and left standing amidst
>   the piercing chill of an Alpine November. [Three minutes of tumult]

Although the theme of being *tired*, spoken with a combination of weariness and rage, is one that would have been known to many black preachers, King adroitly selects the theme of weariness (and impatience) in perfect harmony with the image of a hardworking seamstress whose feet are aching. In years to come, King's own fatigue would give the word *tired* a metaphysical ache that it does not possess in this speech. But here the twenty-six-year-old, who is not tired at all but at the daybreak of an unimagined adventure, manages to intuit the ache in others and to articulate it.

There is more to the speech, but at King's evocation of the race's collective weariness with the everyday humiliations of its existence, the audience takes over. At the conclusion of the set piece, the audience responds with three minutes of sheer joy in hearing its own truth. Out of the deafening applause, shouts of "*King! King!*" can be heard in the church. From this point on, the speech becomes the creature of the speaker *and* his audience.

Midway through the speech King summarizes the grounds on which the boycotters may claim righteousness for this protest. In doing so, he provokes a second eruption of joy. Both the style and content of this summary contain the seeds of all future appeals and arguments for freedom.

> If we are wrong, the Supreme Court of America is wrong.
>   If we are wrong, the Constitution of this nation is wrong.

If we are wrong, Almighty God is wrong.

If we are wrong, Jesus of Nazareth was merely a utopian dreamer and never came to earth.

We are determined to "fight until justice runs down like water and righteousness like a mighty stream."

With each sentence his voice soars into a higher register, reaching its topmost note on the second syllable of al-*might*-y and then descending into the pandemonium of the crowd's response. His summary is arranged in order of authority from the decision of a human court to the Incarnation of the Son of God. The piece ends with his favorite quotation from the prophet Amos, who twenty-eight centuries earlier had stood in the shrine of Bethel and denounced a government that was destroying poor farmers.

Later in his career when his and the Movement's place in history had been assured, King could speak effusively of the judgment of history on the Civil Rights Movement. But in this speech he invokes the meaning of "history" with a way of knowing, a vision, that can only be called prophetic. "We, the disinherited of this land, we who have been oppressed so long, are tired of going through the long night of captivity. And now we are reaching out for *the daybreak of freedom* and equality." The prophet must craft a metaphor to express some new thing that God is doing. With a prescience beyond his years or experience, he finds the right word, "daybreak," with which to capture the hope and fearfulness of the new spirit in Montgomery. The prophet sees it on behalf of the people, and, once again, the people respond with thunderous joy.

In the last quarter of the speech, King appears to recognize the forces he is awakening and attempts to quiet them. He reminds his audience that this is a Christian movement and that love is the pinnacle of Christianity. He shows that when the community works for justice, it is really doing love, for justice is love calculated and distributed. The point he is trying to make is that a calculated action like a boycott does not necessitate a violation of Christian love. Passivity is not a Christian virtue. The speech ends with an appeal for unity and one last promise of the favorable judgment of history upon this "race of people, of black people, . . . who had the moral courage to stand up for their rights." This was the place to end, but because he had not had an opportunity to plan the entire speech, King tacks on three filler sentences that produce an anticlimax and an awkward silence as the young minister concludes and abruptly turns away from the podium. Then the applause thunders forth one last time, and the whole church trembles with its newfound joy.

Members of Dexter who were present that evening were amazed at the transformation of their preacher. The "daybreak of freedom" had proved to be the debut of freedom's eloquent new spokesman. In a single spasm of growth he had shed the categories and confinements of philosophy and made himself one with the people. Here was an artifice more profound and subtle than the homiletician's playing with forms: the integration of civil guarantees and prophetic vision; the intuitive identification with the weariness and rage of the people; the sensitivity to the historical hour; and (not least) the technical skill in manipulating rhetorical formulas—these were the bloodline benefits of an education in the black church. Out of the matrix of Mother Ebenezer, the Sustainers and Reformers, and all his mentors and masters in the Word, King was prepared to become the voice of the Movement. In the Holt Street speech, the Movement was born.

II
PERFORMANCE

4

What He Received:
Units of Tradition

ONE evening in the early days of the Albany campaign, Martin Luther King and his young associate James Bevel visited Mount Olive Baptist Church in "Terrible" Terrell County, Georgia. Only one night earlier, Bevel had spoken at a voter registration meeting at the church. In the middle of the night the church had been torched, one of seven black-church burnings that occurred within a two-week period around Albany. Now King, Bevel, and other volunteers had returned to conduct a service in the charred remains of the sanctuary.

In the service a young woman, a college student and member of SNCC, led the prayers of the community. She spoke with the conviction King had come to expect from the SNCC activists, but also with an innocence and idealism peculiar to the young. The students often spoke of their dream for black people in America, and, as she prayed, the young woman began to intone her own vision of the future with the phrase "I have a dream." That evening the whole church, including its most distinguished visitor, swayed to the phrase, "I have a dream."

The metaphor of the dream is a staple of black religion, as old as the prophet Joel's vision,

> Your old men shall dream dreams,
> And your young men shall see visions,

and as commonplace as the untutored Negro preacher's evocation of a better life across the Jordan. According to one field investigator, an old preacher in Macon County ignited the church with his cry,

> And I *dream*, chill'un!

King himself had been developing the same image in a speech called "The American Dream" that he had been giving at least since 1961. By late 1962 the prototype of "I Have a Dream" had become a fixture in King's repertoire. In the early versions the preacher did little more than complete the phrase with a variety of Bible passages; by August 1963 he had fused the biblical image with the nation's most universal ideals.

There is no such thing as an "original" dream. A powerful instance of the dream was alive that night at Mount Olive, ecstatic as the Holy Spirit and palpable as the hot, damp flesh that kept time with the speaker. King had a genius for absorbing through his pores the power that was all around him in the church and translating it for wider audiences. He *borrowed* nothing from that evening but, as one black preacher says, he *overheard* much. Out of the Bible and the African-American church and the plaintive prayer of a young woman came "I Have a Dream."

I

Like all preachers black and white, Martin Luther King relied on what had been given him. "What do you have," the Apostle Paul asked his congregation, "that you have not received?" For the construction of his sermons, what King received was a body of titles, outlines, and formulas from other preachers. The outlines followed the conventional sermon schemes he had learned in the black church and from his seminary teachers. The formulas were what classical orators would have called *proofs* of the speaker's arguments. The proofs illustrate or substantiate the often unexceptional arguments with a sensual beauty that overshadows the logic of the ideas themselves. Together, the outlines and the proofs constitute what the classical tradition called the *topoi*, or "places," where a culture or religious tradition "stores" its nuggets of wisdom and its basic methods of telling the truth. The homiletical commonplaces are "sure things" in the speaker's arsenal. They can be counted on to produce results because they have been thoroughly field-tested by generations of preachers. King did not have a "canon" of sermons as such. For his day-to-day preaching and speech making he drew on a canon of titles, basic outlines, and thematic formulas, ranging from the epigram to the extended set piece, which he moved from speech to speech and worked into *any* address as the occasion demanded.

The outlines are nothing more than the scaffolding for convictions King had absorbed from sources deep within the African-American tradition. The arguments he used are as venerable as Richard Allen's defense of the integrity of all human beings, Daniel Coker's teachings on redemptive suffering, David Walker's indictment of Christian hypocrites, and even Nat Turner's terrible prophecy of deliverance. The Bible and the black church taught him that you reap what you sow. The sources of his opposition to oppression and violence were as ancient as the Hebrew prophets, who insisted that the poor deserve justice, and the Sermon on the Mount, which forbids hate and counsels peace. These were the convictions. The outlines provided a method of organizing the convictions and applying them to the problem of race in twentieth-century America. Aside from the historic reformers and his contemporaries like Benjamin Mays and Howard Thurman, not even the most liberal of King's sources seriously engaged the problem of race. Thus every homiletical outline King received he necessarily adapted to his own context and used in service of the one and only intellectual passion of his life, the argument for freedom.

The outlines give to the message the appearance of intellectual rigor, but in fact they represent the generic relationships that sustain human life: part to whole, species to genus, cause and effect, up and down, inward and outward, then and now, light and dark, beast and human, lost and found. Thus the Christian life requires us to look (1) inward at our sins, (2) outward toward the neighbor, and (3) upward in worship of God. The human creature is (1) neither pure spirit (like the angels) (2) nor wholly instinctual (like the animals) but (3) a God-created unity of the two, which entails a corresponding Christian witness that is neither purely spiritual nor wholly social but . . . And so on.

The practice of classification and Hegelian triads does not correspond to the narrative or proclamation forms in which the New Testament bears witness to God. In his later sermons, King relaxed his penchant for outlining in favor of a more biblical style of witness. When he used a formal outline, however, King, like most preachers, made it as general as possible for several reasons. Simplicity of outline makes it easy for the hearer to follow the sermon. The generic metaphors do not exclude hearers on the basis of intelligence or experience; they are universal. The more general the main headings, the easier it is for the preacher to include a variety of materials without having to rethink the direction of the whole argument. The outlines of King's sermons are "places" where formulas and more specialized arguments may be stored. The larger (or more intellectually neutral) the dump, the more ammunition may be stored there.

In the early 1950s young King began preaching one of these borrowed, generic outlines around which he fashioned a sermon entitled "The Three Dimensions of a Complete Life." Under a variety of titles, he preached the sermon as a college student, pastor, civil rights leader, and antiwar activist. He preached it in little black Baptist churchs, at baccalaureate services, in New England meetinghouses, at Saint John the Divine in New York, Grace Cathedral in San Francisco, and in many other settings. He published the sermon in two versions.

At no time did he acknowledge his source in Episcopal bishop Phillips Brooks's nineteenth-century masterpiece, "The Symmetry of Life." Brooks's first paragraph reads as follows:

> St. John in his great vision sees the mystic city, "the holy Jerusalem," descending out of heaven from God. It is the picture of glorified humanity, of humanity as it shall be when it is brought to its completeness by being thoroughly filled with God. And one of the glories of the city which he saw was its symmetry. Our cities, our developments and presentations of human life, are partial and one-sided. This city out of heaven was symmetrical. In all its three dimensions it was complete.

King's first paragraph of a 1959 version of the sermon follows Brooks closely:

> Many, many centuries ago, out on a lonely, obscure island called Patmos, a man by the name of John caught a vision of the new Jerusalem descending out of heaven from God. One of the greatest glories of this new city of God that John saw was its completeness. It was not partial and one-sided, but it was complete in all three of its dimensions. And so, in describing the city in the twenty-first chapter of the book of Revelation, John says this: "The length and the breadth and the height of it are equal." In other words, this new city of God, this city of ideal humanity, is not an unbalanced entity but it is complete on all sides.

Brooks continues:

> No man can say what mysteries of the yet unopened future are hidden in the picture of the mystic city; but if that city represents, as I have said, the glorified humanity, then there is much of it that we can understand already. It declares that the perfect life of man will be perfect on every side. One token of its perfectness will be its symmetry. In each of its three dimensions it will be complete.

King's next paragraph closely parallels Brooks's thought but does not reproduce his language:

> But if we look beneath the peculiar jargon of its author and the prevailing apocalyptic symbolism, we will find in this book many eternal truths

which continue to challenge us. One such truth is that of this text. What John is really saying is this: that life as it should be and life at its best is the life that is complete on all sides.

Without pursuing Brooks's questionable interpretation of the heavenly Jerusalem as a picture of "glorified humanity" and without exploring the mystical—and to the black church, foreign—notion of a humanity infused by God, King extracts a more homespun truth, namely, the best life is the well-rounded and complete life. Brooks summarizes:

> There are, then, three directions or dimensions of human life to which we may fitly give these three names, Length and Breadth and Height. The Length of a life, in this meaning of it, is, of course, not its duration. It is rather the reaching on and out of a man, in the line of activity and thought and self-development. . . . It is the push of a life forward to its own personal ends and ambitions.

King counters with:

> There are three dimensions of any complete life to which we can fitly give the words of this text: length, breadth, and height. The length of life as we shall think of it here is not its duration or its longevity, but it is the push of a life forward to achieve its personal ends and ambitions.

At this point, four paragraphs into Brooks's complex and closely argued sermon of twenty-four paragraphs, King parts company with his model and does not return to it. Using the example of a young man who wishes to become a lawyer, Brooks explores the way in which awakening self-consciousness subtly conforms itself to heredity and environment. His is a dissertation that would have been lost on King's audiences and on most contemporary congregations. Instead, King draws on the learned rabbi and pop psychologist, Joshua Liebman, and counsels the importance of self-love. Liebman's book, *Peace of Mind*, was a well of sermon quotations for liberal preachers of the late 1940s and early '50s, and King, whose early sermons are lightly dusted with psychological truisms, had either read Liebman or borrowed the reference from a contemporary sermon. King rounds out the first dimension of the complete life with a set piece, which is a combination of two or more memorized formulas linked by a common theme. The first is the "Street Sweeper," which was popular with black audiences not only because it enjoined Negro pride in ordinary occupations but also because it was a rousing tongue twister. The second was the peppy little jingle of Douglas Mallock:

> If you can't be a highway, just be a trail;
> If you can't be the sun, be a star,

> For it isn't by size that you win or you fail—
> Be the best of whatever you are.

This portion of the sermon is as far removed from Brooks's complex argument as both King's *and* Brooks's sermons are removed from the original meaning of the biblical text!

In Brooks's second dimension, the breadth of life, the preacher describes how a man must become sensitive to the "careers of other men," thereby enlarging his own life. King quickly abandons this line of reasoning and turns instead to a retelling of the parable of the Good Samaritan, which he does with the help of a single, nonstrategic sentence borrowed from the famous preacher George Buttrick. King adds two applications to his treatment of the breadth of life, neither of which appears in Brooks or Buttrick: he admonishes white Americans to quit thinking about themselves and to recognize that their destiny is tied up with that of the Negro. He also suggests that the American government might practice this concern for "breadth" by feeding the starving children of India rather than by establishing military bases.

In his treatment of the final dimension, height, Brooks admonishes his hearers to reach up toward God. He cautions against substituting religion for the transcendent God who is above all religions. In each of the sermons there is but a single passing reference to Jesus. In King's analysis of the final dimension, the sermon falls apart for lack of any meaningful discussion of God. He alludes to a single phrase from Brooks's sermon when he criticizes those who "seek to live life without a sky" but then becomes bogged down in proving the existence of unseen realities in general. He finally closes this section with the admonition to seek God, without whom "life is a meaningless drama with the decisive scenes missing." Both sermons address the importance of humanity's upward reaching toward God. Neither contains a word about God's gracious reaching down toward a fallen race. Although both sermons breathe a spirit of optimism with regard to humanity's ability to reach the divine, neither comes close to the awesome splendor of the Bible's witness to God.

What King borrowed from Brooks, then, was a biblical text, a tone, and a concept for a sermon along with a generic outline consisting of three "places," which for obvious reasons King amplified with proofs unlike those used by Brooks.

He followed Brooks much in the same way in a later sermon entitled "The Death of Evil upon the Seashore," the title and dominant metaphor of which was adapted from the bishop's "The Egyptians Dead upon the Seashore." In Brooks's sermon the striking image of the dead-eyed Egyptians gives way to the sickly theological suggestion that they symbolize the inevitable endings and transitions common to human experience. King will have none of this. In his sermon, the dead Egyptians stand for the death of "inhuman oppression and unjust exploitation." The image from the Exodus provides a cue for King's long rehearsal of colonialism, slavery under "the pharaohs of the South," segregation, and now the stirrings of freedom. In short, beyond the title and the image, King's sermon echoes little more than a sentence or two of Brooks's language and absolutely nothing of his thought.

Why King did not acknowledge the inspiration for the sermons he published in *Strength to Love* is unclear. "The Death of Evil" is a well-known title from one of the nineteenth century's most famous preachers. Perhaps he understood how little of Brooks's substance he had actually borrowed and therefore felt no obligation to give him credit. Perhaps he wished to maintain the illusion of originality—or its obverse, that of commonality with mainstream theological values. In delivery, many preachers would acknowledge their source; many would not. Some would refuse to do so for reasons of artistry and not vanity, insisting that oral "footnoting" is clumsy and interrupts the sense of urgency that should attend the delivery of a sermon. On a few occasions King did in fact interrupt the immediacy of the spoken word by citing the source of his sermon. One Sunday at Ebenezer he acknowledged the inspiration of Harry Emerson Fosdick for "Standing by the Best in an Evil Time" and then proceeded to preach a sermon that bears absolutely no resemblance to Fosdick's original. In another sermon he thanks Fosdick for "Making the Best of a Bad Mess" and accepts its moralistic, life-is-what-you-make-it advice while departing from its specific language and illustrations.

Late in his career King preached a magnificent sermon on the tragic dimension of human achievement entitled "Great . . . But." The idea for the sermon was supplied by one of Fosdick's more pedestrian efforts, "What Does It Really Mean to Be Great?" in which he recalls the biblical appraisal of Naaman: he was a "great man . . . but he was a leper." Fosdick's phrase sparked King's imagination, and he developed from it a powerful statement on the ambiguity that shadows human nature: the great civilizations produced slavery and colonialism; the scientific brilliance that lifts man above the animals also enables him to make napalm and other weapons of destruction. We are great . . . but we are fallen sinners.

The most famous of King's borrowed titles is "The Drum Major Instinct," which he took from the popular Florida Methodist J. Wallace Hamilton's 1949 sermon, "Drum Major Instincts." King's sermon is well known because a tape recording of it was played at his funeral, but, as English Professor Keith D. Miller has shown, the sermon is far from original to King. Save for minor updating of Hamilton's historical references and King's celebrated justification of his own life and career ("Tell them I was a drum major for justice"), most of the sermon comes from Hamilton or other popular preachers of the era.

One could trace King's use of "places," concepts, and titles from many other sermons and preachers. In some cases like "Drum Major" King is heavily dependent on his sources. For the most part, however, King uses an organizing concept as scaffolding for his own race material or enters formulas and longer borrowed pieces into his own concerns. The pattern of his sermons' *effect* is consistent: a spark from Brooks, Fosdick, or Buttrick sets off a conflagration in King. It is more interesting, however, to see how King grew *away* from his homiletical models, and to do that we must return to his flagship sermon and chart its development.

II

In content and style "The Three Dimensions of a Complete Life" (and its preacher) grew decidedly and aggressively *black*. A year and a half after his 1959 version, which contained only four sentences about the problem of race in America, King expanded the section on breadth with an attack on the "cancerous disease" of segregation, using specific examples from the struggle for integration in New Orleans. In this section he also reiterated his famous assurance that the Movement did not intend to humiliate whites. Already in late 1960, he found himself under the obligation of refuting the precursors of black power, and he deftly used the convenient "place" called "breadth" to do so.

In the final section on height, King drops the "life-without-a-sky" echo of Brooks's sermon and replaces it with a personal testimony to God's presence in his life: "There was always something deep down within me that would keep me going, a strange feeling that I was not alone in this struggle." In his own fashion he translates the height metaphor into an idiom more consistent with his training in the theology of Paul Tillich: God is present deep within the struggle. Although King made significant additions to the 1960 version of the sermon, they represent insertions into what is essentially the earlier text.

To appreciate King's development as an African-American preacher, one must hear him preach "The Three Dimensions" one last time on April 9, 1967, in his Ebenezer pulpit. Five days earlier he had denounced the war in Vietnam to an audience of three thousand at Riverside Church in New York. In that speech he had accused his own nation of war crimes and compared American bombing to Nazi atrocities. He had spoken directly from conscience without regard for the political consequences of this stand. As a result, he found himself on the receiving end of criticism from all quarters, including the black civil rights establishment. But he was not, as he would put it later in Memphis, "fearing any man."

"The Three Dimensions" of 1967 is clearly an instance of the prototype he had preached at the turn of the decade. The length, breadth, and height are still there. But the learned allusions have been replaced by examples from King's life and the experiences of the members in his congregation. Into the standard introduction about John's imprisonment on Patmos he injects a reference to his own jailings: "And I've been in prison just enough to know that it's a lonely experience." He continues by explaining the symmetry of the city, but this time in the idiom of his audience: "It was not up on one side and down on the other."

He again refers to Rabbi Liebman's concept of self-love, but in this sermon he translates personal self-esteem into the importance of healthy race consciousness.

> Too many Negroes are ashamed of themselves, ashamed of being black. A Negro gotta rise up and say from the bottom of his soul, "I am somebody. I have a rich, noble, and proud heritage. However exploited and however pained my history has been, I'm black but *I'm black and beautiful.*"

He adds several more examples of self-love not included in previous versions of the sermon. What is significant about them is that each is sketched in the imagery of ordinary black life—no heroic examples, no philosophical name-dropping—but life as it is lived up and down Auburn Avenue. He tells of how he was nearly "flunking out" of a statistics course at Morehouse College, alluding to his limitations with a quotation from *Green Pastures*. "Lord, I ain't much, but I is all I got." He continues in this vein by introducing a new figure from the black Main Street, a shoeshine man from the Gordon Shoe Shop in Montgomery: "It was just an experience to watch this fella shining my shoes. He would get that rag, you know, and he could bring music out of it." The point of this illustration is no longer the psychological importance of self-esteem but the necessity of black people honoring their own gifts.

At every stage of the sermon King introduces new, typically black-consciousness material. In the section on breadth, he not only tells the Good Samaritan story again but this time tries to imagine why the priest and the Levite did not stop to help. Using the black idiom, he surmises that "those brothers" were afraid, just as he was the other night when a fellow standing out on Simpson Road tried to flag him down. "I be honest with you," the preacher confesses, "I kept going." Similarly, in the section on height, which in the earlier versions was the weakest of the three dimensions because of its failure to celebrate the reality of God, King adds set material drawn from the oldest stratum of the Negro church, and suddenly the *God* section becomes the most exciting part of the sermon. After making fun of the trendy, death-of-God theologians of the 1960s, he reminds Ebenezer that God says, "My last name is the same as my first: 'I am that I am.' . . . and God is the only being in the universe that can say *I am* and put a period behind it." He goes on to describe Georgia's "sick governor" Lester Maddox and Alabama's puppet governor Lurleen Wallace, and he proclaims, "[T]he God that I worship is a God that has a way of saying, even the kings and even the governors [should] 'be still and know that I am God.' And God has not yet turned over this universe to Lester Maddox and Lurleen Wallace." This is not the philosopher's God, but the God of Ebenezer who rules the whole chain of being from top to bottom, including petty southern despots. Fittingly, the preacher turns one last time to a set piece on the greatness of this God, which he concludes with a word of personal testimony:

> He's my mother and my father,
> He's my sister and my brother!

The sermon quickly descends from ecstasy to pastoral care, as King, sounding like an oldtime Sustainer, promises his beleaguered people, "When you get all three of these [dimensions] together, your tears of frustration and despair will be wiped away." This is still "The Three Dimensions of a Complete Life," but the preacher has moved light years away from the platitudes of nineteenth-century liberalism and entered the heart of his own black experience.

III

King was not acclaimed as a preacher for his mastery of generic sermon outlines. What moved his audiences were the formulas and set pieces he skillfully inserted into his sermons and speeches. The African-American preacher has at his or her disposal an enormous disassembled inventory of rhetorical parts ready for immediate installation. Some of the set pieces represent King's original composition; some he picked up by ear from the black gospel tradition. His favorite traditional formulas praised the holy Name:

> Sometimes when we've tried to see the meaning
> of Jesus, we said,
> "He's a lily of the valley"
> And that he is a bright . . . morning star.
> At times we've said,
> "He's a rock in a weary land.
> He's a shelter in the time of storm."
> At times we've said,
> "He's a battleaxe in the time of battle."
> At times we've said that somehow he's a mother
> to the motherless and a father to the fatherless.
> At times we've just ended up saying,
> "He's my everything."

Significantly, King repeats the phrase, "*We*'ve said," for the *we* represents the long tradition to which he was heir. The language of this formula is everywhere in the African-American church. King heard it many times from his father at Ebenezer. Field investigators have recorded the formula in a variety of settings from the 1930s to the 1980s—Pipes in rural Georgia, Drake and Cayton in "The World of the Lower Class" in Chicago, Williams in a black pentecostal church in Pittsburgh, Spencer in a southern urban setting. At the *place* in the sermon where the praise of Jesus is appropriate, the preacher calls up the formula and, with individual modifications, passes on what he or she has received. When King, to cite another example, concludes his sermon with, "It will be a glorious day, the morning stars will sing together, and the sons of God will shout for joy," he is "remembering" the old Negro preacher's "Before there was a when or a where, a then or a there, before the morning stars sang together, and all the sons of God shouted for joy, *God* was speaking."

Some of King's formulas he copied whole from the printed sermons of popular preachers. Others he took from books of sermon illustrations and collections he had acquired as early as his Morehouse days. As a young man, he worked from a typed catechism of Shakespeare quotations as well as a thirty-page collection of handwritten aphorisms and quotations. Most of his longer formulas were composed of borrowed phrases that he modified and elaborated to serve his own needs.

The smallest unit of formulaic speech is the Homeric-like epithet. Already in his first oration, the speech at Holt Street Church, he routinely entered the phrase "the iron feet of oppression" to evoke the familiar local tyrannies. Most of his brief metaphoric expressions were his own creations. As a college student he had scribbled such practice phrases on the inside covers of his notebooks. Even as a prisoner in a Georgia jail he wrote by hand, "What is happening in Selma is another flash of the same quest for human dignity piercing the American sky." Homer might have gotten a D in Freshman Comp for his unnecessary repetition of "clever Odysseus" or "wine-dark sea," as would have King for his incessant portrayal of the "dark chambers of pessimism" or his florid description of the stars as "the swinging lanterns of eternity . . . like shining silvery pins sticking in the magnificent blue pincushion." But these were his key-signature phrases by which he transformed the prosaic discouragement of his audiences into the poetry of a Movement. They became the "text" of King's rhetorical authority. His audiences would cheer when he *began* one of his set pieces the way fans respond to the first bars of their favorite song at a rock concert. The formulas not only verified the identity of the speaker, they also guaranteed a collaborative role for the hearer in an important moment of history.

A few of King's short formulas were borrowed: from Howard Thurman's evocation of the slave's life came King's formula, "the rows of cotton, the sizzling heat, the riding overseer with his rawhide whip," which he never failed to recite whenever he spoke of slavery. From Pius Barbour came one of King's favorites, the "paralysis of analysis," with which he regularly chided those who preferred discussion to action. On the basis of Presbyterian Charles Templeton's advocacy of maladjustment to the status quo, King frequently justified his own "prophetic maladjustment" to injustice. But for the most part, the shorter and more self-consciously metaphoric the formula, the more likely it is to be original with King.

In addition to the Homeric epithet, King developed a repertoire of borrowed formulas that included assemblages of poems, paragraphs drawn from popular white preachers, gospel climax formulas absorbed in the black church, and much longer poetic-like pieces of his own composition. When combined around a theme, they formed a set piece. The lines of poems he had been memorizing since childhood:

> *Truth forever on the scaffold*—James Russell Lowell,
>
> *No man is an island*—John Donne,
>
> *God of our weary years*—James Weldon Johnson,
>
> *A tale told by an idiot*—Shakespeare,
>
> *The moving finger writes, and having writ* . . . —Edward Fitzgerald;

the spirituals:

> *I'm so glad trouble don't last always,*
>
> *Go down, Moses;*

songs of the church:

> *Amazing grace,*
> *I have heard the thunder roll,*
> *Where He leads me*

These and many, many others are everywhere in his sermons and speeches. They represent an inventory of classical poetry and inspirational verse filed away in the preacher's memory without regard for ideology or literary merit. He punches them in like numbers on a jukebox, even when they bear only marginal relation to the subject matter he is pursuing. Their beauty both delights the hearer and elevates the tone of the speech, thereby imparting a kind of borrowed grandeur to what was often a sweaty and disorganized business. It is one thing to rehearse the economic deprivations suffered by African-Americans; it is quite another to stand in a sweltering Brown's Chapel and recite in a tremulous voice Paul Laurence Dunbar's,

> A crust of bread and a corner to sleep in,
> A minute to smile and an hour to weep in,
> A pint of joy to a peck of trouble,
> And never a laugh but the moans come double;
> And that is life!

In addition to poetic verses, King memorized a few thematic sections from the sermons of the popular preachers of his day and, usually without attribution, slotted them into his own sermons and speeches. In addition to Harry Emerson Fosdick, a volume of whose sermons he once ordered from a prison cell, King enjoyed the black Baptist Howard Thurman, the Methodist Wallace Hamilton, the Presbyterian George Buttrick, and others whose sermons appeared in collections like *Great Preaching Today* and *Best Sermons*, and in the popular liberal magazine *The Christian Century* and its companion magazine *The Pulpit*. Many of the popular preachers made appearances at the Chicago Sunday Evening Club, a then-prominent showcase for "great" preaching held in Chicago's Orchestra Hall and broadcast nationally. Beginning with his first appearance in 1956, King joined the club of popular liberal preachers in America. He became one of them.

One of his sermons published in *Strength to Love*, "Shattered Dreams," illustrates the complexity of King's literary relationship to this circle of preachers. The shattered dream in the Bible refers to the Apostle Paul's aborted plan to extend the Christian missionary movement to Spain (Romans 15:28). In King's published sermon it refers to anyone who has "faced the agony of blasted hopes" and particularly the Negro, who has long dreamed of freedom but remains "in an oppressive prison of segregation and discrimination." King may have borrowed his title from a sermon by Wallace Hamilton, but there is little else from Hamilton in the sermon. Whether or not King borrowed the idea for the sermon directly from Hamilton is difficult to say since, as one preacher from the period remembers, "During the late 1950s and early 1960s just about every preacher I knew had a sermon about Paul's unfulfilled desires to go to Spain."

Keith Miller has traced this sermon's dependence on Hamilton as well as on Howard Thurman's *Deep River*, a book of reflections on the Negro spirituals that Thurman republished in 1955. Five or six of King's passages echo those of Thurman. For example, Thurman writes of the Negro's "frustrations," which are distilled into a "core of bitterness and disillusionment" that expresses itself in hardness of attitude and total mercilessness. King counters, "One possible reaction is to distill our frustrations into a core of bitterness and resentment. The person who pursues this path is likely to develop a callous attitude, a cold heart, and a bitter hatred toward God, toward those with whom he lives, and toward himself." As in his use of Phillips Brooks, King follows the general line of Thurman's reasoning while varying the language and adding ideas and illustrations of his own.

The intertextual picture becomes more complicated with King's use of a couple of formulaic pieces from George Buttrick (or from Buttrick's source). In Buttrick's sermon "Frustration and Faith," which addresses the same general theme of disappointment, the preacher alludes to George Frederick Watts's portrait of Hope as a forlorn figure seated upon the earth plucking one unbroken harp string. King's "Shattered Dreams" reproduces the brief illustration in the same words. King also quotes the same four lines from *The Rubaiyat of Omar Khayyam* that are found in Buttrick's sermon as well as in the published sermons of many preachers of the period.

Indeed, a survey of the sermon volumes of the 1950s reveals a pronounced lack of originality among the twentieth-century princes of the pulpit not only in the canned illustrations that circulated among them but also with regard to their themes. A comparison of two preachers may give the appearance of direct borrowing of one from the other, but a comparison of *six* of the published preachers will reveal the same tired illustrations—in all six. For example, *everyone*, including King, was using James A. Francis's widely circulated depiction of Jesus as one of the most successful "nobodies" who ever lived: He "was born in an obscure village. . . . He never wrote a book. He never held an office. . . . [Yet] all the armies that ever marched, all the navies that were ever built, all the parliaments that have ever sat, and all the kings that have ever ruled, put together, have not affected the life of man upon this earth like this one solitary personality." Such portraits appear to have been the perfect antidote to the higher criticism of the Bible regnant in European and American seminaries, for many of the preachers shied away from the Jewish eschatological interpretation of Jesus in favor of Greek morality or, if not morality, at least the more circumspect notion of the most influential personality who ever lived.

In that same period, the atomic age (and Paul Tillich) had all the pulpiteers preaching on security and the mastery of anxiety. In *The Pulpit* from 1948 to 1956, there were no fewer than a half dozen sermons on the subject of anxiety, including one that copied from another! Popular psychology inspired many sermons on self-worth and the value of personality. The theology of Reinhold Neibuhr was responsible for a spate of sermons on the paradox of human nature. The Cold War evoked from most of the preachers the requisite sermon on communism, recognizing its aspirations but condemning its conclusions. What is missing in nearly all the published sermons, including those of King, who by 1956 was a member of the homiletical in-group, is a serious exposition of the Bible, which has a great deal to say about the sanctity of God and the worthiness of Jesus but almost nothing on democracy, communism, the sacredness of human personality, and the other favorite topics of midcentury America's most popular preachers.

There is more to King's "Shattered Dreams," however, than a few expressions and illustrations borrowed from other preachers. To appreciate King's shattered dreams, one must hear him preach the sermon one last time, on March 3, 1968, under the title "Unfulfilled Dreams." Although this sermon is not as well known as his "Drum Major Instinct," which he had preached a month earlier and a portion of which was played at his funeral, "Unfulfilled Dreams" is actually the last sermon he preached at Ebenezer. Like "Drum Major," it too is a tortured farewell from one who is obsessed by thoughts of his own death and has come home one last time to "make testimony."

In its structure and some of its illustrations, it is the same "Shattered Dreams" he had preached earlier, but the differences between the two sermons are startling. Instead of preaching from the text in Romans about Paul's postponed journey to Spain, in this last version King preaches from 1 Kings 8:17, which recounts David's failure to build the Lord's temple. David's desire to build the temple was frustrated by his enemies and his own sins, but because it was his heart's intention to serve God, the Lord did not hold his failure against him. In this sermon the shattered dream is not an individual's disappointment or the Negro people's; it is King's own failure to build a kingdom of racial harmony in America. He draws an ominous parallel between his own life and that of Gandhi, who "was assassinated and died with a broken heart. . . ." Alluding to the buoyant hope of his earlier dream, a weary and depressed King cries out, "Life is a continual story of shattered dreams."

In his final sermon to Ebenezer, King announces that the dream is dead. Throughout this last period of his life, his sermons reflect his growing disillusionment with white America. With the urban riots, the white backlash had set in. The war in Vietnam had undercut his earlier reliance on democratic ideals and served to remind him that the ideals themselves were suspect, that white America had been born in genocide and slavery. In those days, King's rage was second to none, neither Stokeley Carmichael's nor Malcolm X's, but his commitment to Christianity offered him no outlet in the rhetoric of violence. During this period, the sermons reflect a gradual relinquishing of personalist themes, fewer assurances from the old homiletical standbys, and an outright disgust for the hypocrisy of American civil religion. The black gospel tradition reasserts itself as the dominant force in his preaching and life. At the end, he is no longer confidently reciting his set piece on the inevitability of justice in America: "The arc of the moral universe is long but it bends toward justice." Instead, he is possessed by the God of judgment and deliverance, the God of Abraham, Isaac, and Jacob, the God of Denmark Vesey, Nat Turner, and his own religious heritage.

In this final sermon the "friendly universe" of his halcyon days at Ebenezer has been replaced with a vision of a frightful battle at the center of reality—"a tension at the heart of the universe between good and evil." Whereas in the earlier version of this and many similar sermons, the conflict would have been resolved by a dialectical sleight of hand, a neat Hegelian synthesis, not so in this tragic farewell. The great hope of a just society, he warns his congregation, may never come true. There are no assurances. In the end, he says, we can only throw our failures and disappointments, which are in fact our very *lives*, upon the mercy of God. Thus what began early in his career as one more entry in the frustrated-hopes genre of popular sermonizing reaches its conclusion in a cry that is more powerful and profound than any of the "originals" among his sources.

IV

What is the creative dynamic at work in King the preacher? Is it tradition or plagiarism? Keith Miller, who has thoroughly documented the sources of King's published sermons and other writings, announces his own position with the comment, "Certainly an awareness of King's plagiarism does little to increase one's admiration for King." Miller is making a legitimate criticism of King's failure to give credit to Brooks, Fosdick, Thurman, Hamilton, Buttrick, and others in *Strength to Love*. But before we begin adjusting our admiration for Dr. King, should we not consider the full force of his preaching and not merely the printed records of some of his early sermons? How did he use what he received and, having used it, was the event that he created a *copy* of someone else's published sermon?

Any appraisal of King's preaching on the basis of his sermons published in *Strength to Love* will reveal something of his citation habits, but it is bound to distort the essence of his preaching, not only because the book does not portray his mature thought or homiletical style but because *no* book can capture oral performance. Even if we had a later volume of his sermons, it could not convey the essence of his preaching, for the sermon's meaning occurs in the voicing of the word. The sermons in *Strength to Love* have been ripped from their context, which was the church's defiant worship in the midst of social and political upheaval in the South. King and his publishers decontextualized his sermons in order to give them a timeless and universal quality, which King should have known is the very antithesis of a sermon. In these sermons, the issue of race appears as an application or addendum to the prudent observations of a midcentury American liberal. The sermons in *Strength to Love* contain no intimate reports of the battle raging over integration, no trace of weariness, defiance, or disillusionment. They tell no stories from the pool halls and barber shops of Auburn Avenue. In them no authoritative voice from his Daddy's chair says, "Make it plain, M. L." The published sermons do not convey the extemporized celebrations of the gospel or the formulaic altar calls characteristic of King's later preaching at Ebenezer and of black preaching in general. They do not get *down* or soar to an ecstatic climax. They do not *deliver*. The printed sermons offer no access to the oral event whose power is real and felt but unrepeatable. They contain scarcely a memory of his voice.

Miller's analysis not only overlooks the eventfulness of King's sermons, it actually exaggerates the extent to which King relied on the words of other preachers. When passages from King's sermons are lined up in parallel columns beside their sources, the configuration on the page conveys the impression of massive borrowing, of a preacher who was neither original nor ethical in his use of secondary sources. But when King's whole sermons are read alongside the whole sermons of the influential preachers, it becomes clear that for the most part King used his peers—Fosdick, Buttrick, Thurman, Hamilton—the way preachers have always used the sermons of others: for an idea, a phrase, an outline.

A survey of the prominent preachers reveals how much homiletical material they held in common and how routinely they all borrowed from one another. To cite a most flagrant instance, it is instructive to notice how many preachers *published* imitations of Fosdick's famous antiwar sermon, "The Unknown Soldier," organizing their sermons around the very same rhetorical device. Or, to cite another example, when one considers how many sermons were "out there," say, on "the nature of man," one is taken with how *little* King borrowed from Claude E. Hill, Lynn Harold Hough, Sidney E. Mead, J. Wallace Hamilton, Harry Emerson Fosdick, and Fulton Sheen, each of whom published a sermon on the paradox of human nature during the period in which King was allegedly scouring the homiletical magazines for ideas. In the 1949-1950 edition of *Best Sermons*, a volume young King would have known, there are no fewer than eight sermons on the paradox of human nature. King's later "What Is Man?" does not so much copy from any of them as *join* them as one more sermonic reflection on the Bible and popular anthropology. King's sermon alludes to the popular piece on the chemical insignificance of the human animal whose ingredients are worth a total of ninety-eight cents. *Many* of the sermons on human nature in that period ring in the same illustration, just as many of them quote Sir James Jeans's comment that "the universe seems to be nearer to a great thought than to a great machine." That is the way preaching works. The anonymous banality as well as Sir James's profoundity were *there* for use—some would say, plundering—and the preachers, who like comedians are always looking for good material, put them to use.

In 1992 one of the most formidable black preachers of our century, Sam Proctor, and, like King, a graduate of "Barbour University," visited Duke University Chapel and, without citation, used as the premise of his sermon Harry Emerson Fosdick's "Making the Best of a Bad Mess." "For this cause left I thee in Crete," Paul says to Titus, "that thou shouldest set in order the things that were wanting." Fosdick's homiletical moral is that according to the Bible, Crete was a terrible place, filled with liars, evil beasts, and gluttons (Titus 1:12), but that it was just the sort of place that Christianity seeks to redeem. Just as Paul left Titus in such a bad mess, so God sends young Christians today to make the best of hopeless situations. Whether Fosdick invented or received his sermon's premise was irrelevant to King, who used the same idea in the 1950s and '60s and to Proctor who repreached it in the 1990s. Aside from the formulas about "Crete," King's sermon bears no relation to Fosdick's, and Proctor's bears none to King's. What is certain is that countless other preachers will continue to drain the swamp that is "Crete" well into a new millennium.

Not all similarities between sermons are the result of conscious imitation. Miller consistently underestimates the body of theological knowledge available to seminary-trained preachers and often mistakes King's allusions to commonly held knowledge for intentional borrowing. For example, it seems pointless to try to trace, as Miller does, King's frequent dissertation on the three Greek meanings of love to a particular source. Anyone who has been to seminary more than a semester knows all about that. (King's handwritten class notes on *agape*, *philia*, and *eros*, taken in systematic Theology I, are available to scholars in the Boston University Library.) Likewise, it is futile to make much of the similarities between four preachers' account of the Prodigal Son as if they copied from one another. More than four preachers have tarried pregnantly over the phrase, "And when he came to himself." Miller should not be surprised that good preachers dwell on phrases like "my father's house" or "outstretched arms," which are a part of the permanent script of Christian consciousness throughout the world.

Even when Miller has overstated his case, he has made his point. King's printed sermons *do* resemble the sermons of others. Preachers tend to echo the work of others—the black expression is "I can hear Taylor *in* you"—because they learn to preach by imitating others. The published sermons that Miller scrutinizes are in fact the polished versions of King's own training sermons. Although in the preface to *Strength to Love* he says that he preached them at Dexter and Ebenezer, most of them originated in his pre-Dexter period of apprenticeship. They are the products of learning by imitation and preaching as rhetorical drill, methods he grew up with and learned from Pius Barbour and others. At Dexter King enjoyed only one year of normal pastoral duties and sermon preparation. The appearance of *Strength to Love* in 1963 effectively "froze" his published homiletical style. He never produced another book of sermons.

King's method of preaching should not be evaluated on the basis of the sermons in this volume. This is to arrest his theology and style at a period in which he was heavily dependent on liberal theology and homiletics and had not publicly revealed his own black voice. Such a reading encourages speculation as to why King tried to sound so "white" in his preaching, as though King were only pretending to be a liberal as a strategic device for ingratiating himself with white audiences. Miller theorizes that King suppressed his black roots, quoted the talismans of Western civilization, and thoroughly immersed himself in the liberal homiletical conventions of his day because he wanted to associate the claims of his Movement with what is best in Western culture. Miller writes, "By adapting and readapting sermonic boilerplate and by refining and retesting his best original material, King successfully placed the strands of his homiletical arguments against segregation into a web of ideas and phrases that the moderate and liberal white Protestant community had already approved." That is only to say he wanted to make his claim with words and ideas that his audience understood and accepted. This is hardly a novel rhetorical strategy on King's part, and it hardly deserves description in as ominous and reproachful a tone as Miller uses.

In a later essay, Miller adds a further spin to his theory by describing King's extensive use of Western philosophy and liberal homiletics as an enormous exercise in "self-making" by which he carefully constructed a public self at variance with his own cultural background and intelligence. In a recent presidential primary, one candidate accused his opponent of "reinventing himself" every few years. The comment was neither given nor received as a compliment. Where does one draw the line between the self-making that occurs in any public speech and self-making as out-and-out fraud? In his analysis of King, Miller leans toward the fraudulency end of the spectrum without, however, meaningfully addressing the *borrowed* quality of many of America's public documents and speeches—whether Jefferson's borrowings in the *Declaration*, Lincoln's reliance on Theodore Parker's "of all, for all, and by all," or Kennedy's well-publicized dependence on Theodore Sorenson. Why figures such as these were not also involved in public "self-making" is not made clear. In point of fact, King followed conventional methods of composition used in politics and homiletics. He immersed himself in the ideas of others and donned the costumes necessary to his cause. King successfully raided liberal culture, which, if one is to accept Miller's account, he could not possibly have digested—despite six years of postgraduate training in it—and has now achieved "iconic status" as an American hero.

It is not only true but a truism that one part of King's many-sided genius was his ability to communicate with white liberals. But the duplicity theory of King's rhetoric simply reduces a complex and accomplished communicator to the sum of his sources. It does not appreciate how fully and unreservedly King at one period in his life joined the circle of liberal preachers in America. The duplicity theory does not do justice to the depth of King's exposure to and appreciation of liberal Protestant theology, and it does not explain why, if he was only mouthing white platitudes in order to ingratiate himself with liberals, he preached portions of these liberal "white" sermons of Brooks, Fosdick, and others in his own black congregations in Montgomery and Atlanta. Nor does it explain why he gave essentially the same liberal speeches at predominantly black colleges and seminaries. If King embraced liberalism as a political ploy, why were its sentiments scattered throughout the sermons of Mays, Johns, Barbour, and other leading black preachers? What was *their* agenda? We are truer to the complexity of King if we admit that the sermons in *Strength to Love* represent his training wheels in theology and homiletics, which, by the time they were published, he had already outgrown.

The duplicity theory focuses exclusively on what it sees as King's self-conscious attempt to copy the style and values of the mid-twentieth-century liberal pulpit. It not only minimizes his appreciation of liberalism, but, by failing to give a *comprehensive* account of King's preaching, it omits its all-important prophetic dimension. Miller's theory leaves the impression that King was so preoccupied with associating himself with political and social liberalism that he never broke with it, never disassociated the Movement from it, and indeed never raged against it. But King's legendary ability to communicate with more than one audience did not compromise his moral vision or quiet his prophetic rage.

The duplicity theory operates with a rigid notion of "self" and "style." Its premise is that there is but one true self that projects one true style. Any variation from this singularity is assigned to fraudulency rather than complexity. If, as one of King's associates once observed, "Dr. King spoke and everybody could understand in his own language," the only possible explanation of this phenomenon, according to the duplicity theory, lies in King's intention to depart from his own true self and style in order to deceive his audience. He fabricates another self. But what if the *self* is known only on its surfaces, and what if *style* bears no obligation to correspond to some metaphysical entity called an inner self, which is unknowable to outsiders anyway, but rather has a responsibility to adapt and mutate according to the demands of a particular audience at a given time? When these questions are addressed to King, one hears a rich variety of stylistic and thematic adaptations tailored to the needs of several sorts of white and black constituencies. When the questions of self and style are related to King's role in American history, one recalls Du Bois's (and King's) moving testimony to the Negro's desire to be *both* fully black and fully American. That anyone could have thought this possible, for however brief a period of time, seems inconceivable to contemporary interpreters of King.

V

Martin Luther King was the product of a preaching tradition that valued originality of effect above originality of composition. The United States is honeycombed with little clapboard Holiness and Baptist churches whose part-time preachers are the last folk poets in the land. They are black and white, literate and semiliterate, rural and urban, at home in the mining towns of Appalachia or the storefront churches of Atlanta. What they have in common is the process of apprenticeship to master preachers and the immediacy of the Spirit's power that inspires their words and moves their congregations. Most of these preachers speak the first part of their sermons, but their real power comes from the rhythmic chant with which they intone the main body and climax. They all religiously depend on the Holy Spirit but, just beneath and behind the Spirit, they celebrate an array of their own rhetorical skills. These include a prodigious memory in which they store thematic set pieces that they have "almost" memorized, the Spirit-given freedom to create new thematic sections on their feet, and the inspired gift of doing it all in poetic meter.

The effectiveness of the folk preachers depends on the dexterity with which they manipulate their sources. Most of their set pieces have been around a long time: "Dry Bones in the Valley," "The Four Horsemen," "Dead (or Live) Cat on the Line," "The Horse Paweth in the Valley," "Jesus Will Make It All Right," and the famous "The Eagle Stirs Her Nest," which Charles Lyell reported hearing in 1868 from a former slave preacher and which continues to be performed today. Ethnographer Bruce Rosenberg recorded a fixed-formula sermon entitled "Deck of Cards," which in one of its performances in Bakersfield, California, ran some two hundred poetic lines. Its message of humanity's sin and God's forgiveness is structured around the numbers and symbols on a deck of cards. For example,

> Mr. Hoyle
> Put a Jack in the deck
> Representin' black death—that death
> That rides from house to house
> Then ya see what he did then
> After a while
> He put an ace in the deck
> Representin' the high card
> The Lord Himself
> Ain't God all right?

Lodged near the conclusion of this sermon, coincidentally, is a portion of the same Jesus-formula that was quoted in King and others earlier in this chapter:

> He's a rock
> In the weary land
> And a shelter
> In the time of storm

An example of a more freely created piece is this Durham, North Carolina, preacher's warning about sin in the city:

> As we drove on in the tax-i,
> We went by the Groov-y Green [bar]
> And the cars were ev-ry-where.
> We came down Down-ing Street
> And we passed two fights,
> And all of the sa-loons were wide o-pen.
> A sleep-ing church and a rov-ing lion
> Is a bad thing for an-y town.
> With the church rest-ing and the dev-il working,
> This is not a safe town.

It would be absurd to pretend that the Nobel laureate King operated like an unschooled folk preacher. He never chanted, he never sacrificed theme to the demands of meter, and he never, or rarely, recited a formula for emotional effect only. But it would be equally absurd to forget that his first teacher, his Daddy, did in fact intone his memorized formulas and that King himself possessed the memory, timing, and mental agility to punch in his formulas when he needed them. Dexterity with near-memorized formulas is the most pronounced similarity between Martin Luther King, Jr. and the folk preachers of America. For example, in the early King one of his set themes was the importance of Negro competence in a white world. Whenever one of his speeches took a turn toward that thematic "place," he never failed to use a set piece that might have been entitled "The Good Negro":

> We must not set out merely to do a good Negro job . . .
> If anyone sets out merely to be
>
> a good Negro doctor,
> a good Negro lawyer,
> a good Negro schoolteacher,
> a good Negro preacher,
> a good Negro skilled laborer,
> a good Negro barber or beautician,

You've already flunked your matriculation exam
For entrance into the university of integration!

The most famous example of King's manipulation of a set piece was his insertion of the "I Have a Dream" sequence into his speech at the Lincoln Memorial. He had already used it to spectacular effect at a mass address in Detroit, but the *Dream* was not included in the original draft of his Washington speech. At Mahalia Jackson's urging, however, he entered it into the moment and thereby produced what one philospher calls "right speech":

> Pseudo speech is speech which externally says the same thing as the right speech. Only, it is not told the right person in the right place and the right time. . . . The world is full of misplaced and mistimed speeches. It lives by the few speeches made at the right time in the right place.

To practice "right speech" means that the speaker will never repeat material verbatim. No matter how identical the folk preacher's renditions of "Deck of Cards" or King's repetitions of his set pieces, time and place will dictate subtle variations. For this reason, folk preachers insist that they do not borrow from anyone and do not repeat their sermons. The latter claim was echoed by Ralph Abernathy, who himself brilliantly exploited elements of the folk style: "See, we always kind of frowned on preachers . . . who repreach their sermons. . . . I have never preached a sermon *in the same way*. I have often repeated a sermon." Both King and Abernathy repeated their set pieces so often that they came to rely on their audiences to energize them for "right speech." The philosopher Rosenstock-Huessy might well have had King and other Movement preachers in mind when he wrote, "[A] word which fully rises to a specific occasion transforms the situation from an accident into a meaningful historical event. . . . [T]he more innocently an utterance is fully dedicated to this occasion and no other, the more original and eternal it may turn out to have been."

Not originality of composition but *timing* creates right speech for preaching. Jesus was not the first Jew to preach the kingdom of God. What was original in him was his announcement of the time of its fulfillment. The sheer unoriginality of Christian preaching has dominated not only the black church, folk preachers, and other products of oral cultures but also the moralizing princes of America's most influential pulpits. Originality implies deviation for the sake of novelty or effect. Christianity prefers faithfulness to the tradition above originality, repetition for the sake of instruction, and sharing of homiletical treasure for the enrichment of the whole church. The author of the church's first homiletics textbook,

Augustine, counseled less gifted preachers to memorize and deliver the words of others. One of Gregory the Great's reforms of the mass included the reading of sermons by the Fathers and other great teachers of the church. In the Middle Ages preachers were swamped with books of homilies, illustrations, outlines, and other helps, leading one historian to observe that "preaching had become an art in the use of borrowed materials." Luther advised his fledgling evangelical preachers to read *his* sermons from their pulpits; Anglicans read and imitated the essaic masterpieces of Bishop Tillotson; Methodists recited Wesley's sermons.

A recent study of the Puritan sermon indicates how closely third-generation Puritan preachers imitated the themes and procedures of Cotton Mather and John Barnard, and "how little regular preaching had changed over seventy-five years." Apart from superficial modifications, "third-generation sermons can hardly be distinguished from their predecessors." Their effect is one of "deliberate repetitiveness and studied unoriginality." Against the massive practice of the church, the modern appeal to individual genius or the invocation of copyright laws trivializes the tradition and makes of it something to be *borrowed* rather than *received.* The tradition is held hostage to the introspections and opinions of the individual. From the perspective of the church's historic practice, to say that King creatively built upon the contributions of others in response to the demands of the historical moment is to say a very good thing about the preacher.

King's *use* of the tradition made all the difference in his preaching style. To the tradition he added his passion for racial freedom, peace, and the eradication of poverty. He shaped it with the full arsenal of his gifts, including his uncanny ability to alchemize everything he touched into rhetorical brilliance.

The arguments he received from figures such as Phillips Brooks were invariably more complicated than King's congregations or *any* TV-benumbed audience could tolerate. Whatever he received he first stripped of intellectual complexity and ambiguity in order to offer plain choices to his hearers. Then he clothed what was left in exaggerated metaphorical langauge in order to elevate the subject matter of his speech. Such language has little appeal to the eye but never fails to work on the ear and the heart.

Finally, in thirteen and one-half years of public ministry, his sermons moved away from dependency on written sources and toward the language of the African-American people. His sermons became less structured and more heartfelt. They became less sophisticated homiletically but more effective from a human point of view. His tendency to choose simple illustrations drawn from the African-American experience corresponded to his growing disillusionment with white liberals and his embrace of a more radical doctrine of black redemption. As his dream of racial harmony began to give way to a more defiant assertion of black sufficiency, his attitude toward his sources changed. The units of tradition were still there, fixed in memory and verbal habit, but he could no longer accept them uncritically and use them for their old purpose.

5
The Strategies of Style

IT had been a long day. The pastor of Atlanta's West Hunter Street Baptist Church, the Reverend Ralph David Abernathy, mopped his brow, did a half-spin in his swivel chair while gazing at the ceiling, and began to reminisce about his old friend's style. As he did so, he slipped into a soft falsetto. "He was gifted, blessed by God. I would often hear him preaching or saying, 'We're *tired.*'" Abernathy tremulously extends the word to three syllables, "*ta-ah-yerd,*" and, now getting into the spirit of it, continues in whispered mimicry of King.

> "We're tired. We are *ta-ah-yerd.* We're tired of being pushed out of the glittering sunlight of life's July and left standin' in the piercing chill of an alpine Novem-bah. We are tired." And often someone would say, "What *is* that boy saying?" And another one would reply, "I don't know *what* he is saying, but I sure like what he said. I like the *way* he's saying it." That's right.

I

According to the old maxim, style is the man, but in King's case style did not mirror a mysterious and inaccessible "inner man," or what King would have called his "personality." Rather, it reflected a strategy for the public presentation of a message, which in turn was related to a larger strategy of social change. He did not preach and speak the way he did because "that is the sort of person he was," but because he had a mission no less calculated or comprehensive than Demosthenes' appeal to Athens or Lincoln's to America.

His mission was, as he put it simply in a 1963 sermon, "to make America a better nation." Paradoxically, he pursued his high and serious purpose with a style whose first principle was the achievement of pleasure. His high style included both unfamiliar and highly resonant words that he sculpted into careful patterns of balance, antithesis, and climax. He played with alliteration, assonance, metaphor, and internal rhyme, and allowed himself and his audiences to drink deeply of the pleasures of repetition until they laughed and clapped in amazement. "I like the *way* he's saying it," his hearers said.

In Africa the rhythms of the epic engulfed the hearer into an exquisite partnership with the poet. By pleasure, and not through rational analysis, the community transmitted its hope and wisdom. In ancient Greece (as well as traditional Africa), pleasure was exploited as "the instrument of cultural continuity." Plato feared the poets just as King's opponents feared his eloquence because pleasure interferes with established patterns of reason and politics. Rhetorical pleasure can move resistant people to unreasonable and destabilizing behavior. It can exalt the spirit and create hope where none has a right to exist.

Like a poet, King took pleasure in the purely labial quality of language. He did not look *through* words or *behind* them for deeper meanings. He enjoyed their surfaces and gloried in pronouncing them. What are faith, hope, and love, he asked, but a "magnificent *tril*-ogy of dura-*bil*-ity"? Why settle for truths that are merely self-evident when with the poet's help we can have them "incan*des*cently clear"? King delighted in euphony, the sweet sound of words and rhythms; it was a pleasure for him to lament the plight of black Americans in three rhymed phrases of equal syllabic length:

> humiliating oppression,
>
> the ungodly exploitation,
>
> the crushing domination.

In a sermon on the Greeks who wished to see Jesus (in John 12), he does not speak of Greek culture in the abstract but evokes the beauty of that culture by reciting the beautiful names of its greatest representatives: Aristophanes, Euripides, Thucydides, Demosthenes, and other "brilliant minds" that "came up around Greece." In the same sermon he powerfully combines a profusion of Latinate words with a series of repetitions ending with the word "Jesus."

> "And Sir," they said, "the radiant lights of philosophy, of poetry, or art, and of all earthly wisdom, hanging resplendent along our pathway are not sufficient to illuminate the way of life.

> Sir, we would see Jesus, the light of the world.
> We know about Plato, but we want to see Jesus.
> We know about Aristotle, but we want to see Jesus.
> We know about Homer, but we want to see Jesus."

His voice stabs at the first syllable of *Pla*-to and *Ho*-mer and drops at the end of each sentence to a gravelly and intimate *Jee*-sus. The sentence structure itself witnesses to the finality of Jesus in a world of culture.

Long ago Christianity accepted Cicero's idea of the three purposes of oratory—to teach, to delight, and to move—and although the church was never comfortable with *delight* as an ingredient of the gospel, many preachers settled for delight, not as a means to an end but as their homiletical end itself. Not so Martin Luther King, who, though tempted by delight, used the pleasure of his language for higher purposes. Broadly stated, he wished to *elevate* the cause he represented to one of noble and historic proportions; he worked to *identify* not only himself with his audience but his oppressed race with its oppressors and vice versa so as to bring them into alignment; and finally, like a prophet, he was forced to *confront* the evil all around him and fight it to the death. King's strategy of elevation is the subject of this chapter. The following chapter will explore his shift from the strategy of identification to the artlessness of rage.

II

From the beginning of his public career King sought to elevate local battles into holy crusades of mythic proportions. In his address at the Holt Street Church he skillfully elevated the significance of Rosa Parks's arrest from an all-too-familiar ugly incident to the test of an entire people's resolve to claim its own dignity—which it was. King exclaims,

> Right here in Montgomery when the history books are written in the future [*Yes, Lord*], somebody will have to say, "There lived a race of people, of black people, fleecy locks and black complexion, of people [*Yes!*] who had the moral courage to stand up for their rights. And thereby they injected a new meaning into the veins of history and of civilization."

He assures an audience in Cincinnati,

> And so we go this way, we can truly be God's children, and we can help America save her soul. Maybe God has called us here to this hour. [*Oh yes!*] Not merely to free ourselves but to free all of our white brothers and save the soul of this nation. . . . We will not ever allow this struggle to become so polarized that it becomes a struggle between *black* and

white men. We must see the tension in this nation is between *injustice* and *justice*, between the forces of *light* and the forces of *darkness*.

The most fundamental characteristic of the poet is the gift of seeing one thing in terms of another or of seeing a greater reality hidden in the squalor of the everyday. This faculty was the truest mark of King's "originality." Before he exercised his gift of speaking, he practiced the poet's— and the prophet's—gift of *seeing*: he discerned the transformation of history taking place in the gray little city of Montgomery, Alabama; he saw the dignity of humanity in striking sanitation workers; he witnessed the army of God in a peaceful march from Selma to Montgomery. Before that march he knelt on the asphalt of Highway 80 and prayed, "Thou hast sent us to fight, not just for ourselves, but to fight for this nation so that democracy might exist here for the whole world." One scholar calls this technique *mythication*, but in this and countless other sermons and speeches King was not so much *magnifying* the importance of the occasion as enabling his audience to *see* its full magnitude. His philosophical background in idealism encouraged him to associate material and political situations with abstract truths. His religious heritage trained him to "see" realities to which others were blind.

King raised the eyes of his congregation by means of a rich assortment of poetic techniques, the most famous of which was metaphoric language. The word *metaphor* has come to signify the language of figures and comparisons, and in that broad sense King dwelt in the land of metaphor all his days. More narrowly, a metaphor is a startling juxtaposition of apparently unlike entities that initially provokes puzzlement, if not denial, in the reader or hearer. *Metaphor* is a "strange" word, said Aristotle. It always says, "It is *and* it is not." Metaphor is a mark of genius and therefore its use is not restricted to adults, the educated, or modern communicators.

Some of the most vivid and arresting metaphors appear in the sermons of unlettered black preachers. A Macon County preacher describes the state of the universe before creation by saying, "The lightning hadn't played a lengthy game across the muddy cloud. Heaven hadn't been burned or hell even warmed up, or even organized." A former slave recounts her conversion as "jus like I had a fiddle in my belly." An emancipated slave preacher named Brother Carper, whose sermons are filled with a richness of images unparalleled in any homiletical era, evokes the River of Life in a sermon from the Book of Revelation:

> Brethren, we all knows what a river am. It am a mighty pretty thing, an' always looks to me like a ribbon danglin' from the bosom of old

mother earth. How often have we stood in the banks of some of these here rivers, an' seed there blue or creamy waters move along dotted an dented with eddies an' ripples, . . . and these eddies, whirling an' gamboling, an' then melting out into each other, like the smile of welcome on the face of a friend, afore he do you a favor, an' seemin' to say. . . . We flow for all, an' flow on, on, forever. . . . But the text speak of the river of life.

Metaphor is primary language based on insights so peculiar that they cannot be cast into formulas and passed on. There is little or no metaphor in this sense to be found in King's sermons. In his sermons he has moved away from the immediacy of pure metaphor toward less peculiar, less obscure, and therefore more reusable picture language: *metonymy*. A metonymy is a predictable metaphor. It is beautiful and it makes sense, both of which attributes are crucial to the orator's success, but its beauty is not drenched in a sensate awareness of nature, and its truth is not startling in its vividness. Metaphor often combines an idea with the image of something material; metonymy does the same, but in metonymy the relationship between the image and the idea is *logical* and suitable for mass reception. The combinations of image and idea in the following phrases make *sense* to the audience and expand its appreciation of the Movement's ideals:

the iron feet of oppression

dark chambers of pessimism

tranquilizing drug of gradualism

dark and desolate valleys of despair
  to sunlit paths of inner peace

sacrifice truth on the altar of self-interest

sagging walls of bus segregation . . . crushed
  by the battering rams of . . . justice

transform it [the jail] from a dungeon of shame
  to a haven of freedom and dignity

Precisely because it lacks shock value, metonymy was useful to King, who was employing his oratory to build a public consensus. It was relatively useless to black nationalists like Malcolm X who were trying to deconstruct the consensus and jump-start a revolution.

King often combines an archetypal image, for example, a *valley* or a *sunlit path*, with an equally universal value, such as *despair* or *peace*. His figures of speech rarely comment directly on his subject matter and never reflect on his addressees. Unlike true metaphor, metonymy does not violate expectations by introducing the "strange" or unexpected word. King remains within the style of his own vocabulary and safely within the worldview of his hearers.

Metonymy engages in the poetry of mythication, for by incessantly associating archetypal images and universal ideas, it subtly produces an effect not unlike that of medieval allegory. It posits the real presence of moral values in the material world and establishes a proportional, almost mechanical, correlation between the two. The mass confusion, brutality, and politics surrounding the march to the Edmund Pettus Bridge in Selma may actually point toward or "stand for" the eternal quest for freedom. The dreariest of southern towns may become the theater of operations for God's righteousness. The spaciousness of King's images serves to *enlarge* the Movement's dimensions and to insure the inevitability of its success. After the darkest night the sun must rise: "I refuse to accept the view that mankind is so tragically bound to the starless night of racism and war that the bright daybreak of peace and brotherhood can never become a reality. . . ."

King's use of metonymy and the language of mythication recalls Garry Wills's interpretation of Lincoln's rhetoric at Gettysburg: "Lincoln eschews all local emphasis. His speech hovers far above the carnage. He lifts the battle to a level of abstraction that purges it of grosser matter. . . ." In remarkably similar fashion, King framed what was an exceedingly mean-spirited conflict between protesters and state troopers in the abstract language of light and darkness, justice and injustice. He thereby prescinded the ugly realities of such matters as strategy, TV coverage, and petty bickering among the demonstrators, and elevated the entire battle to the realm of the ideal. What Lincoln did for the Civil War, King did for the Civil Rights Movement. He made it a moral and religious crusade in which people of goodwill could enlist in spirit and in truth. Of Lincoln's rendering of the battle, Wills writes, "The nightmare realities have been etherealized in the crucible of his language." But such a technique injects a moral dilemma into the rhetorician's strategy of style. In Lincoln's rhetoric at Gettysburg, the horror of war is obscured; in King's early speeches, the viciousness of racism is minimized.

The poetry of elevation *worked* for King in black and white audiences. The images are too grand for the printed page, but their grandeur seemed both to stimulate and then reflect the moral exaltation of packed churches and auditoriums during the Civil Rights years. For whites who were unsure about their involvement in the Movement, here was a voice that spoke from beyond the petty fears that divide us and from above all political motives that compromise us. For African Americans who had been beaten down so long that they doubted their own human value, this was a voice that lifted them up. Some may have believed that the only way they would regain their humanity was by losing it through inhuman acts of violence and hatred. King's style offered a more excellent way.

When his African-American audiences exclaimed, "I like the *way* he's saying it," they were reacting to their preacher in much the same way slave congregations had marveled at theirs. King's hearers were the first to admit that they did not understand his references to Plato, Hegel, the categorical imperative, or ontological anxiety, but his style so thoroughly fused the traditional *what* and *how* of the speech that his meaning came through nonetheless. The point is, said Aristotle long ago, a free man should not talk like a slave. Occasionally King says it bluntly: "You ain't no nigger: but you are God's children," but in fact the entirety of his elevated style could only remind his black audience of its sacred and inestimable worth. His style *was* the message.

King's poetic style also acted as a foil to his realistic appraisal of the human condition. The contrast between his glorious metonymies and the unemployment figures, lynching reports, dropout rates, and poverty statistics with which he peppered his sermons highlighted the contradiction between the Movement's noble goals and the wretched conditions in the South. His poetic language often set the stage for a prophetic denunciation. His exaltations of human nature added poignancy to his matter-of-fact assessments like "It's wrong to live with rats."

Like one of Molière's characters who discovers that all his life he's been talking prose, King went through his public life unself-consciously talking poetry. Although his sermons have been transcribed from audiotapes into prose manuscripts, many of their passages should be arranged in stanza form. In one of his set pieces on the greatness of America he rhapsodizes,

> Just look at what we've done
> We've built gargantuan bridges
> to span the seas
> And gigantic buildings
> to kiss the skies:
> Just look at what we've done.

No one at Ebenezer, including King, would have noticed the density of poetic technique within the brief piece. The passage features a sextain in A B C B C A line arrangement. It employs alliteration in *b* and *g* and *s* sounds, but in lines 2 and 4 the poet inverts the initial consonants in a near-perfect chiasmus. He not only employs the subtlety of near-rhyme (seas = skies) but augments the sibilant sound with a soulful double s at the end of "kiss." Unlike the epic poet, for whom meter, or the measured number of stresses in each line, was all-important, the African-American preacher focuses on theme, imagery, and rhythm. But for good measure, this piece in King exhibits metrical consistency as well.

Later in the sermon he skillfully moves to a longer set piece that might be entitled "The Greatness of Man." This topic King explores by means of a device common to the *exempla* of the medieval pulpit: an extended treatment of the differences between human beings and animals. Although King knows that human greatness rests on the image of God, this piece quickly turns into an exaltation of the human spirit. The congregation joins in the celebration and helps the preacher establish his rhythm by punctuating his calls with well-timed responses. What can "man" do?

> He can climb up the stairs of his concepts
> And enter a world of thought. . . .
> He can think a poem and write it . . .
>   [*Man!*]
> He can think a symphony and compose it.
>
> . . . .
> He has a mind, he has a memory,
> So he can latch onto the past,
>   and he has imagination.
> He can latch on to the future.
> Man is a dreamer—he can dream dreams.
>   [*Man!*]
> Dogs live their days,
> Lions live their days;
> Tigers live their days,
> Foxes live their days,
> But they're all guided by instinct.
> They don't have a way to connect with the future. . . .
> You don't see 'em sittin' 'round working out
>   geometry and algebra.
> You've never seen a group of dogs sittin' round
>   discussing philosophy
> Because they are devoid of speech.
>   [*Man is great!*]
> And just look at him—because of his mind
>   you can't hem him in.
> You just can't hold him in nowhere.
>   [*Preach!*]
> Man is great, but!

The title of this sermon is "Great . . . But." Delivered in the summer of 1967, the sermon is marked by flights of poetic brilliance, and its subject matter is perfectly aligned with the mood of the nation. America is enduring its fourth summer of urban riots. The war in Vietnam is claiming more than one hundred American lives a week, and King's opposition to it has alienated him from Lyndon Johnson and every civil rights organization in the United States except his own. On a recent trip to Cleveland, the white mayor snubbed him and the black mayoral candidate avoided him. The word *Negro* has turned *black* seemingly overnight, and to many militants King appears to be caught somewhere in between. His new book *Where Do We Go from Here: Chaos or Community?* has recently appeared to uniformly hostile notices, including that of one reviewer who observed the irony of its title. The Movement is indeed drifting. Birmingham and Selma are distant victories, the Nobel Prize a dream. "They were great," he seems to say in this sermon, "but . . ."

In this sermon King plies themes that have become familiar to his congregation and the nation. Humanity, he says, was destined for greatness, but it is dogged by its own failures to live up to God's will. He documents that failure by means of his typically visual logic: with the privileged eye of the preacher he takes the long view down "the corridors of history," which, he reminds his hearers, is dotted with the remains of civilizations whose greatness was built on the labor of slaves.

Now he brings his argument home to America, which is a great nation . . . but. At the word "but" it is apparent that the congregation has been waiting in heightening anticipation for the other shoe to fall. A woman's voice croons, "*Whooh,*" in apprehension. The finest scientific minds have used their knowledge to make *n*apalm to burn up villages and little babies. Scientists in the greatest country on earth have used their geometry and algebra to design better bombs. But what goes around comes around. Rapping his knuckles on the pulpit Bible, the preacher recites the black church's theory of history:

> Whatsoever a man soweth,
> That shall he . . .

"Also reap!" the people cry, "Also reap!"

But we are *all* implicated in the evil of our days because we are all sinners. At this point he deftly moves to another component in the set piece, this one with vocabulary and cadences that resonate to the First Epistle of John:

> We know the truth, yet we lie.
> We know we ought to be just, and yet
> we are unjust.

> We know we ought to love, and yet we hate.
> We know we ought to follow the road that is
>   high and noble.
> And yet we walk the low road.
> We know we ought to be pure, and yet we're impure.
>   [*Whooh!* a woman's voice responds an octave higher than before,
>   with a hint of sexuality.]
> Man is great, but he is a sinner.

There ends the set piece. Its lyricism and prophetic candor easily transcend the paradox-of-man genre of sermon so popular in the 1950s.

The echo from 1 John calls attention to the importance of *repetition* in King's preaching. He employs it in the following variations:

Alliteration (the repetition of the first sound of several words in a line):

> not . . . by the color of their skin but by the content of their character

Assonance (the repetition of similar vowel sounds followed by different consonants):

> that mag-ni-ficent tril-ogy of dura-bil-ity

Anaphora (the repetition of the same word or group of words at the beginning of successive clauses):

> How long? Not long, because no lie can live forever.
> How long? Not long, because you still reap what you sow.
> How long? Not long, because the arc of the moral universe is long
>   but it bends toward justice.
> How long? Not long, 'cause mine eyes have seen the glory . . .

The power of this use of anaphora derives not only from the situation, which was the tumultuous conclusion of the march from Selma to Montgomery, but also from the speaker's voice, which by the end of the series has merged with the ecstasy of the crowd. Moreover, the imagery moves from a quotation from Thomas Carlyle to ever-longer quotations until it explodes in the most emotionally and biblically resonant stanza in the English language. King used anaphora to stunning effect in many of his sermons and speeches, most notably his "I Have a Dream" speech in Washington, D.C.

Epistrophe is the repetition of the same word or group of words at the ends of successive clauses:

> In the midst of howling, vicious, snarling police dogs,
>   I'm gonna still sing, We Shall Overcome.
> In the midst of the chilly winds of adversity . . .
>   I'm gonna still say, We Shall Overcome. [tumult]

In the midst of the bombing of our churches, and the burning of our
houses, and the jailing of our children,
 I'm going to still sing, We Shall Overcome
[*Go ahead, sir!*]

King employed epistrophe to great advantage in many of his sermons. In
his last sermon, which he preached at the National Cathedral in Washington, D.C., he ticked off the promises of the Johnson administration,
concluding each with the phrase, "Nothing has been done."

The sermonic leitmotif is a phrase inserted at the beginning of some
sentences and at the end of others. King spots it into a set piece or larger
section of his sermon for its thematic *and* rhythmic value. In a set piece
on Judgment Day he enters the phrase "on that day" throughout the piece.
The phrase establishes a definite rhythm in this portion of the sermon,
which the congregation supplements with its repetition of "*that* day." He
creates the same effect in a sermon on the Rich Man and Dives with the
repetition of the word *hell*, sometimes in the first clause, "This man went
to hell because . . . ," sometimes in the final clause, "and this is why
he went to hell." In this sermon King engages in one of his rare exercises
of *vilification* of an opponent, in this case a self-appointed expert on
genetics and anthropology named Carleton Putnam, who, according to
King, claimed that the Negro is biologically inferior to the Caucasian. He
concludes his appraisal of Putnam's theory with a definitive coda: "If
Mr. Putnam isn't careful and doesn't revise his theory, he too will be a
candidate for *hell*."

In his use of repetition King always engages in *amplification*, either
by saying the same thing in so many different ways that it produces a
cumulative effect (*copiousness*) or, more frequently, by subtly racheting
up the value of the latter phrases in the series (*intensification*). His repetitions characteristically move up the ladder from the profane poets to the
sacred writers or from examples in nature to examples in history. In his
Holt Street address, his repetitions ascend from the ruling of a human
court to the laws of Almighty God. When the good Baptist in him praises
salvation, the final intensification in his series of repetitions often relates
to the *individual soul*, the preacher's own life being *the* case in point.

King also practices what might be called *sacred association*. The preacher
suddenly and illogically slips biblical phrases into the midst of an argument or at its conclusion. A variation of this technique is *heroic association*. The speaker floods his sermons with allusions to the most significant personages and events in the history of the world. The effect is to
elevate his hearers' self-esteeem and to amplify the importance of current events, for he has located his contemporaries in the company of

Socrates, Napoleon, and Churchill, and coupled their cause to the American Revolution and Emancipation.

Two examples of *sacred association*, which is the subtler of the techniques, will suffice. In a sermon on Vietnam, he puts these two sentences back-to-back:

> The shirtless and barefoot people of the world are rising up as never before.
>
> The people who sat in darkness have seen a great light.

This is *parataxis*, the unexpected and syntactically unconnected association of two reports. The biblically literate listener will make an intuitive connection between the Vietnamese and other Asian peoples who are claiming the right to their own destiny, and the long-expectant Jews who in the Christian interpretation have finally seen the Light. Together the sentences produce a minimetaphoric effect. To cite another example, at the conclusion of his final sermon King speaks of justice and brotherhood and then abruptly adds from the Book of Job, "And that day the morning stars will sing together and the sons of God will shout for joy," a phrase that adds an aura of holiness and inevitability to his message. In one sentence he has placed the Movement's agenda into the context of the Judeo-Christian tradition's eschatologial hope.

III

Amplification, intensification, and sacred and heroic association are but a few ways in which King doctored his use of repetition. These techniques are thematic in nature, that is, dependent on the content of the words themselves. The black preacher and audience, however, also have at their disposal a second, nondiscursive, track on which the sermon proceeds. This is the sermon's *sound track*. Its meaning is as theologically rich as that of the theme track, but it is more readily available to experience than reason. King's reputation was that of a cerebral leader, a thinking activist. Anyone around him knew, however, that a significant portion of his message was conveyed on the sound track of his style.

The King sound depends on *rhythm*. Repetition is the father of rhythm in King's sermons, and rhythm is the mother of ecstasy. In the traditional religions of Africa, rhythm represents the most important modality of *nommo*, the word, for it is the source of the word's power and its capacity to give pleasure. We experience our daily lives as an interrelated series of repetitions. We eat, work, chat, tend our children, make love, sleep, and pray in predictable cycles of repetition. The *organization* of these for our well-being and enjoyment constitutes the rhythms of life. Rhythm is all-encompassing. Wyatt Tee Walker writes, "The same beat the Black folks dance to on Saturday night is the same beat they shout to on Sunday morning. . . . If you hear the beat and do not know what the program is, watch the direction of the shout; if the shout is up and down, it is religion; if it is from side to side, it is probably secular."

King's sermons exhibit a masterly organization of repetition into rhythm. Strictly speaking, rhythm is not a sound but the interval *between* sounds that constitutes their organizing principle. The sharper and more pronounced the consonants, the more clearly defined is the interval between them and the more definitely established is the rhythm. African Americans refer to this as "hitting a lick," which implies vocal percussiveness. The repetition of alliterative consonants, multisyllabic words, or whole phrases, for example, "How long?/Not long," quickly establishes the excitement of the rhythm. King, like many musical preachers, often accompanies himself by rapping his knuckles on the side of the pulpit or on the Bible.

In addition to the repetition of sounds, words, and phrases, the predictable use of stress and pitch supports the sermon's rhythm. At regular intervals King will stress a syllable and dramatically raise its pitch and volume. When he moves into this vocal zone, his pattern of stress and pitch does not necessarily follow the *sense* of the words themselves. In the following example the italicized word indicates a sharp rise in stress and pitch:

> You can't *hém* him [mankind] in.
> He has a *mind.*
> Hold John Bunyan
> In Bedford *jáil*
> He set there
> But because he had a *mind*
> His mind leaped *oút* of the bars . . .

Each phrase ascends by degrees to the peak of the accented word, which the speaker does not merely stress but *plays* or bends in a tonal curve. With sufficient repetitions King achieves a hypnotic power that is his to sustain or break.

One of the simplest enrichments of repetition is its alteration. King occasionally varies the pattern of stresses on the consonants and produces *syncopation,* which is a musical version of the black preacher's stylized *stammer* whereby he affects the Old Testament prophet's stuttering failure of speech when confronted with the divine will. The preacher also uses the technique to signify a certain benevolent hesitation before dropping a bombshell on the congregation. The stammer says, in effect, "I, I, I really hate to say this, but I must." It also witnesses to the vulnerability of the gifted Afro-American preacher, whose intellect may on occasion overwhelm an uneducated congregation. King uses the stammer frequently, producing an effect that cannot be duplicated on the printed page.

He also alters the pattern of his repetition by what might be called *run-on rhythm*. In his "Dream" speech, for example, he first establishes the rhythm with the repetition of "I have a dream," but in the third sentence introduced by that phrase he alters the rhythm by ending the sentence, ". . . but by the content of their character-I-have a *dream* today!" He employs a similar technique in a sermon entitled "Guidelines for a Constructive Church." There he comes to a climax with an incredible eighteen consecutive sentences beginning with the phrase, "The acceptable year of the Lord is . . ." Between the seventeenth and the eighteenth he quotes the *Hallelujah Chorus* and cries, "Hallelujah! Hallelujah!" Without a pause or a drop in pitch he adds, "The acceptable year of the Lord is God's year," the effect of which is to shatter his own rhythm. In poetic terms this is a run-on line or *enjambment*. Usually the resolution of the thought and meter occurs at the end of the line; run-on rhythm creates a sense of overflow of emotion. The speaker appears to be overcome with the significance of the "dream" or the "year of the Lord." King uses this technique again in the fourth repetition of "I have a dream" and throughout his sermons and speeches.

While *stress* and *juncture* (or interval) help establish rhythm, *pitch* and *timbre* define the quality of the voice. King had a musical voice but, by most accounts, did not chant the climactic portions of his sermons as his father had before him and as many African-American preachers do. A few of his friends, such as Wyatt Tee Walker, recall that he did "whoop" or "tune" or "moan" on rare occasions away from the glare of the media, while others such as Ralph David Abernathy, Bernard Lee, and Gardner Taylor insist that he never did. In a 1957 sermon he does say, "Preachers can't spend time learning to 'whoop' and 'holler' because they need to study so that we will know the gospel and can help people live right. . . ." And in a later sermon he acknowledges the potential abuses of the sermon's sound track and indulges in some gentle self-mockery: In some churches "the pastor doesn't prepare any sermon to preach; he just depends on his voice, on volume not content. [laughter] And the people leave on Sunday and say, 'You know we have had a great service today, and the preacher just preached this morning!' And somebody says, 'What did he say?' 'I don't know what he said, but he *preached* this morning!' [uproar]." There is no available recording of Martin Luther King whooping, but the gradual ascendency of his pitch from a low growl at the beginning of the sermon to a piercing shout at the upper range of his high baritone, the predictable rhythm of the rise and fall of his voice, and the relentless increase in the *rate* of his speech—all contribute to the melodiousness, the songlike quality, of his voice.

It is a beautiful voice with a breathtaking range. Within a few minutes his voice moves from husky reflection to the peaks of ecstasy, but he always manages to keep both his voice and the ecstasy under control. Like a good singer, he will open his mouth wide to hit the notes but will not reach or strain. His voice never breaks. Its power is such that even in the emotional climax of the sermons, King is usually not letting it out but reining it in. Had he been an incendiary, it would have been the other way around. On a very few occasions, however, King does lose himself in the conflagration of the response. In a mass-meeting address in Birmingham, he concludes with nine consecutive sentences ending in the phrase, "We will still cry 'Freedom!'" As the listeners become ecstatic, King merges his voice with theirs and together he and his audience bring the speech to a frenzied climax:

> We will say Freedom, Freedom, Freedom, Freedom, Freedom, Freedom, Freedom, Freedom, to the world! [tumult]

Martin Luther King, Jr. used the English language in a way that "remembered" his ancestral tongue. West African languages are tonal, as is African-American speech. They convey meaning through the content of the words themselves, in the nuances of their pronunciation, and in the tonal qualities of the voice. Language takes on a melodiousness that allows an African-American singer like Ray Charles to give the word *uh-huh* as many meanings as he chooses. The meaning of words like *uh huh*, *yeah*, *Lord*, or *well* depends on the tone used in their inflection, which in turn depends on the context in which they are spoken or sung. A preacher like King may voice the word *Negro* (*Neg-ro*, *Neeg-ro*, *Nig-rah*) with the connotation that fits his point. Jon Michael Spencer compares African-American preaching, including King's, to jazz performance, both of which improvise by vocal inflection. "Vocal inflections typically used by black preachers and jazz soloists include the bending and lowering (*blue notes*) of pitches, sliding from tone to tone (*glissando*), grace notes, fall-offs, and tremolo."

The blue note, which is characteristic of the blues and black gospel music, recalls the middle pitch of West African tonal languages; it has a modality between sharp and flat that sounds sad to Western ears. When the preacher intensifies the message through a subtly crafted series of repetitions, or colors key words with bluesy intimations of irony, or renders the climax by tuning or chanting, he or she is synthesizing a "meaning" that transcends cognitive analysis.

In his sermon on the greatness of man, King pronounces the word *Lord* in such a way as to demonstrate the Almighty's majesty. *Law-ah-aw-awd* is a glissando. It begins on one note, curves around to two more, and returns to the original. Its very intonation witnesses to a God who is above all human greatness. In addition to these elements of musical style, King also produces a *turn*, which is taking a note slightly up and then partially down though not to its original note, or slightly down and then up but not to its original note:

> Through our airplanes, we've dwarfed *dis*-tance
> and placed time [*t-i-i-i-ahm*] in chains.

As an indication of the intricacy of his vocal stylistics, in one sermon he does a *parallel turn*, which is two turns back-to-back in a syntactically parallel construction. Musical artists such as Aretha Franklin, James Cleveland, and Albert King are more closely associated with these vocal techniques than preachers, especially preachers like King, who do not "tune." But these skills all belonged to King and are present in any of his sermons or speeches before an African-American audience. Their use produces a *feeling* in the audience that offers the possibility of momentary transcendence of the suffering that has produced these blue notes and diminished chords in the first place.

Despite the abuse of his vocal instrument by cigarettes (a pack a day) and overuse (250 speeches a year), King easily ranged from huskiness to trumpetlike clarity. But the visceral response his voice invariably evoked was due to the quality of its natural vibrato. That suggestion of a quaver tells a tale of suffering and hope common to all African Americans. It tells the story of King's jailings and persecutions and thereby certifies his place at the front of the Movement. In the climax of "Great . . . But," he follows the standard practice of reciting a hymn verse in paraphrase, in this case, "Amazing Grace." A general tumult has broken out among his hearers at Ebenezer as the rhythm, range, and tremulous quality of his voice combine to exploit the full power of the stanza. At the line "I once was lost," the preacher roughs his voice like a jazz singer and interjects a throaty "Yes I was." At that point, his identification with Jesus and the congregation is complete.

In the sermon "Great . . . But," the written word cannot adequately convey the pathos of King's voice. On its own track of meaning the voice renders the story of deliverance. It is now ponderous and sad, then happy and defiant, next playful and filled with the wonder of human achievement ("Look at what we've done!"), then choked with the awareness of sin, and finally ecstatic with the hope of salvation—all within the space of twenty-two minutes. One senses that King's voice is only just managing to contain some overflow of suffering whose most natural expression might be a shriek or a chant. The dam is about to burst, wants to burst, but the preacher will not let it. Finally, when he cries, "I once was lost," and adds with a sigh, "Yes I was," that parenthetical admission grounds the old lost-and-found formula in the human frailty of the preacher, a frailty King may hide from everyone in the world but Ebenezer.

The vibrato is present in all King's preaching, as is the hint of his own frailty. The bluesy huskiness, however, appears only at the end of his career. It corresponds to the floundering fortunes of the Movement and his own worsening depression. Nowhere is the tonality of despair more evident than in the directions he gives for his own funeral in "The Drum Major Instinct."

> Yes, if you want to say that I was a drum major,
> Say that I was a drum major for justice [Yes!]
> Say that I was a drum major for peace;
> I was a drum major for righteousness. . . .
>
> I won't have any money to leave behind.
> I won't have the fine and luxurious
>   things of life to leave behind.
> But I just want to leave a committed life behind. [Yes!]

The preacher's imagination of his own funeral is an old device, but the pathos of King's voice, so perfectly attuned to the course of his own life, transcends all issues of style.

IV

The key to any black preacher's style is the responsiveness of the congregation. The call-and-response pattern dates back to the West African ring shout and to the earliest forms of worship among African Americans. Call and response is a metaphor for the organic relationship of the individual to the group in the black church. More specifically, the congregation's response helps establish the sermon's rhythm. While the preacher is catching a breath, the audience hits a lick on his or her behalf. Some-

times the preacher and the congregation are guided by an individual who
acts as a "vocal coach" leading the responses. The coach may be a deacon or mother of the church; at Ebenezer, Daddy King often filled the
role with "Make it plain" or "Make it plain, M. L."

One December evening in Albany, Georgia, King was invited to
preach in the Shiloh Baptist Church, where a great throng was gathered
in preparation for a march. A reporter named Pat Watters remembers that
as King's sermon reached its crescendo, an old man punctuated each of
his remarks with an authoritative *God Almighty!*

> How long will we have to suffer injustices?
> [*God Almighty!*]
> How long will justice be crucified and truth buried?
> [*God Almighty!*]
> But we shall overcome.
> [*Shall overcome*, the crowd choruses back]

King's voice, "full of emotion that flowed into the crowd which poured it
back to him, almost broke, shouting:"

> Don't stop now. Keep moving.
> Walk together, children
> Don't you get weary.
> There's a great camp meeting coming. . . .

The response often provides the preacher a barometer by which to
evaluate his or her performance: "*Preach, Doctor.*" "*Come on up.*" "*Yes.*"
"*All right.*" "*Only the gospel.*" "*Lord help him.*" Often preachers will ask or
"beg" for it, not because they actually want an evaluation but because
they need the emotional encouragement in order to come up to the desired level, for if a vocal congregation is not moved, the preacher has failed.
The legendary Reverend C. L. Franklin begins a sermon, "I hope I can
get somebody to pray with me tonight . . . because you know, I'm a *Negro*
preacher, and I like to talk to people and have people talk *back* to me."
However stylized Franklin's request, it was a point of pride with King
never to ask for help no matter how traditional the ritual might be: "Do
I have a witness?" "Benches can't say 'Amen,'" "Help me, church" (sometimes, "Help me, Holy Ghost!"). Yet he was no less dependent on the
church's vocal response for his effectiveness than any local black preacher.
To verify that dependence, one has only to listen to the remarkably different effect when the same sermon is preached in a responsive church
and a vocally unresponsive church. "The Three Dimensions of a Complete Life" preached at Mount Pisgah Missionary Baptist Church in Chicago and the version preached at Grace Cathedral, San Francisco, have

80 percent of their words in common—but they are different sermons. The responsive congregation appears to pump life into the speaker who, because of a brutal schedule, always sounds bone-tired.

King drew rhetorical energy from lively audiences and in turn energized them in a unique way. In the African-American church—and he preached the vast majority of his sermons in African-American congregations—his message was accompanied first by encouragement, then exhilaration. In the many Shilohs, Friendships, Zion Hills, and Mount Pisgah Baptist churches, he managed to free himself of the heaviness of prepared notes and to preach the gospel of God's deliverance. He would come up, rise, soar, and hit the highest imaginable peaks of defiance, only to rise even higher to an extrarhythmical exaltation of God or freedom, accompanied all the way by responses perfectly timed to his cries. On these occasions he would burn up the church. To vocally unresponsive congregations he preached the same manuscripts, but by the book and without the fire.

The congregation also helped King (and any black preacher) by completing his thought. On many occasions the congregation not only maintained the rhythm but rounded off Bible quotations and hymn verses for the preacher. In the sermon "Great . . . But" King has been building up the greatness of America in anticipation of condemning its sins. The congregation is in on this, so that when King cries, "America is a great nation," the word *but!* can be heard like little firecrackers going off, and one man succinctly finishes the thought, *But she's goin' to hell.* In another sermon at Ebenezer, the congregation's response completes the idea of "the acceptable year of the Lord":

> The acceptable year of the Lord [a single clap is heard
>   like the popping of a paper sack] is that year . . .
> when women will start using the telephone for
> constructive purposes not to spread malicious gossip
> and false rumours on their neighbors.
>   [*oh-oh*, a male voice loud and clear]
> The acceptable year of the Lord is that year
>   [*That year!* one man's voice]
> when men will keep the ends for which they live.

The character of the response often follows lines of social stratification within and among black congregations. Some churches, like King's Dexter Avenue congregation, are what Evans Crawford calls "feel-back" (as opposed to "talk-back") churches that communicate through nods, smiles, and nudges. Bernard Lee notes that sometimes the response is delivered after the service. "Reverend, you were really in my pew this morning." Vocal response ranges from a dignified (if opaque) *well*, the meaning of which falls somewhere between a quizzical "interesting" to a downright *Amen* (not unheard in middle-class white congregations) to *My Lord, Thank you, Lord Jesus, Help yourself, Yes-suh*, and *Preach!* At the more demonstrative end of the spectrum one encounters loud laughing (*ha-ha-ha*), clapping, provocative sounds, and holy dancing. Even at Ebenezer, which falls squarely between a class-church and a mass-church, the response in King's day tended toward the conservative end of the spectrum.

Call and response signifies the communal nature of preaching and biblical interpretation. All are to some degree performers, and none are spectators, though the corporate performance in no way reduces the importance of the individual preacher's virtuosity. Call and response represents the congregation's love for its pastor and the pastor's gift to his or her people. By giving the people an important role in the praise of God, the preacher is reminding them of their importance as God's children.

In the African-American church call and response is an expression of joy. No one has made this point clearer than Henry Mitchell, who caricatures the black preacher as "fiery glad" and the white preacher, whose churches can also be vocally responsive, as "fiery mad." While it is not possible to test his conclusion, a description of the uproarious response to one of Jonathan Edwards's sermons helps to illustrate Mitchell's point:

> As the sermon proceeded many of the people slid forward on the edge of the pews, in their intense interest. The *shrieks* and *outcries* arose from the different parts of the congregation. Some *nervous* individuals actually gripped the pews to keep from sliding into the hell that Edwards had pictured directly underneath them. . . . And there was actually so much confusion that Edwards had to request the people to remain silent so he could finish the sermon.

The response to Martin Luther King's preaching was not "the terrors" that attended a Puritan sermon but the joy of people who wished to be free and to put the terror behind them.

Repetition produces rhythm; rhythm is enriched by the emotional quality of the voice; and rhythm, feeling, and, finally, meaning are sustained by the manifold responses of the congregation. The final element in this cooperative venture is the sermon's climax or series of climactic moments, producing what Du Bois had called the Frenzy. In King's sermons the climax is the "place" (in the classical sense of *topos*) where the *experience* of God replaces *talk about God*. It is the theological culmination of the address, but, stylistically, the climax also represents the most important rhetorical moment in the elevation of a black congregation. Any developed set piece, for example, "the Greatness of Man," will provide a climactic moment in the sermon. The preacher's rate will speed up significantly, and his voice will approach the outer limits of its pitch range and force. King's black-church sermons invariably contained several such climactic plateaus before the final ascent to the climax proper.

The true climax occurs within a few sentences of the end of the sermon. It builds upon the plateaus but contains material not found in them. Wyatt Tee Walker once asked his boss what his top priority was in planning a sermon—"Your three points?" he asked. King replied, "Oh no. First I find my landing strip. It's terrible to be circling around up there without a place to land." The climax is the celebration of the central theme of the whole sermon. From the announcement of its theme in the first sentence, a good sermon relentlessly circles its landing strip.

The climax restates the theme but in other media, sometimes by means of a hymn stanza or an ecstatic elaboration of points made earlier in the sermon. This technique is what older preachers like Sandy Ray referred to as "making gravy." As associate of King's, C. T. Vivian, paraphrases Ray in this way:

> He says there are two ways to make gravy. One of the ways . . . is that you cook meat—prepare your sermon—you take it out and you've got the grease there and you mix some extraneous stuff—flour that wasn't in the original grease, and then you stir that up pretty and you've got gravy. [He says] the other way to make gravy is to put on the pressure cooker . . . and you press all your elements down and you keep turning up the heat . . . you put your meat in . . . so that when you take the top off the gravy has come to the top naturally up out of the meat. That was Martin.

The material of the climax arises out of the ingredients of the sermon itself. The final climax may be built around a set piece, but the preacher's increased rate and run-on rhythms as well as King's extempore interjections and emotional outbursts will overwhelm mere symmetry of expression.

Finally, and most significantly, the climax will be *the* place in the sermon where the gospel of God is celebrated. Even in those sermons that have been heavy with moralism, advice, or admonition, King (in an African-American church) will almost always find a place at the end for a celebration of the identity of God and the salvation of God's people. In "Great . . . But," for example, the only way the preacher can surmount the paradox of human achievement and human sin is by abandoning it to the mercy of God. In King's climaxes, like those of many black Baptist preachers, the solution is often expressed in terms of the speaker's personal witness:

> When I delve into the inner chambers of my own being,
> When I delve into the life of mankind,
> I don't end up saying with the Pharisee,
> "I thank thee Lord that I'm *not* like other men,"
> but I end up saying, "Lord [pronounced *Law-ah-aw-awd*]
>   be *mer*-ciful to me a sinner."
> [*Amen. Go ahead!* great excitement]
> And those sins that I'm able to escape
> I'm just thankful for.
> When I look at dope addicts, I just say to myself
> Law-ah-aw-awd, I'm thankful this morning . . .
>   [*Yes Lord!*]. . . .
> And, oh this morning, this is why I'm thankful to God.
> We must open our lives to him; let him work through
>   Jesus Christ in our being.
> And he can remove that *but* from our lives.

To this he adds a soulful recitation of "Amazing Grace," and abruptly concludes the sermon with the standard invitation to those who wish to come to the altar: "We open the doors of the church now."

The climax is the moment of greatest intimacy in the sermon. Theologically, it is the moment in which preacher and congregation together break through the poverty and prejudice they experience during the week to the joy of that which is to be revealed. Its ecstasy is a promise, a foretaste, of the final victory that one day will be theirs. Like the union of the believer and Christ in medieval mysticism, the spirituality of the sermon's final celebration is often couched in sexual imagery. It is a climax. Its purpose is "rousements." It is intensely pleasurable, and it cannot be faked. One preacher describes this part of the sermon as "making it." William Pipes noticed that Macon County preaching was often accompanied by an undertone of sexuality in the form of comments and suggestive dancing. Although in his use of the climax King maintains extraordinary facial and bodily composure, with no grimacing or posturing—ever—he too participated in a culture of preaching that understood women as receptacles of the Word who are aroused by the eloquence of the speaker. The Movement preachers around him were known to brag about their ability to move "the sisters" to shed their sexual inhibitions, and King himself occasionally evoked highly sensual congregational response in the climax of his sermons.

The joy of style in King's sermons calls attention to the potential for conflict between pleasure and prophecy in all religious speech. In his comments on King's use of repetition, his biographer David Lewis was one of the first to question the social utility of the high style. Repetition was, of course, an essential ingredient in the King sound. With it he established a near-hypnotic rhythm by which he induced pleasure in the audience, won its assent, and, ideally, energized it for action in the community or nation. Lewis argues, however, that far from energizing people for meaningful action, the pleasures of rhetoric dulled blacks to the experience of their own suffering and provided whites with little more than momentary emotional gratification. Like liturgical repetition, King's use of the technique engendered happy expectations of a ritual march or a pleasurable aesthetic sensation, but these turned out to be poor substitutes for real economic or social change.

It is only fair to note that Lewis's comments were not made from the vantage of twenty years of black economic decline and disillusionment but were published within two years of King's death while the battle was still raging. His criticism would be more telling if pleasure had been the only aim or achievement of King's style. But Lewis does not address the prior question, which is What happens if no one *moves* the people in the first place? How are social forces peacefully set in motion? The critique of pleasure is mounted from the territory won by pleasure—and by two additional components in King's strategy of style: identification and confrontation. Beyond the beautiful words and sounds, what was the *rest* of King's rhetorical strategy and how did it work?

6
From Identification to Rage

ONE Sunday morning in 1958 the novelist James Baldwin visited Dexter Avenue Baptist Church to hear its famous pastor preach. A harsh critic of the Negro church, Baldwin admits that he went to services that day anticipating "those stunning, demagogic flights of the imagination which bring an audience cheering to its feet." Instead, what he found was an incredibly humble man whose "secret," he says, lies "in his intimate knowledge of the people he is addressing, be they black or white, and in the forthrightness with which he speaks of those things which hurt and baffle them."

The most complex element in King's strategy of style was identification, which Kenneth Burke calls the rhetorical "principle of courtship." For the first decade of his career King worked incessantly to align the aims of the Movement with the values of moderate-to-liberal white America. His goal was the merger of black aspirations into the American dream. To do this he had to convince black Americans that his methods represented their best interests, and he had to convince white Americans that his vision was consistent with their heritage and in their best interests as well. Due to the growing influence of television, which allowed a Negro unprecedented access to white audiences (*Meet the Press, The Tonight Show*) and his own intellectual background and rhetorical gifts, which granted him unprecedented credibility with white audiences, he carried out his mission of identification before a vast racially mixed audience. Even when he spoke to exclusively black or white audiences, he was in a very real sense addressing the vexingly mixed audience that is America. If that were not complexity enough, he campaigned for identification as a man of dark color in one of the most color-obsessed nations in the world.

I

It was necessary for King to achieve credibility with African-American audiences not because they were insensible to the wrongs he described but because many Negro Christians were skeptical of civil disobedience, and many militant black Americans, however much they admired Martin Luther King, Jr., were more than skeptical of nonviolence. As early as 1955, King was pleading with some Negro audiences to get involved in the struggle and with others to remain nonviolent. As late as 1966 in Chicago he was devoting significant energies to enlisting the support of black leaders in the Daley machine and, sometimes only minutes later, dampening the passion for violence among black urban gangs.

Some of his most potent barbs were reserved for complacent Negro preachers who, as he often said, cared more about the size of the wheelbase on their Cadillac than the size of their service to humanity. He criticized the black church only before black-church audiences, however, and he made fun of its preachers only insofar as they were hindering their people from joining the Movement. Before black-church audiences he engaged in a sophisticated form of what is known on the streets as "loud talking." He assailed the timid, otherworldly preachers with a "you-know-who-you-are" tone while implicitly encouraging those who overheard his scolding to bring them in line. Occasionally, John Lewis recalls, more aggressive lay members resorted to cutting their preacher's salary in order to encourage his militancy. There was never more than a "core" of the black church that made an active commitment to civil rights, reports Gardner Taylor. He adds with a laugh that now it is as difficult to find a black church leader who was not a "close associate of Dr. King's" as it is to find a white person who voted for Richard Nixon. In Taylor's opinion, King overestimated the black church's ability to sustain a high purpose and the white church's ability to discover a conscience. It must be remembered, however, that King's critique of his own church depended upon his identification with it. When he made fun of the preacher "whoopin' his mess," his hearers appreciated it, even if some of them cringed, because everything about his language, personal history, and demeanor indicated an insider's knowledge of the African-American church.

The chief purpose of his identification with African-American audiences, however, was not to scold them or mold them in his own image but to let them know that he understood them and that they now had a voice. When he addressed his own people, he spoke with what *Time* called an "indescribable capacity for empathy." Early in his career he identified with the weariness of the southern Negro, later with the more universal experience of black rage. Both types of identification, though useful and necessary for strategic reasons, were essentially pastoral in function. Even though he had enjoyed a relatively privileged upbringing, King made sure his audiences understood that he had experienced the full effects of growing up black in America. He often told stories of his own childhood experiences—of his father's furious reply to a clerk who called him "Boy," of his own rage at being forced to stand in the back of a bus, of his own children's disappointment when told they could not visit a segregated amusement park.

He often told Ebenezer one of its favorite stories, an interminable and sentimental story about Marian Anderson and her uneducated mother to the effect that behind every Ph.D. is a "No D," behind every Morehouse [Mo'house] College-educated person is a "No House." The happiest day of Marian Anderson's life was not when Toscanini acclaimed her as a "voice that comes only once in a century" [No!], not when she stood before the kings and queens of Europe [No!], but when she was able to tell her mother who had toiled for a lifetime as a domestic, "Mother, you may stop working now." And her mother, with tears of pride running down her cheeks, professed that all her drudgery had been worthwhile. King never tired of telling this story, and Ebenezer, with its generous representation of maids, porters, and day laborers, never failed to sigh with joy.

King was a master of irony in the pulpit. Sometimes he wryly commented on "liberal" whites who wanted to be the Negro's friend but "drew the line" at intermarriage. He invariably added his own punchline. "I want to be the white man's brother not his brother-in-law," a joke that is about as close as Martin Luther King comes to a put-down in the pulpit. He usually omitted his dialect material when speaking before predominantly white audiences, and even when he tried to tell stories to such groups, he often came off stiff and uncomfortable. Among his own people, however, he role-played dialogue with a freedom he did not enjoy in other settings. In his own congregation he mimicked the genteel southern white who was so used to saying "nigger" that whenever he tried to say "Negro" it came out "Nig-ra":

> A man told me. . . . "You know, I grew up with so much affection and love for, for, Nigras" (He couldn't say "Negro") . . . "that I always did nice things for Nigras. . . . But over the last few years since, ah, you Nigras have been demonstrating and, ah, you got others shouting 'black power' and all of this, we just don't feel the same kind of lu-uv we once had."

In a more serious vein, he assured a black audience in 1962 that he understood the anger of those who have been "trampled over" so long.

He can identify with those who have seen lynch mobs with their own eyes, who have been on the receiving end of police brutality, who are still the last hired and the first fired. "So many doors have been closed in our faces." At Ebenezer he said, "You've seen racial injustice in an old tired Negro woman shopping all day in a downtown store and coming to the point of wanting a cup of coffee and a hamburger to keep going, only to be told that 'we don't serve niggers here.'"

King ocasionally told stories in the pulpit but he was more gifted in the use of the *focal instance*, which is not characterized by a plot and punchline but captures in irreducible form an image or a slice of life from the black experience in America. In a 1966 sermon, "Who Are We?" he invites his congregation into the world of a Chicago tenement where "you will see people hovered up in rooms without heat, without electric light. There you will see eight and ten families using the same bathroom. Can't even get in there." In another focal instance he captures the abject poverty of rural blacks who live in Marks, Mississippi, in the poorest county in America:

> And I saw mothers and fathers who said to me not only were they un-employed, they didn't get any kind of income—no old age pension, no welfare check, nor anything. I said, "How do you live?" And they say, "Well, we go around—go around to the neighbors and ask them for a little something. When the berry season comes, we pick berries; when the rabbit season comes, we hunt and catch a few rabbits, and that's about it."

Before African-American audiences King unaffectedly used black colloquialisms. He could say "he don't" when the occasion seemed appropriate or "get it over to them" for "teach" or "people are going to lie on you" for "tell lies about you." In speaking of church-sponsored colonialism he spits out his judgment in disgust. "And they messed up the world with that kind of mess." In the same sermon preached shortly before his death, he recalls a particular morning with the black preacher's phrase, "The sun got up that morning." And always and everywhere he ritualized the lament, "We're *tired*."

Like many black preachers before him, King had a set piece on the futility of highfalutin' education. Its point is the superiority of faithful discipleship above formal education. That a Ph.D. in theology is reciting this piece only adds to the congregation's delight. Out in the hinterlands of Georgia, a rural congregation has little choice but to agree with its preacher who says, "[N]ever hear a man gwine to school to git his diploma on God." A middle-class Chicago congregation of professionals most likely does not agree with these sentiments, but it *admires* its preacher's learned condescension when he hits the same notes in explaining Jesus' conflict with the Pharisees: "That fixed 'em—all those Ph.D.'s in their long robes and mortarboard hats, all puffed up with their education. With all their degrees and learning, they couldn't trick the Son of the Living God!" Significantly, King's version of this piece subordinates the traditional anti-school sentiments to the importance of social ministry:

> Oh there will be a day.
> The question won't be how many awards did you get
> in life.
> Not *that* day.
> It won't be how popular were ya in your social setting.
> That won't be the question *that* day.
> It will not ask how many degrees you've been able to get.
> The question that day will not be concerned whether
> you are a Ph.D. or a No.D.,
> Will not be concerned whether you went to Morehouse . . .
> or No House. . . .
> On that day the question will be "What did you do
> for others?
> Now, I can hear somebody saying,
> "Lord, uh, I did a lot of things in life.
> I did my job well. . . . Lord, I went to school and
> studied hard. I accumulated a lot of money.
> Lord, that's what I did."
> Seems as if I can hear the Lord of Light saying,
> "But I was hungry, and you fed me not
> [rhythm is now building]
> I was sick and you visited me not.
> I was naked in the cold, and I was in prison
> And you weren't concerned about me,
> *so get out of my face!*" [uproar]

The congregation loves the playfulness of the language, the drama of what the preacher "hears" on its behalf, and the shocking dysphemism at the end of the piece. "Get out of my face" reflects the unexpected judgment of God, who sits *high* but looks *low*. (It's also a pleasure to hear a *doctor* talk like that.) The style is a vehicle for conveying the cleverness of God who, after all, has chosen the foolish things of this world to confound the wise. God is playing by a secret set of rules known only to this preacher here and, thanks to him, also to us. At Ebenezer the bond between preacher and congregation is evident. When King performs this piece at the National Cathedral in Washington, not surprisingly the

Almighty does not use the pool-hall vernacular, "Get out of my face," but the more Episcopalian "That was not enough!"

Before white congregations he dispenses with several other dramatic techniques popular in the black pulpit, the most prominent being the first-person dramatic testimony that places the preacher on the scene of revelation. He or she then "reports," "Seems as I can hear the Lord saying . . ." This technique was popularized by no less an evangelist than George Whitefield, who rhapsodized on the Apostle Peter as follows:

> Methinks I see him wringing his hands, rending his garments, stamping on the ground and, with the self- condemned publican, smiting his breast. See how it heaves!

Nevertheless, King omits this technique, as well as the stammer, the understatement ("'Thou shalt not kill.' I understand it's in the Ten Commandments"), and the gospel climax from his sermons in white congregations.

Some have argued that King was a kind of media phenomenon whose message was packaged for moderate-to-liberal white audiences. But this interpretation overlooks both the strategic and the pastoral dimensions of his identification with the people of his own race. Strategically, he needed to build a coalition of African Americans who would commit themselves to the Movement nonviolently. Pastorally, he supplied a voice that understood their hurts and was unafraid to rebuke their sins. The result was King's unprecedented impact on African Americans. Thousands of anonymous and forgotten African Americans wrote to tell him about their employment difficulties, marital conflicts, and money problems (one from Greenville, South Carolina, asked for ten dollars and a mattress). But most wrote simply to praise him for voicing their frustrations and hopes. "THANK YOU, THANK YOU, THANK YOU," wrote a young woman from the Bronx. "You are truly a remarkable person; a great leader and a courageous fellow citizen. We are all very proud of you as our leader in the fight for human dignity." A bishop in the African Methodist Episcopal Church introduced him to a black-church conference with these words:

> [W]hat's more important is that he is identified with the average man:
>> the man of the mill,
>> the man of the valley,
>> the man of the hill,
>> the man at the throttle,
>> the man at the fire,
>> the man with the fruit of his toil on his brow.

II

When it came to identifying with white audiences, King was faced with a cultural dilemma. How could he be true to his African roots without denying his ties to the great themes of Western culture? The obverse of the dilemma was equally perplexing. How could he remain true to his most cherished convictions about humanity without denying the black particularity of his own people?

Was King ill-advised to pursue a strategy of identification with white audiences? Franz Fanon counseled protesters to use *different* symbols, myths, and sounds from those of the established order. The oppressed, he said, can never use the language of the established order with as much skill as the oppressor. The oppressed must introduce a new language and another sound. Fanon's alternative was never an option for King. He himself had drawn deeply from the well of the oppressor's values and internalized them, with the result that his rhetoric effortlessly identified the aspirations of African Americans with mainline Christianity and liberal political ideals.

His theologial language closely mirrors the dominant theological and homiletical movements of his day. When analyzed line by line, his sermonic material, which also found its way into his public addresses and thence into the media, reflects the same worldview and ethos to which white Baptists, Methodists, and Presbyterians had long been accustomed. The Brotherhood of Man and the Fatherhood of God (or what one of the Movement's most generous supporters, Governor Nelson Rockefeller, derided as BOMFOG) was not a King invention. In fact, he had not received these liberal sentiments directly from white thinkers. Much of the liberal theological and political tradition was mediated to him through his sophisticated mentors, such as Benjamin Mays, Pius Barbour, Howard Thurman, and Vernon Johns, whose sermons also reflect the dominant theology of their era. What they all added, of course, was the other *sound* that Fanon had insisted upon.

Martin Luther King was the last of the great liberals in America to identify the purposes of social reform with those of Christianity. White Christians *had* to listen to King because he was speaking their language. He routinely cast the struggle for civil rights in terms of light and darkness, good and evil, and the two kingdoms. He effortlessly merged his voice with the prophet's cry "Let justice roll down like waters and righteousness like an everflowing stream." Even in his *public* addresses he interpreted race relations by means of the biblical Parable of the Good Samaritan, and the Vietnam War in terms of the Prodigal Son. He assumed that his secular audiences would understand and respond to *agape* love, and so thoroughly absorbed the imagery of death and resurrection into his political addresses that even when he failed to mention the name *Jesus*, his audience considered itself no less a congregation. He preached this consensus so persuasively that even when the biblical God failed to appear in the address, the audience was no less convinced that history is guided, to paraphrase Matthew Arnold, by an enduring Power not ourselves that makes for freedom. When King's publicist deleted references to the name of God from his tapes, the speeches still sounded like sermons.

King intuitively exploited the cultural hegemony of the Bible in what G. K. Chesterton called "the nation with the soul of a church." In the wider culture he wielded the Bible with as much rhetorical as theological authority. His manner of speech reflected the cadences and phrasing of the King James Version of the Bible. He thereby created resonances to a shared sacred worldview without making explicit doctrinal demands. His *style* communicated a subliminal message to black *and* white audiences, though each was moved by the precise elements it needed and wanted to hear, and each heard what its level of religious commitments and biblical literacy allowed. When he began his set piece "There will be a day," the fundamentalist heard it as a prediction of a literal Second Coming. The evangelical black Christian discerned in it the triumph of God's righteousness over racism. The liberal Christian extracted from the same phrase a poetic evocation of an eternal principle of balance.

King *sounded* like the Bible. His language favored structural parallelisms such as

> to suffer and sacrifice
>
> with dignity and discipline
>
> for truth and justice
>
> with goodwill and strong moral sensitivity

The second term often subtly advances or intensifies the thought as, for example, the term "moral sensitivity" adds definition to the more general "goodwill." Critic Robert Alter suggests that poetry resists complete parallelism. Careful examination of any parallel construction usually reveals a small wedge of difference between similar terms. The effect of this verbal habit in King is to add a sense of completion and authoritative competence to his utterances. Less tangibly, the structural parallelisms echo the poetics of the Psalms and Wisdom literature:

> Why art thou cast down, O my soul,
> And why art thou disquieted within me? (Psalm 42:11)

> Hide thy face from my sins,
> And blot out all mine iniquities (Psalm 51:9)

> My son, attend to my words:
> incline thine ear unto my sayings. (Proverbs 4:20)

King sprinkled his sermons with *sententiae*, brief, pithy, and balanced maxims reminiscent of the Book of Proverbs and other Hebrew poetry. They too are characterized by parallel construction, but the clauses function antithetically. He often complained,

> We've ended up building guided missiles
> and leaving misguided men.

Of America he said,

> She has allowed her technology to out-distance
> her theology.
> She has allowed her mentality
> to outrun her morality.

He occasionally warned,

> If a man hasn't found a cause worth dying for,
> he's not fit to live.

Like parallel constructions in general, these also convey a sense of completeness and authoritative competence. A truly *public* orator like King does not present shaded options but clear choices whose mode of presentation leaves no room for indecision. These antitheses do not *suggest* a truth, they nail it down. In doing so, they echo a biblical "sound" with which most of his audiences would have been familiar.

The pointed phrase is actually an intrusion into King's usual "running style," which is characterized by ponderous successions of subject-verb-object sentences in which the thing or value, for instance, "freedom" or "righteousness," often overpowers a lackluster verb. For example, in one sermon he begins eight consecutive sentences with the phrase "We see" to introduce a variety of objects and values. He uses the phrase sixteen times in all in the first three minutes of the sermon! His sentences are connected by a staggering number of "ands," with as many as *seventy-five* sentences in one sermon beginning with the word "and." His use of *and* may be the preacher's habitual way of buying time, but the "additive style" is so common in the King James Version that its use has come to be recognized as a mark of biblical style. For King it was one more unconscious means of stylizing his rhetoric into the language of the Bible and thereby strengthening the bonds of identification between speaker and audience.

On a political level, King's sermons and speeches mined out the unassailable bedrock of civil religion. Before he could confront white America with its betrayals of its own ideals, he first had to establish that "these truths" are woven into God's truth and therefore belong to all people. They are transcendent truths and therefore apply even to those who have been traditionally excluded from their blessings. For an oppressed people, the principles embedded in the *Declaration of Independence* and the Constitution as well as the mythology surrounding Abraham Lincoln function as a moral sanction that transcends the bigotry of local, state, and national administrations. As a twenty-six-year-old orator, King cried out to the Holt Street throng:

> This is the glory of our democracy. . . .
> If we are wrong, the Constitution of the United States
>   is wrong.
> If we are wrong, God almighty is wrong!

A little over a year later the 1957 Prayer Pilgrimage at the Lincoln Memorial concluded with this pledge:

> We believe in the Fatherhood of God and the brotherhood
>   of mankind.
> We believe that democracy is the noblest political expression
>   of this religious faith.

At the same event, King closed his speech with a couplet from James Weldon Johnson that expresses the powerful link between patriotism and religion.

> Shadowed beneath thy hand, may we forever stand
> True to our God, true to our native land.

In these early expressions of civil religion, always behind the transcendent principle lay the authority of an equally transcendent God.

> And behind the dim unknown
> Standeth God within the shadow
> Keeping watch over his own.

King would later step away from the shelter of transcendent principles that merely *protect* marginal people and boldly *ground* his critique of America in the biblical authority of a God who does not merely protect the oppressed but *chooses* them to be his instrument. But King never wholly ceased to support the liberating claims of the Bible with the protective claims of the Constitution and vice versa. In a 1966 sermon on the Georgia State Legislature's refusal to seat black activist Julian Bond, King dramatically appeals to Lincoln, Jefferson, and John Kennedy ("come here, brother Kennedy") and asks for their witness to this moral issue (quoting "Brother" Kennedy), "as old as the Bible and as modern as the Constitution."

Even in the sphere of civil religious discourse, King's argument for freedom rests on a theological foundation, *not* on the easy equation of God and the American way of life that constituted civil religion in the Eisenhower era but on the *discrepancy* between God's eternal will and America's sorry performance. God has called America, King could say with Lincoln, but the Deity does not *smile* on America. God does *not* bless America because America has not been good. Rather, God works indirectly in the formation of political ideals and correctives, sponsoring what Sidney Mead called the "transcendent universal religion of the nation." God's law supports certain transcendental principles enshrined in America's foundational documents, and together they both supersede local "laws" against integration. The Constitution protects all people and, lest African Americans not be included in the legal definition of "people" (as historically they were not), King reminds his listeners that the Bible contains the true definition of "people." The Bible says a person is a creature who is capable of a relationship with God. *All* people were created for this relationship in the image of God.

Where the Constitution fails to keep faith with the divine intention, the Declaration of Independence succeeds. Just as Lincoln had exploited the transcendent truths of the Declaration to undermine actual *laws* regarding slavery, so King attacked segregation by means of the same strategy. This is the backdrop for his set piece on inclusiveness in his 1967 sermon "Great . . . But." America is such a great nation that it has written into its charter the God-ordained equality of all people.

> It didn't say "some men."
> [*No!*]
> It said *all* [*aw-a-wal-ll*] men.
> [*Amen. Make it plain!*]
> It didn't say, "all white men."
> It said "all men."
> [*Men!* the congregation replies]
> Which includes black men.

The word "all" is a glissando. Not only is it given four distinct syllables but it describes an arc of four distinct notes. King makes of it a long and soulful word to insure that the whole church knows that it is sheltered under its wings.

Theologically, King's allusions to the image of God or other Christian doctrines helped sustain the self-respect of his black audiences. Politically, the same language reminded white audiences in no uncertain words: "We are Americans like you. We want what is coming to us." Thus James Baldwin was only half right in declaring that unlike earlier black leaders, King always said the same thing to black and white audiences. The words were often the same, but his audiences heard them in radically different tones.

III

In the identification phase of King's rhetorical strategy, his sermons and speeches are marked by a nonconfrontive style of *inclusion*. In the early years of his career, he regularly addressed white audiences with the pronoun *we*. His profuse quotations from the standard white poets and philosophers—Donne, Lowell, Carlyle, Hugo, Kant, Hegel, Emerson, and many others—as well as his quotations from the Bible and the Constitution reinforced the commonality of culture that allows an educated American Christian Negro to address the AFL-CIO or the B'nai B'rith with the pronoun *we*. The speaker is the go-between, the transformer, by which the audience comes to associate the values of the Movement with its own. If *freedom* is truly the essence of human nature, as King and the traditions of Western idealism and romanticism insist, then these poor black creatures who are being beaten and arrested are doing only what any civilized human being would do. They are struggling for their (and our) common birthright. Their fate is our own.

In addition to his use of inclusion, the early King tends to reduce the distance between potential antagonists by explaining the evil of segregators in psychological or anthropological categories. To a group of white Southern Baptists, whose institutional strength rested on the rock of segregation, King soft-pedaled the sinfulness of segregation by explaining that racists really aren't evil but only afraid. Race prejudice is a "communication problem." The authority he cites to refute racial supremicism is not the divisive claims of the biblical prophets but "the best evidence of the anthropological scientists." Before the same group of Southern Baptists he portrays the Negro's suffering in spiritual terms rather than the more threatening political or economic categories. Not only has the Negro been subjected to a spiritually debilitating "I-It relationship" but, what is equally distressing, segregation has damaged the spiritual and mental health of the *segregator*.

A close reading of the speeches of Martin Luther King and Malcolm

X sheds light on the technique of inclusion. King's starting point was the unity of the races; Malcolm's was the evil of the white race. Malcolm spoke for and with black Americans but made no attempt to identify with even the most liberal whites. Unlike King, who made compounds of black and white, Jew and Gentile, Catholic and Protestant, Malcolm never placed *black* and *white* in a coordinated structure within a sentence. King said "we." Malcolm invariably said "you" and made fun of the "house Negro" who spoke of America as "we":

> If the master said, "We got a good house here," the house Negro would say, "Yeah, we got a good house here." Whenever the master said "we," he said "we." That's how you can tell a house Negro.

In his famous "Deaf, Dumb, and Blind" speech, Malcolm attacks his audience.

> You are *deaf, dumb,* and *blind*. You are lost in the wilderness of North America, and the black man's new day has been delayed because of you. Now I am here to get you ready.

King spoke incessantly of "man," that is, humanity; Malcolm appears not to have believed in "man" or a corporate humanity that is not divided and barricaded according to race and religion. He and King both knew there were *enemies* out there, but only Malcolm would acknowledge them. Before any audience King would begin, "I need not pause to say how honored I am to be here." Before *any* audience Malcolm would begin, "Mr Moderator, . . . brothers and sisters, friends and enemies: I just can't believe everyone in here is a friend and I don't want to leave anybody out." Malcolm's language was peppered with imperatives and sarcastic rhetorical questions. More than 90 percent of the sentences in King's published sermons are declarative; only 2.5 percent are openly imperative.

Malcolm's use of metaphor was startling, in the original sense of metaphor. He often used animal imagery to make his point, as when he chided a black audience, "You didn't come here on the Mayflower. You came here on a slave ship. In chains, like a horse, or a cow, or a chicken." Whites he compared to the more rapacious animals: the wolf, the fox, but most of all, the snake. "If I were to go home and find some blood on the leg of one of my little girls, and my wife told me that a snake bit the child, I'd go looking for the snake. And if I found the snake, I wouldn't necessarily take time to see if it had blood on its jaws. As far as I'm concerned the snake is the snake." On the other hand, King employed metaphors like "the quicksands of hatred," which, according to his detractors, actually softened the concept of hate by poeticizing it. In King's rhetoric (until the mid-1960s), the future is always a promise; in Malcolm's it is always a threat. The difference between King and Malcolm is the difference between the rhetoric of absorption and the rhetoric of attack.

Yet there is in King's strategy of identification an aggressive element, perhaps a passive-aggressive element, that is rooted in a preacher trick as old as Augustine, who said, "I need not go over all the other things that can be done by powerful eloquence to move the minds of the hearers, not telling them what they ought to do, but urging them to do *what they already know ought to be done.*" King so thoroughly associated the goals of African Americans with the values of mainstream America that the latter group was left with the choice of taking action on behalf of the oppressed or accepting its own guilt. In an interview with psychologist Kenneth B. Clark, King said, "I think it [love for the oppressor] arouses a sense of shame within them often—in many instances. I think it does something to touch the conscience and establish a sense of guilt." The use of shame for political purposes was Gandhian *satyagraha*, but, as a psychological weapon, guilt was a natural outgrowth of King's childhood experiences. At an early age he was torn between hatred for racists, which he learned from his father's example, and his own need to conciliate powerful opponents, which he first experienced as fear of a powerful father. That conflict within him produced a sense of guilt from which he never escaped. The leader who used guilt as a rhetorico-political strategy was described by one of his closest associates as "an intensely guilt-ridden man."

King was operating with a theory of social change as old as Richard Allen and Daniel Coker. The strategy of identification inevitably created invidious comparisons that laid the groundwork for the moral and eventual social defeat of white segregationists. For a black person to turn the other cheek and forgive seventy times seven could only expose the hypocrisy of good churchgoing racists, though there is little evidence that King's hard-core opponents ever felt the stirrings of shame. Moderate white Christians, on the other hand, as well as liberals susceptible to BOMFOG, were already self-convicted. In their hearts they knew they were wrong, guilty as sin. Thus the first, and to King's mind, the most important battle in *any* contest—the moral one—had already gone to the Negro. King also believed that the Movement offered whites the opportunity to redeem their sins and purge their guilt by embracing their black brothers and sisters. "Somehow the white man needs the Negro to get rid of his guilt," he said. The worst tactic imaginable, he insisted, was black-on-white violence, for black violence relieves white guilt by providing an excuse for hate and retaliation. His more militant critics, on the other hand, ridiculed the emotional catharsis whites experienced when they heard King speak of love and brotherhood. The *feeling*, they said, had become a substitute for committed social and political action. A good soaking in guilt was not enough.

Depending on how one interprets King's strategy of identification, King was either caught between cultures in a no-win situation or he was adroitly playing both ends against the middle. By identifying the Movement with mainstream religious and cultural values, he was effectively cutting off two unwanted ends: the ultraconservative whites and the militant blacks. The trick was to maintain his position on a constantly shifting field without alienating his supporters or compromising his bottom-line values.

The word *trick* recalls the myth of the "signifying monkey," which originates in Africa. The scholar of African literature, Henry Louis Gates, has drawn attention to the figure of Esu-Elegbara, a divine trickster who specializes in playing off superior forces against one another to his own advantage. In the jungle he shuttles between the fierce lion (the white racist?) and the powerful elephant (the white liberal?), tricking each to do his will. The signifying monkey's discourse, metaphorically, is double voiced. *Signifying* among African Americans has come to mean the artful use of language in the forms of innuendo, needling, evading an issue, indirection, loud talking, exaggeration, and many others, in order to go one up on the other person or group. In the history of American race relations, signifying was the only mode of communication, other than suicidal confrontation, that was available to an oppressed people. It was the language of slave preachers who coded their sermons with double signification: one for the Massa, one for a knowing congregation.

King did not engage in signifying in the verbal manner of H. Rap Brown or other nationalists. There was no jiving, trickery, or race machoism in his speech. In Movement mass-meeting speeches, Ralph Abernathy was a master of the playful put-down and the double entendre, but even before these friendly audiences King was comparatively restrained. In a larger sense, however, he was thrown into a historic signifying contest. His entire rhetorical strategy of identification can be interpreted as a form of signifying. He is the double-voiced spokesman shuttling between innumerable audiences. He inspires whites and blacks, mediates between two factions of blacks, and deftly manipulates two factions of whites: the racist lion and the liberal elephant. His job is to make teammates of black and white moderates and to get the liberal elephant to stomp the racist lion. He does so by identifying with the elephant's values, by mimicking them. We know he is *unconsciously* mimicking because his particular version of the signifying game is to articulate the values of white liberalism, which he had thoroughly internalized, but in his own black voice. The black voice, imbued with the irony, judgment, and genuine hope of the black gospel, modifies liberal values even as it articulates them. It exposes their provisional nature. And like the trickster Esu, King gets his way but is smashed in the process.

IV

For all its unconscious cunning, King's strategy of identification led ineluctably to the language of *confrontation* and to the eventual abandonment of rhetorical strategy. The Movement's identification with mainstream values created a temporary and fragile consensus The by-product of that consensus was a high-definition portrait of the enemies of freedom. The isolation of evil was a perennial Movement strategy, one that was frustrated in Albany, Georgia, where Police Chief Pritchett and other officials resisted mythication and politely refused their assigned roles as villains. The strategy worked in Birmingham, where the actors enthusiastically coveted their parts. King understood his role as crucial, for what he and his followers did enabled the villains to play their part and so expose themselves by means of the mass media to the whole world. "Just let them get their dogs," an angry King shouted, "and let them get the hose, and we will leave them standing before their God and the world spattered with the blood and reeking with the stench of their Negro brothers." It is necessary "to bring these issues to the surface, to bring them out into the open where everybody can see them. . . ." King and his people marched into the teeth of the opposition and allowed themselves to be slathered with the abuse of racists in order to define and localize the evil. Then they drew it to a head and, by means of public opinion and the federal government, neutralized it.

White violence was absolutely necessary to the success of this strategy. There had to be snarling dogs and fire hoses. "Bull" Connor had to be the star. Without them, the TV cameras, which were bored by peace, brotherhood, black religion, and the most eloquent rhetoric of the twentieth century, would have packed up and returned to New York. Only gradually did King and his aides realize that media coverage depended on telegenic acts of violence. By the Birmingham campaign, Wyatt Tee Walker, for one, thoroughly understood the principle. He later bragged, "There never was any more skillful manipulation of the news media than there was in Birmingham." David Garrow has documented the irony of the shift in strategy. The nonviolent movement turned from winning the hearts of the oppressors to provoking violent reactions among them in order to insure TV coverage and federal intervention.

On the one hand, confrontation represents a media strategy as current as the TV era; on the other hand, it echoes a homiletical method as venerable as that of Jonathan Edwards, whose proclamation of grace only highlighted the sins of his people. It is striking that King employed the same metaphor to explain his strategy of confrontation as Jonathan Edwards used two hundred years earlier to describe his: that of lancing an overripe boil.

Confrontation was the theological flip side of identification. The God who wholly identifies with us in Jesus Christ is the same God who thunders from Sinai and seethes through the mouths of the social prophets. The prophet whose divinely inspired *pathos* enables him to understand the suffering of his people is the same person who must threaten their oppressors. The civil religionist who celebrates the transcendent values that created the Republic is the same one who warns that the Republic will be judged by them.

The philosopher R. G. Collingwood once observed that the artist "tells his audience, at the risk of their displeasure, the secrets of their own hearts" for "no community altogether knows its own heart." The terrible secret that the prophet began to tell in his last desperate years was this: it doesn't matter if America has or has not lived up to its principles, because the principles themselves are a lie. King would have agreed with the colonial Quaker Dr. Benjamin Rush who laid open the heart of his community when he said, "Remember that national crimes require national punishments. . . ." The orator who once climaxed his speeches with renditions of *The Battle Hymn of the Republic* and *My Country 'Tis of Thee* issued the following indictment of his country:

> Our nation was born in genocide when it embraced the doctrine that the original American, the Indian, was an inferior race. Even before there were large numbers of Negroes on our shores, the scar of racial hatred had already disfigured colonial society. From the sixteenth century forward, blood flowed in battles over racial supremacy. We are perhaps the only nation which tried as a matter of national policy to wipe out its indigenous population.

King essentially took on the project of demythologizing American civil religion. Although the foundational documents should apply to all, he told a Montgomery congregation in 1968, they were written by men who owned slaves. "Now a nation that got started like that . . . has a lot of repenting to do." The history of oppression has produced a "terrible ambivalence in the soul of white America." "Yes it is true," he said two weeks before his death, ". . . America is a racist country." In a 1968 sermon to Ebenezer he wonders aloud if black Americans will be able to celebrate the Bicentennial. "You know why? Because it [the Declaration of Independence] has never had any real meaning in terms of implementation in our lives." He is publicly admitting, with Langston Hughes,

> O, yes
> I say it plain,
> America never was America to me.

Even Emancipation, the holiest day on the calendar of African-American civil religion, was a lie. White America uses it only to foist a bootstrap ideology onto blacks, but it is a myth to say that white immigrants pulled themselves up by their own bootstraps.

> Remember that nobody in this nation has done that. While they refused to give the black man any land, don't forget this, America at that same moment, through an act of Congress was giving away millions of acres of land in the West and the Midwest to white peasants from Europe. Never forget it. What else did they do? They built land grant colleges for them long before they built them for us, in order to teach them how to farm. They provided county agents long before they provided them for us, in order to give them greater expertise in farming. And then they provided low interest rates for them so that they could mechanize their farms. And now, through federal subsidies, they are paying many of these people millions of dollars not to farm. And these are the very same folk telling Negroes that they ought to lift themselves by their own bootstraps, but it is a cruel jest to say to a bootless man that he ought to lift himself by his own bootstraps. . . . Emancipation for the black man was freedom to hunger.

In these later sermons, racial prejudice is no longer a psychosocial projection of the white person's inner fears or ignorance; it is no longer a sin that can be redeemed through contrition, repentance, and goodwill. Racism is written into the heart of white America and into the documents to which African Americans futilely appeal for protection. You can't trust your own country, he tells his people, because it isn't really yours. Before a group in Montgomery in 1968 he makes this astonishing comparison:

> And you know what, a nation that put as many Japanese in a concentration camp as they did in the forties . . . will put black people in a concentration camp. And I'm not interested in being in any concentration camp. I been on the reservation too long now.

This is the secret of your own heart, the black prophet tells America. The moral victories of conscience won through the strategy of identification are a sham. A moral victory is a contradiction in terms, since practical American morality will only continue to defeat and enslave people of color. Radical legislative and economic surgery is needed in the form of guaranteed jobs, equality of health care, the distribution of wealth, and other measures.

The strategy of confrontation was never absent from any phase of King's public career. In every city from Montgomery in 1955 to Selma in 1965 King's rhetoric included negotiations, accusations, recriminations, and lists of demands. His strategy of identification never failed to meet with resistance. Even in the early and middle years of his career, rank-and-file unaligned whites regarded King as a threat to social stability. During those years the majority of whites did *not* respect him or resonate to the biblical and philosophical basis of his demands. Neither were they shamed into repentance. They feared him—and hated him. After Selma, when the Johnson administration failed to implement the Voting Rights Act, and with the nation's deepening involvement in Vietnam, King simply dropped the guise of inclusiveness, quit talking about America with the pronoun *we*, and no longer rhetorically masked the true nature of the conflict. His predictions of riots were interpreted by whites as instigations of violence. Many whites began to make a *post hoc ergo propter hoc* connection between a King visit to their city and later urban unrest, this despite his steadfast and unwavering support of nonviolence.

With his increasing focus on economic issues, and with the burgeoning crises of war and urban unrest, the old metaphors seemed overmatched by events. By 1968 "I Have a Dream" and other stylistic flights, perfect for their day, had taken on a hollow sound. His sermons of this period do not dispense with metaphoric language, but the soaring images of hope have disappeared. In a 1967 sermon he warns, "And what we must see is that our nation's summers of riots are caused by our nation's winters of delay. And as long as justice is postponed, we're going to have these summers. . . . I'm worried about America because it's sick with racism still." In this sermon he introduces the Poor People's Campaign and adds, "I can only say God help America if the response doesn't come." In Mississippi he throws down the gauntlet: "America has a choice. Either you give the Negro his God-given rights and his freedom or you face the fact of continual social disruption and chaos. America, which will you choose?" In Chicago he gives notice: "But if these agreements aren't carried out, Chicago hasn't *seen* a demonstration." Before the Poor People's Campaign he promises that he will remain nonviolent but adds that if progress isn't made, "the discussion of guerrilla warfare will be more extensive."

His most confrontational language, however, he trains not on the issues of poverty or race but on the Vietnam War. He ruefully notes that public opinion applauded nonviolence so long as it served white interest, but when it comes to the children of Vietnam, America wants nothing to do with nonviolence. His graphic description of what American firepower is doing to the Vietnamese peasants exceeds his angriest denunciations of racial injustice.

> They wander into the towns and see thousands of children homeless, without clothes, running in packs on the streets like animals; they see the children degraded by our soldiers as they beg for food. They see the children selling their sisters to our soldiers, soliciting for their mothers. We have destroyed their two most cherished institutions—the family and the village. We have destroyed their land and their crops. . . . We have corrupted their women and children and killed their men. What strange liberators we are!

In the sermons of 1967 and 1968 new rhetorical elements begin to make their appearance. Many of the old formulas remain, but they now appear incongruous beside the new language of social and economic analysis. King was self-consciously crossing the boundary from reform to revolution, liberalism to liberationism, announcing a more profound critique of the American system. He alluded to a "class struggle" that transcended the questions of race. He was both "tired of race," as he confessed to Ebenezer in 1968, but also deeply aware that white racism was the "destructive cutting edge" that was splitting America into "two hostile societies." To one audience he said, "We're dealing in a sense with class issues, we're dealing with the problem of the gulf between the haves and the have-nots." Before Ebenezer he criticized the increasing disparity between the poor and the wealthy and called for a redistribution of wealth.

In earlier sermons he might have condemned the individual behavior of those who engaged in rioting or looting, but his later sermons condemn the system that makes riots inevitable. Like the Hebrew prophets who assailed the structures of enslavement in Israel, King shows how the woes of the fathers are visited upon their children in urban America: the disturbances in Watts are voices of anger from children who have grown up in fatherless homes, because the fathers were unable to find work and often had to leave home so that their families could qualify for Aid to Dependent Children. Watts continues to seethe with the bitterness of people who share in none of the benefits of our great society and who are reminded of that fact daily by the humiliation they receive from police, welfare workers, and city councils. Not only are the riots immediately destructive to African Americans, they lead to greater repression. They intensify the fears of the white majority while relieving its guilt. They give "permission" to the oppressor to take sterner measures.

When it comes to drugs and other vices, King takes a similar approach. He cautions young people against drugs and gives thanks to God that he has never used them, but he also criticizes a society that creates the need for drugs and then crushes the individual kid who is "just out there selling a little dope." In a 1967 sermon appropriately entitled, "Judging Others," he asks, Why don't the police pay more attention to the "total sin" that begins in the syndicates rather than arresting "the Negro who is just trying to make him a little money?" The same goes, he says, for the numbers racket and prostitution. The government should be more effective at attacking the causes of these ills than their often-underprivileged perpetrators.

In these late sermons he is not as successful in masking his personal mood from the congregation. He speaks of the Movement—and his own life—as history, and unconsciously adopts a retrospective tone that is accompanied by flights of maudlin reminiscences and sad stories of people who failed in their endeavors. When he is not sad in these sermons, he is uncharacteristically angry with those who have obstructed his programs or deserted him. His rhetoric becomes more confrontational, more prophetic, in nature. The multiple "authorities" upon which his earlier sermons rested have been reduced to One. In a variety of ways he says, in effect: This is God's will for America. You can take it or leave it.

The unmasked reality of confrontation now entails a cleaner, more direct mode of address to a nation that has accommodated itself to the bloodless rhetoric of Lyndon Johnson and the generals. If every rhetorical strategy implies self-interest and a careful weighing of costs and benefits to the audience, then King's practice of confrontation has taken him to the outer boundary of strategy. His style may not represent the complete abandonment of rhetorical strategy, but in its leveling of the complexities of identification, it comes close. He is no longer cleverly shuttling between the lion and the elephant, no longer the double-voiced one. He is past signifying. He knows that the prophet is answerable only to God and will not be denied a word befitting his vision.

7
The Masks of Character

IN a 1961 article in the *New York Times Magazine*, Ved Mehta observed that Gandhi and Martin Luther King shared a certain dramatistic genius. Both directed demonstrations as if they were theatrical performances, and both would therefore be remembered as "imaginative artists who knew how to use world politics as their stage." Several years later, David Halberstam described the Civil Rights Movement as "a great televised morality play, white hats and black hats; lift up the black and there would be the white face of Bull Connor; lift up the white hat and there would be the solemn black face of Martin King." Mehta's and Halberstam's comments—like those of historian Lerone Bennett, who insisted King's greatest gift was his "instinct for symbolic action"—are a reminder that Martin Luther King was, in the fullest and most positive sense of the word, an actor. He perfected his craft in the black church, where he learned how to assume the appropriate role and to perform according to the expectations of his public.

Implicit in King's strategies of style were a few personae by which he communicated his purposes for America. These roles he accepted and played with absolute fidelity. When speaking prophetically, for example, no unseemly aside, unmeant gesture, or hint of backstage behavior ever detracted from his role or diminished the high ground he had chosen for himself. Like a Greek actor, he moved across the stage speaking his lines with a passion appropriate to his mask. He never broke character. What sociologist Erving Goffman calls the "front," which is a performer's setting, appearance, and manner, remained in King utterly consistent.

The most important dimension of oratory, said Aristotle in his *Rhetoric*, is the character of the orator, by which he meant not only the speaker's

morality but also his competence, attitude, and the "front" he presents
to his audience. Those who followed Aristotle were quick to seize upon
his first point, morality, and formulated the ideal of "the good man speaking well." But Aristotle, who also wrote the rules for actors in the *Poetics*,
had craftily qualified his praise of character to such an extent that the
character projected by the orator cannot be distinguished from the masks
worn by the actor. The speaker, he said, must *appear* to be competent.
The rhetorical argument proceeds not by the strict logic of syllogisms
but by a line of reasoning that is *approximately* true—which certainly describes King's orderly if often associative methods of argumentation. The
speaker, he advised, "must make his own character *look* right." In classical
Latin *persona* was "mask": *per-sonar* literally means that which is *sounded
through*. The classical view of "person" is the very opposite of the hidden
and irreducible *me* of churchly piety, psychoanalysis, or modern individualism. It is a role. In other words, orators have more in common with
actors than the orators—or the preachers—like to admit.

The Christian church has never quit worrying about the relation of
moral character and effective preaching. Although the theologians debated
endlessly among themselves concerning the efficacy of the Word of God
when it is spoken by a reprobate, that the preacher *ought* to be a good person
whose life is steeped in prayer and charity was never in question. All Christians, including King and the African-American church, have unquestion-
ingly accepted Jesus' doctrine of the good tree that produces good fruit
and applied it to the moral character of the clergy. The church has done
so, however, without coming to an explicit appreciation of the masks of
character necessarily worn by any person in the act of preaching.

I

As a result of his African heritage of the "little me" that dwells in each of
us, as well as his extensive training in the philosophy of personalism,
King never seriously doubted that there is something called a *person*
behind the mask at the very center of everything we say and do. This
person is known to the speaker and accountable to God. But more clearly
than any twentieth-century orator, King also understood that that person is unknowable and inaccessible to others. In theory, he would have
rejected the sociologist's claim that the self is only the *product* of the many
scenes we play, an imaginary peg we hang our actions on, but not their
cause. In practice, however, the demands of his public-appearance sched-

ule, which had a monster-life of its own, drove him to the instinctive practice of roles. At times he appears to have felt impoverished in his "true" self by the many roles thrust upon him. The inner person, he reminds the Ebenezer congregation that "knew" him so well, is always hidden from view. You think you see "a fellow named Martin Luther King?" he asked his congregation. Oh, no! "You can never see that something that the psychologists call my personality." How can you see the "'me' that makes me *me*?" In this passage King indirectly reflects the tension between the core of identity in our heart of hearts, in which he never ceased to believe, and the many roles we all play in the world. The hidden *I* exists all right, but it is available only to God. For all others, there are only masks.

In his published sermons King rarely discussed his own life or feelings. He occasionally prefaced the sermons he delivered at Ebenezer with reports from the battles raging on various fronts, confessing at times to his own feelings of fatigue or discouragement, but for the most part even these sermons do not convey explicit revelations of his "personality." They do contain set pieces, formula stories in which he plays a central role, for example, the story of his visit to India or the formulaic account of Mother Pollard, who had encouraged him in the early days of the Movement, but these always serve a purpose in doctrine or ethics other than simple autobiography. Increasingly, these Ebenezer sermons come to reflect his appreciation of ordinary black experiences in America, including his own, which he shared with a genuine and artless narrative style. But on the whole, his entire canon of set pieces and personal accounts offers comparatively little in the way of self-revelation.

Indirectly, however, the sermons at Ebenezer are massively self-referencing. They subtly reflect the normal needs of the human ego: the tendency to defend oneself against accusations, to justify one's decisions, to celebrate one's own accomplishments, to confess one's sins, and otherwise to establish a reliable, predictable, public character. He communicates by means of indirection because the African-American preaching tradition cares little about soul-searching and introspection in the pulpit. It features the proclamation of a church's and a race's deliverance, not the feelings of the preacher. "Be yourself" is advice the young black preacher jettisons early in his or her apprenticeship. Paradoxically, King found intellectual nurture in the therapeutic traditions of personalism and, like many of his generation, he was at home in the religious-neutral language of "the self." He was schooled in Paul Tillich's attempt to correlate the concepts of theology and psychotherapy, and he frequently alluded to the lessons of psychology. The result of this clash of traditions was a style of veiled self-referencing in which he spoke about other people and events but was often indirectly meditating on his own experience and the state of his own soul.

Like most preachers, he told stories about people who had overcome seemingly insurmountable obstacles and made something of themselves, the ostensible purpose of which was to encourage his hearers to persevere in hope and hard work. The stories, like those about Gandhi or Nkrumah, emphasized the enormous disparity between the humble origins of these men and their lofty accomplishments. His early "Sermon on Gandhi" might well be entitled "Sermon on King," for, like a *Bildungsroman*, it sketches the rise of Gandhi in terms chosen for their obvious parallel to the young preacher. Indeed, the most intriguing feature of the sermon is its implied role assignment to King himself. Both King and Gandhi insisted on nonviolence, both were jailed, both sought to reconcile sociopolitical antagonists, both were impelled by absolute self-discipline, including voluntary poverty, and both were targets of hate. "And here was the man of nonviolence falling at the hands of a man of violence. Here was a man of love falling at the hands of a man with [= of] hate. This seems the way of history."

It is equally difficult to read King's sermon on the heroics of Kwame Nkrumah and the birth of Ghana without hearing echoes of his own heroic struggle against segregation in Montgomery. When he recounts Nkrumah's imprisonments and advocacy of peaceful revolution, the listener cannot help but associate the twenty-nine-year-old preacher with these aspirations. And just as Nkrumah's success was symbolized by dancing with the duchess of Kent at the Independence celebration, so King's new stature is signaled (in the sermon) by his conversations at the same party with Vice President Nixon, Ambassador Ralph Bunche, and other dignitaries. In an indirect but unsubtle fashion, the sermon announces the heightened trajectory King has set for himself.

King's ambition created something of a problem for him. Good rhetorical technique dictates that even the most ambitious of speakers must mask that ambition from his audience by presenting it under the guise of other qualities and forces. As an alternative to the open celebration of his own importance, which would have been distasteful to his natural humility, King spoke of his role in the early Movement as one that had been fated by history or God, and presented himself as a person who had been called to a special task. For example, in describing the difficult circumstances surrounding his Holt Street address, he ascends to the language of mythication: "Now I was faced with the inescapable task of preparing, in almost no time at all, a speech that was expected to give a sense of direction to a people imbued with a new and still unplumbed passion for justice." Seven weeks later, after his parsonage was bombed, his speech from the front porch quieted the passions of an angry crowd. He describes its effect, again, in the grandiose imagery of mythication: "This could well have been the darkest night in Montgomery's history. But something happened to avert it: The spirit of God was in our hearts; and a night that seemed destined to end in unleashed chaos came to a close in a majestic group demonstration of nonviolence." The editors and cowriters of *Stride Toward Freedom* noticed his tendency to promote his own heroics above the contributions of others like E. D. Nixon and Jo Ann Robinson and advised a rewrite of several chapters. Despite his revisions, King's real or perceived self-aggrandizement provoked hard feelings and jealousy among his early associates in Montgomery.

More than his accomplishments, King's parish sermons indirectly reflect the fragmentation of his own life. In August 1967, he confides to Ebenezer, ". . . I'm tired of all this traveling I have to do. I'm killing myself and killing my health, and always away from my children and my family. . . ." In a sermon entitled "Ingratitude," he voices his usual complaint, but with a greater-than-usual bitterness, against the ingratitude of successful blacks who quickly forget the sacrifices of others. "I say to you this morning they are ingrates. . . . [They say,] 'Oh, I'm the first Negro here.' And they think it just happened. They talk about the jobs they have and, and, they think it happened out of the benevolence of industry." In a sermon entitled "Interruptions," he advises his listeners how to deal with life's disappointments, while reflecting at a deeper level on the harried and fragmented nature of his own existence. In the sermon he offers some transparent if dubious psychological wisdom: "Daydreaming is dangerous precisely because the more you daydream, the thinner and thinner your personality becomes until, ultimately, it splits. . . . And you should never do anything that makes that personality so thin that it ends up splitting." Yet it is precisely the "splitting" of his own existence that now, in early 1968, preoccupies him. Although he is preaching to others the value of rising above the forces that threaten to destroy "our personalities," it is clear that the preacher, beleaguered by criticism of his antiwar activities and his plans for the Poor People's Campaign, is ministering to his own spirit. "The question is whether you're going to stretch your wings forth and then go on above the storm. That's the question. It's life's question." He tells about a man who committed suicide by jumping off a bridge and another who considered it but resolved to live and went on to write the popular song *Good Night, Irene*. As King desperately exhorts his congregation to choose life over death, it is himself he is urging to persevere. He promises his congregation that he himself will put his faith not in things or good health but in God. He concludes by addressing his congregation much the way Daddy King did, as a single entity. "And I close this morning, Ebenezer, by urging you not to jump. When the interruptions of life come, reach down into the deepest bottoms of your soul, and you'll find something that you didn't realize was there."

A few scholars have combed the sermons looking for confessions of marital infidelity or other specific sins. They are not to be found except in his coded and more general reflections on the fragmentation of life. Even the somewhat stylized allusions to his own moral failures, which represent little more than a common preacher technique, are present only within the welcoming circle of Ebenezer. His favorite approach to sin is to distinguish between the intention of the heart and sinful "habits" whose repetition can lead to destruction. "Habits are easy to develop, but they are hard to break up," he says in "The God of the Lost." "Are you grappling with some bad habit that's destroying your real moral fiber, some tragic habit that has gripped you?" he asks in "Is the Universe Friendly?" "It keeps you from being really integrated with yourself, it's embarrassing your family, it's embarrassing your community, it's embarrassing your church. Christ is the answer." The habit he has in mind is sexual impurity, as he makes clear in subsequent remarks on Augustine. In "Answer to a Perplexing Question" he exhibits a pastoral understanding of the persistence of such sins:

> Many of you here know something of what it is to struggle with sin. Year by year you became aware of the terrible sin that was taking possession of your life. It may have been slavery to drink, untruthfulness, impurity of selfishness or sexual promiscuity. And as the years unfolded the vice grew bolder and bolder. You knew all along that it was an unnatural intrusion. Never could you adjust to the fact; you knew all along that it was wrong and that it had invaded your life as an unnatural intruder. You said to yourself, "One day I'm going to rise up and drive this evil out. I know it is wrong. It is destroying my character and embarrassing my family." At last the day came and you made a New Year's resolution that you would get rid of the whole base evil. And then the next year came around and you were doing the same old evil thing. . . . [T]he old habit was still there.

With these words King is ministering to the needs of a typical congregation. His comments do not constitute a personal revelation, but, by the same token, the preacher would not have missed their significance for his own life or exempted himself from his own message of pastoral care.

Sources as disreputable as Hoover's FBI and as reliable as King's closest confidants have alleged a pattern of sexual "habits" in King that eventually threatened to compromise his effectiveness and became the object of containment efforts by concerned aides. One SCLC staffer confided that it was King's sex life that stymied his associates from writing their own account of his leadership in the Movement. They did not know how to deal with the contradiction between the public mask and the private person. It was the secret they all shared. During his lifetime its revelation would have shattered his mask and with it the awe that hung over each of his performances. As Erving Goffman has noted, all great performers present "a basic social coin, with awe on one side and shame on the other." This is in some measure true of all preachers as well, who must make peace between the holiness of their message and the failures of their personal lives.

King's activities might have been troublesome to him on two scores, as a source of shame and a source of fear. If he was laboring under a burden of shame because of his secret sins and the contradiction between his public and private life, the sermons do not indicate it. With the exception of a few oblique references to "habits" that destroy personality and relationships, the sermons give little evidence that he was engaged in a moral struggle with his own infidelity or the casual sex practiced by many of the Movement's leaders, including the preachers. It is clear from his sermons and other comments that he did fear the exposure of his sinful habits. Exposure was but one danger in the pack of wolves gathering about him. As the FBI bore down upon his private life, he complained more frequently of how people will "tell lies on you" trying to destroy one's reputation. It appears that it was more the threat of exposure than the moral dilemma itself that helped precipitate his later depression.

By referring to impurity or drunkenness as a "habit," King relegates personal behavior (as opposed to social policy) to a position of second-echelon importance. Although sinful habits have potential for destructiveness, they do not touch the sacrosanct region of the heart, or personality, where the true disposition of the soul resides. The source of evil lies in the heart, but its most destructive effect occurs in society. He reminds his congregation that sins of the spirit, not lesser sins of the flesh, lead to lynching and bigotry. Besides, those who condemn specific personal sin often do so "because deep down within they really want to do it themselves." King sets aside a place of good in each person, including himself, that cannot be effaced by even the worst of personal habits. In his final sermon to Ebenezer, "Unfulfilled Dreams," in which he transparently associates his own failures and disappointments with those of

King David, the decisive criterion by which God judges King David (and Martin Luther King) is not the king's failure to perform his appointed tasks but the intention of his heart. "It was within thine heart; you had the desire to do it. . . ." To paraphrase another Martin Luther, grace has no receptacle save the human heart.

Beyond unfaithfulness or any other personal "habit" lay the menacing threat of despair. In his later years King preached more frequently on the topic of despair without, however, openly or confessionally confronting his own demons. His later sermons nonetheless corroborate the mounting depression that his friends and aides have variously ascribed to conflicts with other leaders, his fear of marginalization within the Movement, death threats, fatigue, harassment by the FBI, and the daily pressures of simply *being* Martin Luther King Jr.

In recent years many have commented on King's melancholy and its causes. SCLC aide William Rutherford asserts that his malaise was based more on "intellectual and spiritual" considerations than personal ones. King was desperately seeking to integrate all that he and his philosophy of nonviolence had achieved into the new world order of the late 1960s. But the world he had helped create in Montgomery, Atlanta, Birmingham, and Selma was already changing, and King was beginning to see with amazed eyes the enormity of the economic and class struggle that awaited blacks in America. The problem he had cast in terms of moral absolutes was proving intractable to the symbolism of moral solutions. Neither "Burn, baby, burn" nor white backlash fit into his dream for America. When touring the burned-out devastation of Watts, he had been heckled by a solitary black man, in itself an insignificant occurrence, yet also a portent of new tribulations awaiting him among his own constituency.

King's encounter with despair was given focus by his personal transition from the role of "central prophet" to that of "peripheral prophet" in the Republic. In ancient Israel the central prophet operated from within the power structure. He recalled the nation to its originating covenant and served as a consultant to kings on military alliances and other matters of national importance. The peripheral prophet, like Jeremiah, was an outsider who identified with the poor, taunted the monarch, and railed against war. As a central prophet, Martin Luther King had been deeply involved with the Johnson administration's efforts to pass civil rights legislation. Beginning already in the Eisenhower administration and increasing steadily throughout the turbulent Kennedy years, he had been regularly consulted on matters of interest to the Negro in America. After the March on Washington, Kennedy had scrambled to align himself with King's beautifully articulated ideals. In some of Lyndon Johnson's early civil rights speeches, King was gratified to hear echoes of his own ideas. No black leader had ever enjoyed comparable access to the Oval Office and the power it represented.

With his opposition to the war in Vietnam, however, King began his prophetic trek from the halls of power to the periphery. He now played Jeremiah to LBJ's tragic Zedekiah (see Jeremiah 38 and 39). All privileges of access were forfeited. Ostracized by Johnson and his administration, bedeviled by opposition from the increasingly powerful militant factions of the Civil Rights Movement, King appears to have felt that the road was closing before him. He confided his fear to Bernard Lee, "I'm going to be left out there alone." Contributions to SCLC were down. King was written off by the NAACP, the Urban League, and young black nationalists. Roy Innis denounced both his methods and his goal of integration. From New York came word that Adam Clayton Powell, always something of a thorn in King's side, was referring to him as "Martin 'Loser' King." His biographer David Lewis writes, "The verdict was that Martin was finished."

He began to fixate on one last Big Win in civil rights and, more worrisome to his aides, his already morbid personality gave way to full-blown obsession with his own death. If anyone had cause to contemplate death, it was King. By 1968, the FBI had investigated fifty plots to kill him. He was also the recipient of an enormous volume of hate mail that poured into SCLC headquarters on a daily basis. For example, after Malcolm X was assassinated, King received a grisly newspaper photo of Malcolm's body being carried from the hall with the following: "You damn nigger this is what going to happen to you some day. . . . Somebody sure as hell is going to kill you some day. . . ." He confided to Ebenezer, "Morning after morning, you will get up and look into the faces of your children and your wife, not knowing whether you will get back to them because you know that you are living every day under the threat of death." Staffer John Gibson remembers that toward the end of his life King was able to relax only when surrounded by friends in rooms without windows. In public he let his eyes unconsciously dart from face to face, looking for his assassin. Gibson concludes,

> I talked with Andy [Young] the week after Dr. King died—no, it was the Saturday before his funeral—and he said he was sure that Dr. King welcomed it because it had gotten to the point where the agony of wondering when it was coming was overpowering. But it never deterred him; it never deterred him.

King addressed his fears in his usual indirect mode of expression by generalizing them. He told stories about *other* people who collapsed or who did not collapse under pressure, encouraged his congregation to remain strong, and vowed that he himself would continue to trust in God. His longtime favorite hymn assumed a newfound sense of urgency:

> Why should I feel discouraged,
> Why should the shadows come,
> Why should my heart be lonely
> And long for heaven and home,
>
> When Jesus is my portion?
> My constant friend is He:
> His eye is on the sparrow,
> And I know He watches me.

The husky emotion of his voice and the ardor of his protestations of faith indicate the terrible emotional price his courage was exacting. The *feeling*, even the violence, with which he promises not to give in to despair tells the story. In "Mastering Our Fears," a favorite topic among liberal preachers, the intensity and honesty of his counsel to the congregation transcends the usual liberal advice. As we get older, he says, we begin to fear that "our personalities may collapse." "[A] sense of impending failure" encourages some people to waste their lives in excessive drinking and sexual promiscuity. Their "sunrise of love and peace" gives way to the "sunset of inner depression." They attempt "to drown the guilt [their persistent sin produces] by engaging more in the guilt-provoking act." He quickly summarizes what Freud, Tillich, and Karen Horney have to say about the multitude of phobias spawned by a capitalist society, culminating in the ultimate anxiety, the "ontological fear" of non-being itself. "It's the fear of death, really. . . ." It "seems that you are burdened down with the greatest trials, and you wake up crying sometime." His sermon on the words of Jeremiah, "This is my grief and I must bear it," reflects his own dark night of abandonment at this period in his life. In his last speech in Memphis he recited words he had been using since 1965, but at the phrase "I'm not fearing any man," his eyes rolled up in his head, and as he left the podium, he moved, almost fell, directly into the arms of Ralph Abernathy. He was living on the edge.

II

King organized the achievements, failures, fears, anger, depression, and indeed his whole sense of private and public identity in the only way he knew how: he adopted a series of biblical personae, masks, that captured the several roles he understood himself to be playing in American life. These roles coincided—and continue to coincide—with the most profound expectations that America, "the nation with a soul of a church," lays upon its public servants. King not only wore these masks but spent his life in an effort to live into them. They became his truest self. The first of these was the prophet.

Martin Luther King, Jr. was received by his followers as a prophet. Ralph Abernathy, whose willingness to play the role of jester enhanced his friend's nobility of "front," routinely introduced him as the "Moses of the twentieth century." SCLC organization kits contained similar phrases, reminding sponsoring groups of King's selflessness, historic importance, and other personal qualities. At his eulogy, King's mentor, Benjamin Mays, offered the most extended comparison of King and the biblical prophets:

> If Amos and Micah were prophets in the eighth century B.C., Martin Luther King, Jr. was a prophet in the twentieth century. If Isaiah was called of God to prophesy in his day, Martin Luther was called of God to prophesy in his time. If Hosea was sent to preach love and forgiveness centuries ago, Martin Luther was sent to expound the doctrine of non-violence and forgiveness in the third quarter of the twentieth century. If Jesus was called to preach the gospel to the poor, Martin Luther was called to give dignity to the common man. If a prophet is one who interprets in clear and intelligible language the will of God, Martin Luther King, Jr. fits that designation. If a prophet is one who does not seek popular causes to espouse, but rather the causes he thinks are right, Martin Luther qualifies on that score.

In reflecting on King, Fred Shuttlesworth understood him to be a man designated by God, "even as Moses was," to do God's work on earth. His father put it simply, "Martin is just a twentieth-century prophet." C. T. Vivian considers King to be "the prophet of our time." A black insurance executive from Montgomery remembers that young Martin "put a plumb line on the Cradle of the Confederacy. . . ." A retired beautician and member of Ebenezer voices the opinion of many when she says, "He was like a Moses to the black race."

Moses was the prophet-of-prophets who led his people from captivity. African slaves in America often blended the figures of Moses and Jesus into one. The lyrics of the spiritual recall both the open tomb and the Exodus:

> O Mary, doan you weep, doan you moan,
> Pharaoh's army got drownded.

Moses had liberated the slaves from their earthly bondage. Jesus, by his suffering, had freed them from their spiritual bondage. Together they symbolized one comprehensive agent of salvation. Perhaps out of respect for Jesus, African Americans tended to identify their leaders and hoped-for messianic figures with Moses. Not that Moses was the Messiah, but in Deuteronomy 18:15–18 God promises to raise up for Israel a prophet like Moses, a promise that has animated the messianic imagination of subsequent religious communities. African Americans have traditionally decorated their leaders with messianic imagery and have given the name "black Moses" to such figures as Harriet Tubman, Booker T. Washington, Marcus Garvey, Joe Louis, Martin Luther King, Jr., Jesse Jackson, and many others. Paul Laurence Dunbar gave poetic voice to the hope in *Joggin' Erlong*:

> Dey kin fo'ge yo' chains an' shackles
> F'om de mountains to de sea;
> But de Lawd will sen' some Moses
> Fu' to set his chillun free.

From the beginning of his career King embraced his prophetic role, attaching it both to his work for racial equality and his broader advocacy of peace and economic justice. In 1958 he wrote, "Any discussion of the role of the Christian ministry today must ultimately emphasize the need for prophecy. Not every minister can be a prophet, but some must be prepared for the ordeals of this high calling and be willing to suffer courageously for righteousness." In his 1963 "Letter from Birmingham City Jail" he compares his traveling campaigns to those of the eighth-century (B.C.E.) prophets who left their villages to speak on behalf of the Lord. Although he offered no systematic treatise on the prophet's role in society, his many allusions create an impressionistic profile of the prophet's vocation. The prophet is a spokesman for the Lord whose ministry is not restricted to a single locale or to local issues. He travels with the freedom of the Spirit. Whether a central or a peripheral figure, the prophet's most important audience is not the church but the nation, before whom he will speak words of judgment. No prophet is honored in his own country. He must steel himself with courage and be anointed with the willingness to suffer for the truth.

King substantiated his prophetic vocation with an account of the very thing he had lacked as an apprentice preacher: a *call*, which in the African-American tradition "is deemed more important than any type of preparation." Israeli scholar Mechal Sobel, who has read a great volume of conversion literature, including the narratives of former slaves included in *God Struck Me Dead*, notes that the African-American experience of conversion follows a pattern of eight steps. With several modifications, King's "kitchen experience" in which the voice of God called him to a prophetic ministry reproduces the traditional pattern outlined by Sobel. The individual is brought before God, if only, as one slave remembered, "in imagination," and God addresses the seeker by his very own "private name," signifying that God knows this person in the depths of the soul. The individual then "dies" and recounts the experience of his or her own death. The "little me" is divided from the "big me" and brought to the brink of hell, but a little white man leads the "little me" from hell to heaven. God makes his presence known to the saved person, and promises to be with his child forever. Now reborn, the saved person is sent back to earth by God. Now, no matter what dangers or temptations arise, the redeemed is free of them all and will never again be confounded. God promises the converted, "Be not afraid, for I will be on your right hand and on your left." In his call narrative King reports that a voice promised, "And lo, I will be with you, even until the end of the world."

The individual seeker may be burdened by the weight of an unconverted life or cut off from family and loved ones. In the conversion narratives the words "lonely" and "alone" appear frequently. On that winter night in the parsonage kitchen in Montgomery, King's burden was the loneliness and sense of embattlement so characteristic of the prophets in the Old Testament. In this respect, King's call narrative resonates to the biblical story of the call of Elijah in 1 Kings 19. Like Elijah who confessed, "I, even I only, am left, and they seek my life," King also acknowledges his loneliness and his utter dependency on God. Bowing his head over a cup of coffee, he prays, "But Lord, I must confess that I'm weak now, I'm faltering. . . ." And like Elijah, King is addressed by a "still small voice" that assures him of the divine presence. In traditional African-American narratives, some of the seekers report their death and rebirth in visionary imagery. "It was like lightning." "Like a flash, the power of God struck me." King is given a promise that he hears in an "inner voice," the content of which he reports by means of the lightning-and-thunder imagery from one of his favorite hymns, *Never Alone*:

> I've seen the lightning flashing
> And heard the thunder roll.
> I've felt sin's breakers dashing,
>   which tried to conquer my soul.
> I've heard the voice of my Savior,
> He bid me still fight on—
> He promised never to leave me,

> Never to leave me alone.
>   never to leave me,
>   No, never to leave me alone.

King preached the story of his call at Mount Pisgah Missionary Baptist Church in Chicago, where he elaborated its details considerably over the more staid account published in *Strength to Love*. At Mount Pisgah, he emphasized the importance of his *name* by repeating it as God (or his mother or father) would have spoken it: "Martin Luther, stand up for righteousness. Stand up for justice. Stand up for truth." *Martin Luther* signifies what Sobel would identify as the "private name" often celebrated in the traditional accounts, but it also represents what someone steeped in personalism like King would affirm as the essence of his identity. Only God can penetrate to the "fundamentum" of the person represented by one's name. Only God knows what makes "me" me. Although King does not pictoralize the division of his two selves and the transmission of his "little me" into heaven, the repetition of his narrative serves to produce the same effect, for in describing the fearfulness of his former state and contrasting it with his present courage, he is in effect revealing two Martin Luther Kings to his audiences and telling how one was delivered from the other.

In the traditional version of the call, the person who has been saved now witnesses to God's salvation and speaks of being "free at last." "Thank God almighty, I am free at last." King too rejoices in God's promise never to leave him and adds, "And I'm going on in believing in him." The slave conversion often coincided with the call to preach. For King, who had been preaching for seven years out of a general desire to serve humanity but without a dramatic call, the kitchen experience provided the long-awaited spiritual focus for his work and filled him with personal courage. It also helped confirm his public identity as a prophet. At last, he had his call.

In his sermons and civic addresses King executed a ritualized series of prophetic functions. In the middle years of his career King produced an imaginative picture of a better America. The America he evoked did not yet exist, but it had *preexisted* in the ideals of the founding fathers and in the mind of God. Only because of its preexistence was his ideal of the American covenant an attainable goal. King's prophetic imagination enabled America to envision a society in which skin color was incidental to friendship, goodness, and achievement. Many white Americans could not "imagine" eating with Negroes, sending their children to the same schools, living in the same neighborhood, or working as equals with those of another race. We are accustomed at least to such concepts and images today, but in the America of the 1950s and '60s these ideas defied imaginative representation. They were unthinkable because they were unpicturable. Segregation and discrimination were equated with the God-ordained order of nature and society. How could it be otherwise?

One of the most important tasks of the prophet, then, to whom God has given sharper eyes than others, is to perceive the true nature of the evil around him. What others accept as "doing what comes naturally" the prophet sees in all its grotesque horror. It is an abomination to the Lord. Citing Amos 8:4–6, Abraham Heschel writes, "The world is a proud place, full of beauty, but the prophets are scandalized, and rave as if the whole world were a slum." King frequently preached on the necessity of being "maladjusted" to the evils of segregation and discrimination, a notion he probably picked up from Heschel, who taught that the prophets in Israel were "morally maladjusted" to society's "conventional lies." King cautioned preachers against their easy assimilation into the worldview of the society in which they serve. Like Amos, who despised the religious feasts of Bethel, King warned against substituting "eloquent sermons," "glad outpourings," "the beauty of your anthems," and "your long prayers" for genuine acts of justice.

What God expects of the prophet is not flowery ritual but a kind of divine madness that shatters the complacency of religious people. The origins of prophecy are in ecstasy. The root meaning of to prophesy may be "to slaver," "to foam at the mouth," hence the utterances of one whose sensibilities the spirit has completely alienated from civilized life and discourse. Hebrew scholars contend that prophets are "maladjusted" figures whose "pathological" visions are given utterance in tones "one octave too high for our ears."

Once the prophet sees the true nature of the evil around him, he names it. In his prophetic ministry, King gave names to what he saw: sin, racism, genocide, doom, cowardice, expediency, idolatry of nation, militarism, religious hypocrisy. In his 1966 sermon on Julian Bond he cries, "And there is a time when prophecy must speak out. 'Your hands are full of blood,' says Isaiah. 'Get out of my face. Don't pray your long prayers to me, don't come to me with your eloquent speeches. Don't talk to me about your patriotism. Your hands are full of blood.' Somebody must tell the nation this."

Only after the evil has been named does the reconstructive effort begin. King is undoubtedly best known for this phase of his prophetic activity, namely, his imaginative alternative to a racist society. Just as Isaiah metaphorically transported the exiles into a lush oasis of peace and exaltation, King's prophecy of an ideal America in which "the sons of former slaves and sons of former slave-owners" will sit at table together, gave a rich visual content to the old American dream. What makes this metaphor *prophetic* is precisely its metaphoric quality. If he were merely painting a nice picture of racial harmony, as most Americans still believe he was, the "I Have a Dream" speech would not qualify as prophecy. But the image of the shared table *stands for* God's justice in the world. The modern prophet is claiming that God's covenant with the world, which entails the restoration of relationships between people and with God, will *look like* white kids and black kids eating at the same table and treating one another as kin.

The dreamer King is actually the *mediator* of a religious covenant with the American people. The covenant is an amalgam of the biblical God's covenantal pledge to Israel, the Puritans' self-assurance of God's special favor, and the Constitution's guarantee of mutually protected rights in an ordered society. The complexity of the American covenant is appropriately symbolized by the deist Jefferson's suggestion that the Great Seal of the United States should contain a picture of Moses leading Israel across the Red Sea or by Lincoln's characterization of America as "the almost-chosen people." In his second inaugural address Jefferson invoked "that Being in whose hands we are, who led our forefathers, as Israel of old, from their native land. . . ." King perpetuated this thoroughly civil, yet explicitly biblical, tradition by functioning as an intermediary between the noblest features of the sacred American covenant and a racially torn society at war with itself. Like many of the biblical prophets who exhorted Israel to *return* to the provisions of the covenant, the early and midcareer King fulfilled a conservative function, for his imaginative vision of what America would become was derived from his vision of what America had once been. With Langston Hughes (again) he could sing,

> Let America be America again,
> Let it be the dream it used to be.

He therefore called America *back* to its moral and religious foundations. Sometimes he did so by adroitly juxtaposing quotations from the Scripture with the lyrics from patriotic anthems.

> I have a dream that one day
>   every valley shall be exalted,
> every hill and mountain shall be made low,
> the rough places shall be made plain,

and the crooked places shall be made straight
and the glory of the Lord will be revealed
and all flesh shall see it together. . . .

This will be the day when all of God's children will be able to sing with new meaning "my country 'tis of thee; sweet land of liberty; of thee I sing."

On many other occasions he exploited the sacred character of America's greatest civil hymn, *The Battle Hymn of the Republic*. The hymn draws on the apocalyptic imagery of the Book of Revelation in order to portray a time of testing for America: "He is sifting out the hearts of men before His judgment seat." In the hymn (and in King's use of it), according to Robert Bellah, "Christian holiness and republican liberty are finally conjoined: 'as he died to make men holy, let us die to make men free.'" No less than Lincoln, who interpreted God's covenant and judgment to the nation, King mediated the salient features of divine revelation to the almost-chosen people of America. Like the Israelites who could look upon Moses only after his face was veiled, Americans preferred to confront King's splendid dream of the American covenant, which perpetuated the myth of America's chosenness, rather than the terrible facts of the nation's origins. They did not understand that when the dream is a *prophet's* dream, it activates God's will and *must come to pass* (". . . *knowing* that we *will* be free one day," he said). The twentieth-century Moses, then, mediated a covenant with which many Americans felt comfortable, although—or perhaps *because*—they did not grasp the eschatological inevitability that underlay its rhetoric.

During this midperiod of his career, some of King's utterances reflect the typical American conflation of the two covenents. Occasionally, his biblical-prophetic posture is accompanied by civil-prophetic content, with the result that the biblical prophets' absolute reliance on *God* as the sole criterion of judgment is so thoroughly sifted through the imagery of the Republic that it is devalued. The biblical prophets do not comfort Israel or remonstrate with her on the basis of the moral law or any virtue that is separable from the nature of God. God is in favor of justice not because it is a profound moral idea but because it is a part of God's nature. For *all* his ways are justice, the Bible says. Thus according to Abraham Heschel, "To identify God with the moral idea would be contrary to the very meaning of prophetic theology. God is not the mere guardian of the moral order." In a 1963 sermon King conflates the prophetic style and the civil idea in a way Jeremiah (or Lincoln) would have never done: "American is doomed . . . because she has failed to live up to *the great dream of America and to her great ideas.*" To an early audience he defines a prophet as one who fearlessly declares the doctrine of the fatherhood of God and the brotherhood of man. He sometimes equates the mission of the prophets with his self-described "commission" to apply the "ethical insights of our Judeo-Christian heritage" to contemporary social problems.

That this job description collapses prophetism into social ethics is probably owing to the influence of Walter Rauschenbusch, who did the same thing fifty years earlier. He wrote in *Christianity and the Social Crisis,* "The prophets were the heralds of the fundamental truth that religion and ethics are inseparable, and that ethical conduct is the supreme and sufficient religious act." The biblical prophets, however, based their demands on the holiness of God. They recalled the community to obedience to the covenant and purity of worship. Rauschenbusch and King understood that, but the circumstances in which they ministered led them to articulate their prophecies in terms of social improvement. King's focus on ethics is more readily understandable than Rauschenbusch's, for unlike his predecessor, King enjoyed a truly national platform. King understood prophecy as a means of transforming the nation, but only secondarily as a method of correcting and empowering the church. He knew that no matter how dyed-in-the-wool its Christian soul, this nation, born in racism and nurtured in pragmatism, would not be persuaded by appeals to the holiness of God.

For much the same reason he intentionally did not appeal to or in any way reprise the roles of the black Reformers who had preceded him. The roles created by Allen, Walker, Garnet, Vesey, Turner, and Douglass were available to him, and he apparently knew them. But King understood prophecy as a word addressed to the *whole* nation because, due to unprecedented historical circumstances, the whole nation was accessible to him. His first tactic was to identify the Movement with mainstream Western values, thereby creating a symbolic consensus between white and black people of goodwill. The corollary of this tactic was the isolation of what he initially believed to be a small knot of racists and unsympathetic white opponents. Finally, by means of his visionary portrayal of the "beloved community," he would persuade the new majority to embrace equality and brotherhood.

III

Only when the majority stiffened in its resistance and the beloved community failed to materialize, did King begin to sound like his radical predecessors and embrace the more dangerous function of prophecy, namely, the exercise of speech that is received directly from God and hurled against God's opponents. King understood that the greatness of Moses lay not only in his leadership of the Exodus but in his unique privilege of communicating immediately, that is, without the necessity of a mediator, with almighty God. Only a prophet like Moses can say, "Thus says the Lord." The prophet shouts from the housetops what others will only whisper in their closets. King intuitively seized upon the immediacy of the prophet's relationship to God as a nonnegotiable source of his authority. This led him to two prophetic moves.

First, like the prophets in ancient Israel who debunked their own nation's false religiosity, King turned against the American covenant and began to demythologize it. The so-called Christian America in which he had first believed, then brilliantly exploited, he now bitterly attacked. He pointed out that "liberty and justice for all" had never meant *all*. The so-called divine covenant by which the nation was founded suffered from serious flaws. In a 1968 address he said, "[T]he problem is America has never lived up to it [its own covenent] and the ultimate contradiction is that the men who wrote it owned slaves at the same time." This has produced a "terrible ambivalence" in the soul of white America. It was the war in Vietnam, however, that helped him rediscover the racist basis of American history. As he suddenly saw it, once again the United States was treating people of color as nonpersons; once again it was obliterating entire cultures in the name of God in pursuit of national vanity. In the late period of his life, he preached as if he had just stumbled upon the dirty secret of America's misery.

King's second move follows closely the first. Along with his demythologization of the American covenant comes the prophetic privilege of speaking univocally for the Lord. He has eliminated the middle man of civil religion and is now witnessing as if from God's heart. Individual phrases that simply do not exist in his earlier sermons and speeches now appear everywhere.

> The judgment of God is on America now.

> I don't know what Jesus had as his demands other than "Repent, for the kingdom is at hand." My demand in Washington is "Repent, America."

> Our ultimate allegiance is not to this nation. Our ultimate allegiance is to the Almighty God, and this is where we get our authority.

> Our government and the press generally won't tell us these things, *but God told me to tell you this morning.*

> When God speaks, who can but prophesy?

America's on the way to hell.

[In Memphis the night before his death:] God sent us here to say to you that you're not treating his children right.

In this period he consistently grounded his right to speak for God in the office of the ministry. The first reason he usually gave for his opposition to the war in Vietnam was his "calling" as a minister of the gospel, by which he meant his prophetic vocation. To a packed house at Victory Baptist Church in Los Angeles he explained his position in this way: "Now there are those who say, 'You are a civil rights leader. What are you doing speaking out? You should stay in your field.' Well, I wish you would go back and tell them for me that before I became a civil rights leader, I was a preacher of the Gospel. . . ."

King reinforced his prophetic persona not only by quoting the biblical prophets but also by creating resonances to their style. With Isaiah he can be tender: "My people, my people," he cries from the Capitol steps in Montgomery, "listen. The battle is in our hands." With Jeremiah, who mourns for Israel's sin,

> My grief is beyond healing,
> my heart is sick within me.

King, too, weeps for America's hardness of heart. "This is a grief," he says in unison with Jeremiah, "and I must bear it." Like the prophets, King's most characteristic emotion is not rage but sorrow at the intractable sin of the people. The racism of his fellow Christians in the white denominations has moved him "to weep tears of love."

King also replicates the prophet's sense of urgency. Because the prophet experiences the world from God's perspective, the prophet's sense of timing does not correspond to the world's. All that matters is that sinful persons should turn from their evil "today" as soon as they hear God's command. King frequently merged his appeal with that of the biblical prophets, reminding his hearers that there will never be a convenient or acceptable time for racial justice. His insistent mode of speech corresponds to the prophet's call to action: "Today" is not a program for self-improvement or a blueprint for a Great Society but a cry in the wilderness. Although King's demands for immediate action grew more radical in his later years, they did not become more specific. In his last SCLC presidential address, he calls for "restructuring the whole of American society," but beyond the questions he raises about capitalism's ability to distribute wealth, the speech makes no concrete proposals. Indeed, he appears to shy away from specifics by retreating into prophetic language, concluding that the economic structure of America must be "born again."

Many of the Old Testament prophets engaged in parabolic actions that foreshadowed coming events. Isaiah went naked in the streets as a portent of Egypt's imminent humiliation. Ezekiel refused to mourn his wife in order to prophesy a time when Israel's dead would not be mourned. In ancient Israel, says Gerhard von Rad, these symbolic actions not only predicted events but began the process of their realization. When a modern prophet enacted some fragment of the future, say, an integrated lunch counter, the act itself represented the first phase of the prophecy's fulfillment. It was the prophet's way of saying, "It's going to happen. It's already happening!" What many assume to have been tactical moves on the part of the leaders of the Civil Rights Movement were in fact prophetic actions. Of course, it may appear that to interpret ordinary actions like walking or eating as prophetic signs of an imminent reality is to overinflate them with symbolic or theological significance. But how *does* one explain the irrational fear and anger that a sit-in or a march engendered in white southerners? Outsiders have occasionally wondered why the authorities did not simply permit a group of Negroes to walk a few blocks to city hall without offering violent interference. What provoked these distorted grimaces and snarls on the faces of spectators, now frozen for history in old photographs and museum exhibits? Perhaps these Bible-believing southerners suspected what ancient Israel knew, that the actions of the prophets, just as surely as their words, are the signs of a new order that is rapidly approaching.

Prophetic actions are harbingers of the future. Regardless of how humble the march, it signified movement from one condition to another, sometimes from a church located in a poor section of town to the commercial or political center of the town. "We're goin' *downtown*" meant more than "We are going for a walk." When the march was undertaken in the name of God, it participated in all those other movements, most notably the Exodus, by which God's people attained their goal. Before the march from Selma to Montgomery, King's prayer on Highway 80 announced the alignment between the march and the biblical journey of God's people through the wilderness. Once again, the implicit presence of Moses, who guided his people from captivity into freedom, hovers over the assembly: "Almighty God, Thou hast called us to walk for freedom, even as Thou did the children of Israel. We pray, dear God, as we go through a wilderness of state troopers that Thou wilt hold our hand."

Other prophetic actions conveyed similar symbolic messages. When King and others went to jail without acrimony and in good spirit, that act signified the Movement's embrace of suffering as a means of redemption. It was as if the gospel was being imprisoned by a hateful law. Since the jail terms were relatively brief, the release signified the ultimate triumph of the good over injustice. When King and his associates put off their preacher suits and donned overalls for a march, it was a way of demonstrating the humility of the Movement. When they did not resist personal abuse and scorn, their nonresistance symbolized the way of suffering over violence, a way that had already been validated by Jesus and the prophets. When King preached, as he sometimes did, in the smoldering foundation of a firebombed Negro church, his sermon constituted a prophetic act symbolizing the victory of the gospel over every attempt to destroy it.

IV

King overlaid his prophetic vision with imagery and verbal expressions suggestive of the Apostle Paul. Paul too preached prophetically, for he courageously proclaimed Christ in prisons and before the rulers and emperors of his world. The pictures of Paul recorded in Acts, Philippians, and Ephesians portray an apostle whose passion for the truth cannot be dampened by arrests, imprisonments, beatings, controversies, and the threats of the authorities. They reveal a man who, because of his surpassing knowledge of the truth in Jesus Christ, is at peace with himself. So powerful is the truth of Paul's witness that when he and his associates are imprisoned, their very presence in the jail is enough to convert their fellow prisoners and the jailer, an occurrence that King and his associates also report on several occasions. The Movement loved to sing about its jailed heroes in its adaptation of a spiritual that first appeared shortly after Nat Turner's insurrection:

> Paul and Silas *bound* in jail
> Got no money for to . . . go to bail
> Keep your eyes
> On the prize—
> Hold on . . .

While Paul was in prison he wrote much of the canonical literature of the New Testament. When Martin Luther King, Jr. was imprisoned in the Birmingham City Jail, he produced a canonical classic of prison literature. Like Paul, King was occasionally reverenced above his message. At Lystra and Derbe, Paul was forced to decline acclamations of his own deity. When a woman literally fell at King's feet and worshipped him, one of his asssistants remembers that King was "mortified." Like Paul, who preached a "philosophical" sermon in Athens in order to appeal to the "cultured despisers" of the faith (see Acts 17), King crafted his message to the capacities of his varied audiences. He too tried to be "all things to all people." Like Paul's, some of King's actions produced controversies, even riots, and like Paul, King was forced to explain these upheavals before the highest authorities in the government—with little success.

It was in Paul's role as *apostle* (one who is sent), however, that King found echoes in his own peripatetic ministry. King repeatedly justified his crusades in cities other than his home base by alluding to the "Macedonian call," which summoned Paul beyond the bounds of Tarsus. In going to Birmingham or Chicago, where he was branded an "outsider," he invoked the same *mission* that propelled Paul throughout the Mediterranean world. In his "Letter from Birmingham City Jail," he wrote, "[J]ust as the Apostle Paul left his little village of Tarsus and carried the gospel of Jesus Christ to practically every hamlet and city of the Graeco-Roman world, I too am compelled to carry the gospel of freedom beyond my particular hometown." The truth of the gospel justifies Paul's and Martin Luther King's freedom of movement. Justice knows no boundaries. The Spirit blows where it wills. Just as Paul was usually greeted on arrival by a small cell of believers already present in the city, so King came by invitation only and was reverently received as the chief Apostle of Freedom. His arrivals at packed churches in Albany, Selma, or Memphis touchingly recreate a Pauline arrival at one of his house-churches. There is a powerful scene in Acts 20 in which Paul kneels on the beach in prayer with his associates in Ephesus, then embraces them before bidding them farewell. The spirit of that bond between apostle and coworkers is captured by the reporter Pat Watters's description of the students welcoming King after his release from Reidsville State Prison. "They stood in a long line. It was dusk, and a full moon glowed on their silhouetted forms. Dr. King got out of his car and waved to them, and they responded by beginning to sing *We Shall Overcome*.

It is always the bittersweet suffering, however, to which Paul and King return as to an old and trusted friend for the most profound evocations of their ministries. They both count the cost and complain of their sufferings: Paul of his frequent imprisonments and beatings—five times whipped, three times beaten with rods, once stoned, three times shipwrecked—all this in addition to his "daily pressure" and "anxiety" for the churches. King also enumerates the trials and indignities he has undergone on behalf of freedom:

> Since that time [in Montgomery],
>> I've been in more than 18 jail cells.
>
> Since that time
>> I've come perilously close to death at the hands
>> of a demented Negro woman.
>
> Since that time
>> I've seen my home bombed three times.
>
> Since that time
>> I've had to live every day under the threat of death.

All these he endured over and above his daily and unremitting anxiety for the Movement. Paul's sufferings helped to substantiate his credentials as an Apostle. King recited his trials for the purpose of assuring his hearers that just as God has preserved his life, God will take care of them too. But at a deeper level, and especially when away from Atlanta, the catalogue of suffering also functioned for King as a proof of his apostolic authority.

In his most profound comments on suffering, the Apostle linked his own weaknesses and failures to the suffering and death of Jesus. His ministry is not *like* Christ's; it is organically joined to it. We carry in our bodies the death of Jesus, Paul says. Therefore just as Jesus' diminishment produced an ultimate victory, so all the suffering that attends the apostolic ministry will be the agent of the church's final vindication. "So death is at work in us, but life in you." This is a far deeper exploration of suffering than mere recital of one's own trials and inconveniences. While King resonates to Paul's language of apostleship and suffering, it will be clear from the following section that he also shares Paul's mystical participation in Christ's suffering and death.

King's unity with Christ provokes several verbal expressions in the Pauline style, most notably the aggressiveness of one who is free in Christ and unafraid to offer himself as an example to others. On more than one occasion Paul speaks of "my gospel" and tells his fledgling Christians to "do as I do." Unity with the crucified One is the seal of apostolic authority. The apostle is the authority to be emulated by others. As King's ministry began to look more and more like a failed project, his own sense of personal authority as an apostolic partner of Jesus Christ gathered strength. He became more uninhibited in telling his congregation to "do as I do." He assured Ebenezer that in Hitler's Germany he would "have openly broken the law. I would have practiced civil disobedience." If he were of draft age, he tells them, he would go to jail before fighting in Vietnam. He is no longer fearing any man or government. Like Paul, who finally began living altogether from a divine rather than human perspective, King distanced himself from the judgments of politicians and the opinions of the newspapers. His final insouciance—"and I'm happy to-night. I'm not worried about anything. I'm not fearing any man"—echoes a weary but defiant Apostle who said, "But with me is it a very small thing that I should be judged by you or by any human court." And again, "I know how to be abased, and I know how to abound; in any and all circumstances I have learned the secret of facing plenty and hunger, abundance and want. I can do all things in him who strengthens me."

V

Early on, when King began casting himself in a prophetic role, he alluded to the necessity of suffering. In his 1958 pronouncement on prophecy he anticipates the final role of the prophet: "[B]ut someone must be prepared for the ordeals of this high calling and be willing to *suffer* courageously for righteousness." He would later characterize "the calling to speak" as "a vocation of agony." What appears in his early comments as a few thinly veiled hints of the dangers he and all prophets face erupts at the end as a frank association of his own suffering with that of God's suffering servant, Jesus.

For his part, King's identification with Jesus was less calculated than his association with Moses and the prophets. At the beginning of his career, his supporters occasionally compared him to Jesus, as after his conviction in a Montgomery court when he was introduced to a mass meeting with the words, "Here is the man who today was nailed to the cross for you and me." King's early biographer reports that after the introduction, singing, weeping, and general pandemonium broke loose until King rose and uttered words of forgiveness for the judge. At another mass meeting in Montgomery, when King entered the church to speak, the waiting crowd shuddered with excitement and began to clap in "nervous, broken rhythms." The first hymn produced an emotional double entendre: *Come, Thou Almighty King!* During the same crisis a reporter for the Los Angeles *Tribune* wrote, "Undoubtedly King is the 'king,' this movement's leader . . . its crucified symbol, its blood and body-of-Christ. . . ."

A Methodist Sunday school guide in February 1957 identified Martin Luther King as a "saint," for which a tiny Methodist Church in Selma, Alabama, voted to withhold funds from Emory University because the author of the article was a professor at Emory's school of theology. Eventually the author, Claude Thompson, replied, "Any other candidates care to compete with the Rev. Martin Luther King for the saintly crown in Montgomery? In my way of thinking this is amazingly similar to the way in which Jesus of Nazareth faced the enemies of His day. This is saint-hood in man's church. . . ." King indirectly contributed to the Jesus identification when under enormous stress during public prayer he cried out, "Lord, I hope no one will have to die as a result of our struggle for freedom in Montgomery. Certainly I don't want to die. But if anyone has to die, let it be me." From 1956 forward, both church and world routinely identified Martin Luther King, Jr. with Jesus of Nazareth in his piety and his crucifixion. King had to cope with that symbolism, as well as his premature canonization in the Methodist Church, for as long as he lived.

King's measured theological appraisal of Jesus focused more on his life and teaching than his death and resurrection, which is one of the paradoxes of King's religious thinking, for the martyr who touched a primal religious sensibility in his identification with the death of Jesus actually venerated the Lord's teaching and example above his atoning death. True to his graduate training in liberal theology, the young pastor King found the greatest appeal in Jesus' Sermon on the Mount. "This is," King declared to his first biographer, L. D. Reddick, "a wonderful statement of the practical solution of the major problems that man must face in any generation. In it, answers to life's great questions are given in terms of the love ethic." A decade later, at Riverside Church in New York City, he held up the nonviolent ministry of Jesus Christ and his love for his enemies as the model for his own opposition to the Vietnam War. He frequently quoted the words of Jesus in the Sermon on the Mount, "Love your enemy," as the inspiration for his entire public ministry. In a sermon about Jesus to Ebenezer, King is clearly thinking of his own ministry when he says, "He [Jesus] always went into the temple, but the temple gave him the kind of inspiration to go out and do something for others." King often confided to Ralph Abernathy his worry that seminarians were being trained in a theology of the cross and resurrection, which he took to be Germanic and otherworldly, but were not being taught to emulate the deeds of Jesus' ministry.

Despite his reservations with regard to a theology of the cross, King's public character increasingly came to be shaped by resonances to the Crucified One. He had long taught that "unmerited suffering" was redemptive, a principle drawn from his tradition's (and Abraham Heschel's) reading of the Servant Songs in the prophet Isaiah. Heschel, whose writings King greatly admired, noted that the Servant's suffering is greatly disproportionate (= unmerited) to his offense and that his suffering is redemptive not only for himself or even Israel but for all people. King substituted "the Negro race" for Israel and "all Americans" for all people, and thereby produced a particularly African-American version of Israel's mission.

Scholars have long disagreed over the precise identity of the Servant in Isaiah. Some identify him with corporate Israel, others with the individual prophet. King's own theory of redemption through suffering also vacillated between the corporate (in which case the Servant is signified by the black race) and the individual (with himself in the role of the suffering prophet) whose death will in some unspecified way bring not only redemption to the black race but reconciliation to all people. With this persona, he fitted to his own face the most powerful rhetorical mask available to him, both because it united his psychological tendencies with his religious heritage and because it is the most recognizable and evocative symbol in Western culture.

Like the central figure in a passion play, King anticipated his own death as a redemptive occurrence. In the summer of 1966 he assured Ebenezer that he was ready to go the whole way for the sake of the disinherited: "If it means suffering a little bit, I'm going that way. If it means sacrificing, I'm going that way. If it means dying for them, I'm going that way."

In an earlier sermon he makes the same promise: "Some people are hungry this morning. Some people are still living with segregation and discrimination this morning. I'm going to preach about it. I'm going to fight for them. I'll die for them if necessary because I got my guidelines clear. . . ." A year later King drew the clearest distinction between the reformism of the Movement, which until 1965 was characterized by the advocacy of old-fashioned American values, and the radicality of the Movement post 1965, which would necessitate "a radical redistribution of economic and political power." He also confessed that the suffering of little children in Vietnam had led him to a radical new position on the war. In the context of this abandonment of safe positions on war and social justice, he speaks of the cross: "When I took up the cross, I recognized its meaning. . . . The cross is something that you bear and ultimately that you die on."

Like Nat Turner—who, when he was asked shortly before he was executed, "Do you find yourself mistaken now?" replied, "Was not Christ crucified?"—Martin Luther King, Jr. used the crucified Savior as the organizing symbol for his own depression, failures, rejection, and impending death. Rarely in his last sermons and speeches did he speak about Jesus without indirectly referring to himself, and rarely did his comments about himself fail to convey overtones of the Lord's betrayal and death. In "Standing by the Best in an Evil Time," he puts words into Jesus' mouth that more accurately reflect his own problems: "I want to thank you," Jesus says to his disciples, "for sticking with me when men were lying about me and saying evil things about me and spreading false rumors on me." In a January 1968 sermon entitled "Interruptions," he characterizes the life of Jesus as one of fatigue and constant distractions from his mission. He adds, "We see so many tragic interruptions these days. They have come to transform the buoyancy of hope to the fatigue of despair. The lights have gone out in so many of our lives."

In his famous "The Drum Major Instinct" preached in February of 1968, King reminds his congregation that Jesus tells those who want to be great that they must suffer and die with him. Toward the end of the sermon, just before giving the directions for his own funeral, he recites his set piece based on James A. Francis's portrait of Jesus in "A Solitary Life," but he revises it in such a way that it clearly reflects his own troubles. Concerning Jesus' reputation, he says, "They called him a rabble-rouser. They called him a troublemaker. They said he was an agitator. *He practiced civil disobedience; he broke injunctions.*" Later in the sermon he advises the preacher what to say about him in his funeral sermon. What he would have the world remember about Martin Luther King, Jr. he phrases in messianic language: "I want you to say that I tried to love and serve humanity." In this sermon, above all others, the words by which he associates himself with the rejected Messiah are overpowered by the husky pathos of his voice. Its quality of emotional desperation reproduces King's suffering more eloquently than the content of his *apologia*, which, when read instead of listened to, conveys only a fraction of the sermon's power. It was the agitation with which he preached that sermon, one of his parishioners remembers, that made so many in the congregation cry. The following month he went on to Detroit where he told a Methodist congregation not to give up, no matter how hopeless the situation seems. The sermon is unusual in that it contains no fewer than three stories about suicide.

No one will ever fully understand the psychodynamics of King's depression. That he was, in Stanley Levison's words, "an intensely guilt-ridden man"—and a depressed man—has been meticulously documented by his biographers Stephen Oates and David Garrow. Levison's explanation for King's guilt, that he felt that others also deserved tribute, does not resolve the matter. His explanation does not account for King's early struggle with his own feelings of aggression not only toward white people but also toward his overbearing father. Young King channeled these feelings into a variety of activities designed to please his father, and finally suppressed them through a philosophically and theologically cogent theory of nonaggression. Levison's remarks about King's mental state echo the Reverend Gardner C. Taylor's allusion to "some darkly brooding element in his makeup." His friend was "a peculiar man," said Taylor, and "a deeply troubled person. I don't think anybody will ever know the interior torture that Martin King went through."

Levison alludes to King's feelings of unworthiness, and Taylor places King in the company of many great preachers, including Fosdick, who suffered from depression, but neither adequately addresses the general morbidity of King's personality that led him to cast himself in the role of victim from the beginning of his prophetic ministry. Of this tendency Allison Davis asserts that "the typical response by King to his own angry and aggressive feelings was to feel guilty, and to *turn his anger against himself as punishment for his guilt.* . . . There was some deep compulsion in King to suffer, to sacrifice himself." Even without a satisfactory acount of King's mental state, which, of course, not even the most compelling psychohistory can claim to present, it is clear in the record of his sermons how thoroughly he associated the symbol of the crucified Messiah with the mission of his people and his own impending sacrifice.

VI

At the March on Washington, Martin Luther King, Jr. was introduced by A. Philip Randolph as "the moral leader of our nation," and everyone stood. By that time, the relation of character, message, and historical context, as it might be analyzed by Aristotelian or other criteria, had become a moot point. He was not only closely identified with the Movement, he had become its chief symbol. The print and television media had seized upon the oddity of a black man who could articulate the white man's philosophy in a black man's voice and, in the absence of other contenders, had equated Martin Luther King with the Civil Rights Movement. In a 1957 article in *Time*, Lee Griggs wrote, "Personally humble, articulate, and of high educational attainment, Martin Luther King, Jr. is, in fact, what many a Negro—and, were it not for his color, many a white— would like to be." King's personal history of suffering on behalf of freedom, including his Pauline-like résumé of arrests, imprisonments, and bombings conferred upon him automatic moral stature. He carefully wove allusions to his ministerial office into his public speeches, all of which he delivered in a dignified homiletical style. His personal history of courage and integrity, his careful management of the roles assigned to him, and his vocation to preach prophetic religion along with the gospel of the Republic—all converged to make of him *the* symbol of the sacred American covenant.

A true symbol confers the qualities of the thing it symbolizes. If King's *personality* was no longer a critical factor in his effectiveness as a speaker, it was because when he spoke he depended on something larger than personality, namely, the public presentation of his *character*. His character was now intimately related to the new history that was breaking out like flash fires everywhere. In the language of the 1960s, the new history was a "happening" named Freedom or the Movement in which an audience might be chosen to participate. His admirers validated their participation in the events, which, he assured them, were world-historic, by assembling in one of his audiences, marching with him, or otherwise establishing a link with him. He became the sacrament of the Movement. To participate in the Movement, one had to have a piece of Dr. King, or, after his death, be able to demonstrate one's connection to him. Authority in the Civil Rights Movement began to function on the model of apostolic succession. When an interviewer suggested to former aide C. T. Vivian that at the end of his life King was searching for his place in the Movement, Vivian responded incredulously, "Man, Dr. King *was* the Movement."

As the symbol of the Movement, it was necessary that King's personal life remain above reproach, free not only from sexual or financial scandal but also from any of the telltale signs of pride, pettiness, or venality. By the time of his death, the immense responsibility of being Martin Luther King had become a heavy burden. He found himself trapped by his own symbolic function. The pastoral liking for people that he had cultivated in his Dexter days now coexisted with a growing sense of his own importance. The legendary humility that was rooted in the acknowledgment of his own limitations gradually gave way to pomposity. According to one of his closest aides, he began to receive the opposition of fellow African Americans as a personal affront, as an act of lese majesty. On one occasion he responded to the criticism of an opponent with a condescending, "I shall *pray* for him." "King is a frustrating man," David Halberstam wrote in 1967:

> Ten years ago *Time* found him humble, but few would find him that way today, though the average reporter coming into contact with him is not exactly sure why; he suspects King's vanity. One senses that he is a shy and sensitive man thrown into a prominence which he did not seek but which he has come to accept, rather likes, and intends to perpetuate. . . . He has finally come to believe his myth. . . .

He believed it and was trapped by it. Since Aristotle, rhetorical theory has separated persuasiveness from logic and linked it to *virtue*—or the appearance of it. What Aristotle began, NBC completed with a succession of sound bites and images that could make or break a man (and a Movement) within seconds. King responded to the challenge with magnifient fidelity to the roles that had been assigned him. He carefully cultivated the patience and nobility necessary to maintain the moral high ground. Drawing on his training in the black preaching tradition, he was ever the prepared and public man. So much so that the necessity of the measured response, along with a brutal schedule of appearances and speeches, discouraged originality of expression and stifled spontananeity of argument. Even worse, his myth allowed him no public expression of his inner self, in which, despite his several masks, he fiercely believed. Although he frequently complained of the repetitiousness of his existence, what he truly chafed under was the necessity of his own *virtue* for the sustaining of the Movement. His repeated indiscretions, despite the warnings of friends and the threats of enemies, may have been his attempt to reclaim his "personality" and demythologize his own virtue.

He understood very well the sort of virtue the Movement required, and he knew that he had satisfied its requirements. He had not capitulated to the greater sins of cowardice and greed. The downright stubbornness with which he resisted co-optation would make politicians blush. Like Jeremiah, he had burned his bridges to the centers of power, making angry theological judgments and radical economic demands that middle-class Americans would never forgive. He played his public role so faithfully that he, knowingly, canceled his own future as a public person. After the war, no second incarnation would have found him in the Ivy League or running for president. Jeremiah never reappears in the Old Testament as the beloved doyen of the prophets, and neither would have King.

King appropriated the personae that were proven "performers" in Western culture. These personae—prophet, missionary, victim—also enabled him to take his place in the long line of black reformers and preachers who had worn the same masks and fought the same fights. Finally, it was to the symbolism of the masks that the American people responded.

White American liberalism embraced the victim but rejected the prophet. It accepted its own guilt but not the radical changes necessary for its liberation. Black America found it increasingly difficult to identify itself as a vessel of redemptive, or unmerited, suffering. It rejected such sentiments as these, delivered by King in sworn testimony:

> [I]n order to redeem the soul of the situation, of the nation, you must be willing to suffer and have violence inflicted upon you, but you don't

inflict it upon your opponent, and this is why I say that maybe there will be some blood in the State of Alabama before we get freedom, but *it must be our blood and not the blood of our white brothers.*

Black America embraced the threatening prophet who predicted long hot summers in the ghettos, condemned the war in Vietnam, and demanded a radical redistribution of wealth, but it grew weary with the priestly victim who in his latter years clung ever more tenaciously to suffering and death as the only means of redemption. In negotiating the immense social distance between the prophet's wrath and the messiah's suffering, King became in C. Eric Lincoln's suggestive phrase, "the unbearable symbol." His prophetic edge alienated his liberal white audience; intimations of corporate self-sacrifice worried and puzzled his own people. The split in King's sense of role (and America's religious consciousness) signaled the end of the Civil Rights Movement as a Christian phenomenon.

III

THEOLOGY AND BEYOND

8

In the Mirror of the Bible

KARL Marx once said that every great historical movement occurs in costume. Luther played Saint Paul, and the French Revolutionaries wore the Roman togas of the Republic and the Empire. The proletarian aside, of course, there has never been a revolution without borrowed imagery. From the beginning in 1955, King's leadership of the Movement was a calculated act of interpretation carried out in the mirror of the Bible's imagery, stories, and characters: the morning star of freedom illumined the darkness of an ordinary Southern city. Baptist and Methodist Rotarians were assigned and grudgingly assumed the role of "the pharaohs of the South." A festive parade of thousands along a state highway symbolized the Exodus from Egypt. Blood-spattered Negroes enacted the mystery of unmerited suffering. All of it was presided over by the black Moses who was willing to die for his people.

Some, like Ella Baker, assert, "The movement made Martin," alleging that the raw materials of this drama were there and available to the first opportunistic actor to come along. King's now-mythic stature, they imply, rests on a scaffolding of historical accidents. Even the most compelling rhetorical performance depends on the situation that evokes it, much in the way an answer is beholden to its question. Didn't King himself speak of being pursued by the furies of history?

But the Civil Rights Movement did not "make" King any more than the Civil War "made" Lincoln. Admittedly, like Lincoln, King was summoned by events he did not initiate and exposed to conditions he did not create, but his response was so powerful an *interpretation* of events that it reshaped the conditions in which they originated. His answer was so true that it reframed the question. Martin Luther King, Jr. was not the first Negro to champion the cause of civil rights in the twentieth century. He was merely the first to name the struggle and to declare its meaning. To divine that meaning, the young preacher turned instinctively to the Bible.

In his first public address at the packed Holt Street Baptist Church, the twenty-six-year-old King responded to the events in Montgomery by interpreting their meaning in relation to American history, the Constitution, and the Bible, but most of all the Bible. "We, the disinherited of this land," he cried, "we who have been oppressed so long, are tired of going through the long night of captivity." In one sentence the young preacher evoked five biblical images, each one ripe with the suffering and hope of Israel and the Negro people. From that night forward, King and the black church community forged an interpretive partnership in which they read the Bible, recited it, sang it, performed it, Amen-ed it, and otherwise celebrated the birth of Freedom by its sacred light.

I

At Crozer Seminary King could not have received a biblical education more at odds with his own tradition than the training given him by his Old Testament professor James B. Pritchard and his New Testament teacher Morton Scott Enslin. By the time King arrived at Crozer, the historical-critical method, or "higher criticism" as it was called, had been a fixture in liberal seminaries for more than a quarter of a century. Higher criticism analyzed the biblical text "scientifically" (as they still say in Germany) in order to arrive at the original or most primitive expressions of Israel's religion or the church's faith. In New Testament studies, the method did not claim access to the historical Jesus (who, said Albert Schweitzer in *The Quest of the Historical Jesus*, "will be to our time a stranger and an enigma"). Its more rigorous practitioners, therefore, did not necessarily adhere to the liberal agenda of imitating the moral example of Jesus, since all that can be known of the "real" Jesus is that he was an apocalyptic figure who had no ethics because he believed that his death would usher in the end of the world. From Pritchard, the future Moses of the twentieth century learned that "Moses" was a legendary code word for the tribe. From Enslin he learned that the life of Jesus is out of range to modern historians.

According to historical criticism, the text is a window through which the scholar catches a glimpse of the social and religious situation that lay behind it. King and his classmates were required to write essays explaining how cultural and psychological factors helped shape key Christian doctrines such as the Virgin Birth and the bodily resurrection. The seminarian King was so preoccupied with the supposed alien nature of the Bible that he did not notice that the historical criticism he was learning stifled the Christian impulse to get close to the Scripture, to live it, and did nothing to bridge the gap between cultures. In fact, higher criticism only magnified the distance between the Book and its modern reader. It exposed the cultural dissonance between the two and questioned the possibility or even the desirability of the modern reader's participation in the alien world of the Bible.

For students of that generation the presenting problem was, in the words of one of King's student papers, how to reconcile modern life and the "unscientific views in the Bible." The modern reader "comes to see that the science of the bible is quite contrary to the science that he has learned in school. He is unable to find the sun standing still in his modern astronomy. His knowledge of biology will not permit him to conceive of saints long deceased arising from their graves." It will not do to practice allegory, as the ancient church (and the black church) did, for "allegory is empty speculation." With a scientific flourish that would have startled his father, the seminarian concludes, "The interpretation of any portion of the Bible must be both operative and disinterested."

King seems to have gotten help in resolving the impasse between modern science and the ancient text by reading Harry Emerson Fosdick and immersing himself in the "evangelical liberalism" of his favorite professor, George Washington Davis. In a paper for Davis he outlined several ingredients in the proper use of the Bible in modern theology. The most important is the notion of progressive revelation. The anthropomorphic qualities of God found in the earlier books of the Old Testament gradually give way to the more humane and universal truths that are revealed in the later phases of the Bible's chronological development. The primitive tradition describes Yahweh as "a man of war," but Jesus reminds us that "God so loved the world." Although the African-American tradition had long celebrated the unity of a people, King saw progress in Israel's movement away from the "tribal solidarity" of Sinai to the personal religion of the prophets and its consummation in "Jesus who found the center of all spiritual values in earth in personal lives and their possibilities." Between the precritical tradition of the black church and the higher criticism of the exegetes, the theologian Davis mediated to King a progressive method of interpretation that moved from the particular to the universal and the corporate to the personal. "Here the phrase 'progressive revelation' becomes a reality," King wrote. "We can start with the major ideas of the scripture and follow them as they develope [sic] from the acorns of immaturity to the oaks of maturity, and see them as they reach their culmination in Christ and his Gospel."

King's student papers show that he entertained the claims of higher criticism with proper intellectual respect. He then responded to them in a manner common to an entire generation of preachers trained in the method: he ignored the method's reductionist conclusions but embraced its implicit notion of progressive revelation. He searched the Scripture for its enduring, universal truths. All that is precious in the Bible he had already discovered without the aid of sophisticated historical methods. King ignored higher criticism not because its results offended his piety but because its disinterested objectivity was as useless in his crusade to the nation as it was in the ministry of the African-American church.

When preachers trained in the method quietly abandoned it, they tended to retreat from higher criticism to morals or applications drawn from the text. Both historical criticism and moralistic interpretation imply an ability on the part of the interpreter to step back from the text intellectually and to appraise the difference between the world of the Bible and contemporary existence. That ability is a legacy of the Enlightenment, whose philologists and philosophers discovered the difference between historical events and the Bible's history-like narratives. If the biblical stories were not historically verifiable, they at least contained moral truths for religious edification. The Enlightenment's theologians were responsible for exchanging the great stories of salvation whose themes and characters had *enveloped* the medieval world for what eventually would become lessons in morality used to *decorate* the modern world.

The African-American tradition of interpretation, however, did not pass through the Enlightenment, never enjoyed the leisure of disinterested analysis, and therefore did not distance itself from the Bible or settle for moral applications. The cruelties of slavery made it imperative that African Americans not step *back* but step *into* the Book and its storied world of God's personal relations with those in trouble. Bereft of a remembered history of their own in a culture that valued historical consciousness, the enslaved Africans listened to the Bible and adopted a new history. Their leaders were Moses, Joshua, Samson, and Jesus. Their only hope was to recognize their own suffering and captivity in the Bible stories. The freed slave Brother Carper, whose sermon "The River of the Water of Life" is quoted in Chapter 5, understood that Old Testment history and the events of his own day were not identical; yet his sermon is not content to draw applications or lessons from one to the other. The world that he (and the African-American tradition) experiences does not merely correspond to the Bible; it is *enrolled* in the world of the Bible. It reprises the ancient story in a different costume while at the same time honoring the priority of the Bible's personages and events. The method of interpretation that accompanies this worldview is appropriately called "figural," for the Bible and contemporary experience take the shape of a single, enormous tapestry whose figures are repeated in many locations with a variety of significations. The interpreter (preacher + community) decodes the pattern as it occurs in contemporary events by means of a divinely given spiritual discernment.

Brother Carper begins another of his sermons, "The Shadow of a Great Rock in a Weary Land," with a a brief colloquy on figural interpretation:

> There be two kinds of language, the literal and the figurative. The one expresses the thought plainly but not passionately; the other passionately, but not always so plainly. The bible abounds with both these mode of talk. . . . It am as full of Christ as the body of heaven am of God.

African Americans were no friends to a strict historical literalism, for by that method the white preachers had "proved" the godliness of slavery. The "figurative" reading employed by slave preachers was a conscious strategy to enliven the spirits of a people held in literal chains. But even when white preachers switched techniques and read what they held to be figurative or "spiritual" texts to black congregations, their audiences instinctively did a double-reverse by *hearing* the text literally, that is, in a manner to their advantage. When, for example, a white Methodist minister addressed a black gathering on Psalm 68, "Let God arise, let his enemies be scattered," he was dismayed by the congregation's tumultuous response. "What was figurative," he ruefully observed, "they interpreted literally."

That interpretive reversal participated in one of the oldest metaphors for the Bible's relation to human life: the book as mirror, *liber et speculum*. The Bible mirrors or contains all of life, and life mirrors or replicates the figures and stories of the Bible. Paul Ricoeur writes, "Scripture appears here as an inexhaustible treasure which stimulates thought about everything. . . . In this way, the understanding of Scripture somehow enrolls all the instruments of culture" on its own pages. Henry Mitchell puts it this way: The black preacher reads the Bible as an inexhaustible encyclopedia of life. "He sees it as full of insights—warm and wise and relevant to the everyday problems of a black man." There is nothing in the Bible that one does not meet here in the world. And there is nothing in this world—no cruelty, stupidity, struggle, or marvel—that one cannot find in the pages of Scripture.

Typology discerns a continuity between figures as they are repeated in the Bible and as they reoccur in history. For example, Christ is a new Adam (see Romans 5:14); the eucharist is a latter-day manna (John 6); baptism "corresponds" to Noah's deliverance in the Flood (1 Peter 3:2). But typology is not limited to the pages of the Bible. In a recent sermon by Gardner C. Taylor, John the Baptist reappears as Martin Luther King, Jr. in the wilderness of America: "Dwight D. Eisenhower being President of the United States, and John Patterson the Governor of Alabama, J. Edgar Hoover, the omnipotent autocrat of the FBI, and Billy Graham and Norman Vincent Peale the high priests of middle America [*wild applause*], the Word of God came to Martin Luther King in the wilderness of America" [*tumult*]. Typology is the most important form of figural interpretation for it allows for the fullest participation of one reality in another. Allegory, which is the assignment of external values to textual figures, plays a less significant role in the same tradition. Both methods are employed in Brother Carper's (and the African-American church's) rich and allusive reading of Scripture.

Figural interpretation was in the atmosphere King breathed at Ebenezer. It was also mediated to him by his learned mentors, such as Benjamin Mays, Vernon Johns, and Pius Barbour, whose intellectual sermons rested on interpretive assumptions much older than the African-American church. In his 1957 sermon "The Birth of a New Nation," King engages in his own version of figural interpretation. In the first few minutes he establishes the two poles of his typology. First, there is the *type*, which is the Bible's story of the Exodus from bondage, the beauty of which was recently reinforced in the preacher's enjoyment of Cecil B. DeMille's *The Ten Commandments*. Since Emancipation, Negroes in America had been reading the Exodus as the story of their own deliverance from bondage. Before that, they had read and sung it in the hope of future redemption. In his sermon, King assumes that figural tradition without mentioning it. He moves quickly from type to the *antitype*, which is no one political program but an ontological condition of existence, "man's explicit quest for freedom."

"The Birth of a New Nation" is important to King's development because it marks a reversion from the philosophical method he learned in graduate school to his native theological assumptions. Like the idealists and the Marxists, King believed that freedom is the essence of the human spirit, but, unlike the philosophers, he confessed that the universal quest for freedom owes its existence to a biblical precedent. The pattern of freedom movements now appearing throughout the world is a direct result of God's commitment to freedom. King's use of typology follows the pattern of the New Testament, which spiritualizes Old Testament events, and it is consistent with the African-American tradition, which, except for periods of mass migration, interpreted the Exodus as a change of political and social *status* rather than geographical location.

He cites two instances of this typology of freedom in "The Birth of a New Nation": the emergence of Ghana from the British Empire and the struggle for freedom among American Negroes. In his discussion of the latter, he makes it clear that the American Negro's exodus will not be to a *place*—Africa—but to an enhanced *quality* of life: "We find ourselves breakin' a-loose from an evil Egypt, trying to move through the wilderness towards the promised land of cultural integration." Not incidental to his typology is the figure of Moses, who reappears as Kwame Nkrumah and Martin Luther King, Jr., both of whom, like Moses, lived as "exiles" in their own countries and acted as agents of liberation but will not live to see the fullness of the promised land. That the agent of deliverance *never* lives to enjoy the benefits of freedom is a law of history.

He now moves to the signification of Ghana's nationhood for the Negro's quest for integration into his own nation of America. The birth of Ghana means "that the forces of the universe are on the side of justice." The new African nation symbolizes the birth of a new universal order. After a long series of reflections on British colonialism, he returns to the biblical promise of deliverance, though this time not to the original Exodus from Egypt but to yet another layer of the tradition's use of the Exodus, the deliverance from Exile in Mesopotamia. This Isaiah 40 celebrates in the hills made low, crooked places straight, rough places plain; and all flesh shall see the glory of the Lord, which means, King exclaims, that they shall see it from Montgomery, from New York, and from Ghana!

The Exodus from Egypt reoccurs in the return from Babylonian exile, and both events, now fused in the nature of God and the essential aspirations of humankind, are available to freedom movements around the world. King's use of the Exodus pattern to describe the American struggle once again releases the power of the most liberative "type" in the Old Testament. At the same time he is careful not to grant "secular" events a life of their own outside the terms and framework of the biblical world. In such an atmosphere of interpretation, unbiblical ideologies such as revolutionary violence or racial separatism have no ground because they have no root or history in anything larger than themselves. They are not nourished by the ancient spring. Some of his opponents criticized King's focus on the New Testament concept of love as contrary to black self-expression, but more than love or any single virtue, it was his maintenance of the larger biblical world, the *figural imperative*, that both set the

Movement on fire and contained it within the boundaries of a sacred worldview.

He deftly enrolled the racial conflicts of midcentury America into the world of the Bible. In his famous "A Knock at Midnight," he admits that "one of the interesting things about all the parables of Jesus is that over and over again" (figure after figure?) "they reveal many truths . . . , and you can often use them as an outline to many things that we face in the contemporary world." He agrees with the historical critics that a parable has only one main point but then inserts his rationale for figural interpretation by allowing that one can find many other good things "not intended as the main point." Thus in a sermon on the Good Samaritan he focuses on the racial rather than religious differences between the Jews and Samaritans and compares the mugging on the Jericho Road to price gouging in the ghettos of Chicago. In a later rendition of the same parable, the man in the ditch is the Memphis sanitation worker. Who will stop to help him? he asks.

One Sunday morning at Ebenezer, he looked out upon a congregation that included his brother A. D. King, Andy Young, Dorothy Cotton, and Wyatt Tee Walker. They, along with King, had just come from jail for refusing to honor a court injunction. In a rambling twenty-minute preface to his sermon, he refers to them as "other young men" and then proceeds to preach a fire-breathing message about Shadrach, Meshach, and Abednego (Daniel 3). The relation of type and antitype is obvious but effective: Nebuchadnezzar's "sheriffs" issued an "injunction" and three young men "practiced civil disobedience" and refused to obey it. They were tossed into a fired-up furnace and delivered by God's intervention. The same scenario is now being played out in the South. The deliverance of the three young men is reoccurring among those who are jailed for civil disobedience. "Go to jail if necessary," King cries, "but you never go alone." "So I'm not sorry that we broke that injunction in Birmingham . . . I'm happy that in breaking it I have some good company. I have Shadrach, Meshach, and Abednego. I have Jesus and Socrates. And I have all the early Christians, who refused to bow." He concludes the sermon with an allusion to his vision in the kitchen, which he implicitly recognizes as a contemporary instance of God's deliverance of the defiantly righteous.

In his 1963 sermon "Lazarus and Dives," which he prefaces with the comment "Its symbols *are* symbols," the biblical figures reappear in the tapestry of history and race relations. Echoing a phrase of Albert Schweitzer's, he compares Africa to a beggar lying at Europe's doorstep. Moving closer to home, he identifies the gap between Lazarus and Dives

as "segregation" and the stupidity of white supremacy. He then proceeds
to universalize the parable by interpreting it as a lesson on the dangers
of materialism and inordinate self-love. If a parable is meant to pique the
hearer's contentment, King makes it do just that, for, having established
one rather predictable typological frame of reference in a black congregation, he lulls his people into a state of complacency. Ebenezer is comfortable with the concept of its own victimization at the hands of white
oppressors. Just when Ebenezer expects its preacher to reinforce its identification with Lazarus, King says to his congregation, "Each of us is a
potential *Dives*," who may be rich in education or social influence. Even
the oppressed, he insists, are potential oppressors. This odd coupling of
a politically specific and an a-political interpretive perspective surfaces
in many of his early and midcareer sermons. In his 1964 "Discerning the
Signs of History," which is largely devoted to the enslaving ideologies of
colonialism and militarism, he also includes a scathing attack on black
materialism. He does not, as he does in later sermons, focus exclusively
on white capitalism's manipulation of blacks to buy things they don't
need—a form of exploitation thoroughly scorched by Malcolm X—but
lashes out at brothers and sisters who put their faith in "beautiful auto-
mobiles." The interpretive method of these sermons represents King's
unwillingness to write off the pastoral implications of the Christian message. Typology for King, at least in his early and midcareer sermons, is
not a static political stencil to overlay every text but a dynamic instrument by means of which the pastor may probe the lives of his hearers. As
a small and quiet strategy—like a parable—these sermons also define his
distance from those whose appreciation of the Bible is limited to its political function.

II

King's best-known use of the allegorical method is his flagship sermon,
"The Three Dimensions of a Complete Life." The perfect proportions of
the heavenly city *stand* for something else, namely, the complete life of
the Christian. Unlike typology whose "type" is real and of intrinsic importance, in the allegory of Revelation 21 the actual measurements of the
heavenly Jerusalem are irrelevant. What matters is the spiritual values
they represent. Typology is more important than allegory in black biblical interpretation because it sustains its hearers at a deeper level. The
disinherited participate more readily in an event or person than in a spiritual value. A downtrodden community can participate as a body in the

Exodus or the cross of Christ, but the dimensions of the heavenly city or some other textual detail might mean anything. The community is left utterly dependent on the preacher's ability to read between the lines. Allegory permits clever interpretations of difficult texts and often elicits admiration from the congregation, but it can leave the text and congregation subject to manipulation.

King used allegory for the same reason the church has always used allegory: to permit a richness of expression not obtainable from historical literalism, which is another way of saying that he wanted to *use* the text in his own world. The Bible is more than a witness to a few key doctrines; it is an almanac one can live by every day. As Henry Mitchell says, the black preacher mines the Bible as an "inexhaustible source of good preaching material," not an "inert doctrinal and ethical authority." What good is the Book if it is so Israel-specific that it can't help black people survive? The medieval church allegorized texts to make them speak explicitly of Christ, salvation, and the church. Allegory was preparation for a *worldly* reading of the text, which was nothing other than the application of allegorically received doctrine to the lives and morals of the faithful. The moral reading grants the interpreter entree into contemporary existence. He or she reads the *world* by means of the Bible. Even the most arcane texts yield moral advice—and in King's reading—relevant political insights as well.

The progression from the allegorical to the moral interpretation of the Bible reached its conclusion in nineteenth- and twentieth-century topical preaching. In a topical sermon, a text is cited and briefly explained, but it serves as little more than a pretext for the exposition of a variety of moral, psychological, or political topics. In "Discerning the Signs of History," King's text is Luke 12, in which Jesus admonishes his followers to discern "the signs of the time" of God's visitation in his own person (v. 56). King bypasses the text's focus on the imminent revelation of Jesus and stresses instead the importance of understanding "the face of the past." The only way America will move forward in race relations is by reading the lessons of history. These turn out to be the moral laws that will inevitably consign colonialism, militarism, racism, and materialism to the ashbin of history. In the course of the sermon he supports these laws with no fewer than eleven historical examples ranging from Marie Antoinette to the Berlin Wall.

Jesus' parable of the Good Samaritan (Luke 10) has a checkered history of allegorization. The simplicity of King's treatment of it is restrained by comparison to the wider tradition. He begins by identifying the point of the parable: a neighbor is "any man who lies needy at life's

roadside. This is all the parable is saying." Having paid his dues to the
one-point approach of modern criticism, he opens the parable as if it were
an almanac of social behavior. Its characters represent three philosophies
of life. The robber lives by a philosophy of acquisitiveness. "What's yours
is mine. . . ." This is as true of the mugger as the slum landlord. The
Levite lives by a philosophy of indifference. "What is mine is mine, and
what is yours is yours. I won't bother with you." Both whites and middle-
class Negroes, he says, follow this philosophy. The Samaritan lives by
the philosophy of generosity or "other-preservation." "What is mine is
yours, and I'll give it to you." These are the three philosophies. The
preacher gives testimony to the third, and the sermon ends. He had begun it with the admission, "I'm very tired this morning, and I know if I
preach a long time, I'll be even more tired." The "Good Samaritan" he
preaches at Ebenezer is a far cry from the elaborately articulated allegory
on three kinds of altruism that appears in *Strength to Love*, but the rudiments of the method are the same. In both versions allegory opens the
gates to moral admonition.

In one of his favorites, "A Knock at Midnight," King allegorizes Jesus'
parable on persistence in prayer in order to make it yield a luxuriance of
moral and churchly lessons. Luke 11:5-13 tells the story of a person who
beats on his friend's door at midnight in an effort to borrow three loaves
of bread. The friend resists the request but, Jesus adds, he will finally
give in because of the supplicant's persistence. As with "The Good Samaritan," King begins the sermon giving a one-sentence summary of its meaning: "Now this is a parable dealing with the power of persistent prayer."
He promises that he will find within it, however, "a basic guide in dealing with many of the problems that we confront in our nation and in the
world today. . . ." Then, in a strategy worthy of the master allegorist
Origen, he proceeds to assign meanings to each of the parable's details.
"Midnight" signifies the deep darkness America confronts on a social,
personal, and moral level. The "knock" is the world's need of material
and spiritual help. The "bread" stands for the spiritual nourishment that
only the church can give. The man's initial disappointment reflects the
world's disillusion with the church's moral failures. The sermon is one
King loved to preach in the Mount Zions, Shilohs, Victory Baptists, and
others on his circuit. It is explicitly oriented to the mission of the black
church in America. In a little church in North Carolina, for example, he
warns the people not to get carried away by emotionalism or preoccupied with status ("we have *so* many lawyers!") to the neglect of its mission. Returning to the terms of his allegory one last time, he exhorts the
black church to "keep the bread fresh," that is, to retain the integrity of

the message for a world that needs it. "Keep the bread fresh" becomes the motif for a miniclimax in the sermon, which reaches its thunderous conclusion with a set piece on discouragement and the preacher's final allusion to his own call. In this sermon King's use of allegory permits him to associate his own themes with several details in the text, leaving the congregation with the impression that he has miraculously discovered the mother lode of these riches in a simple story about prayer.

The further King strayed from typology to allegory and moralizing, the more he underwrote the Enlightenment's "step back" from the text—and the more susceptible he became to the banality of religious lessons. Hans Frei writes, "The Enlightenment is known for many great accomplishments, but religious profundity is not usually among them." The religious stepchild of the Enlightenment, mainline liberalism, misread the great narratives of the Bible as the raw material for morals, truths, or lessons on the human condition. It cannibalized them for their spare parts. The story of the Exodus, to give the most prominent example, which in the Bible reveals nothing less than the identity of Yahweh and the deliverance of Israel, was taken by King's teachers to be an illustration of several laws of human nature and history that are ontologically *prior* to God's self-revelation. Liberal theology discovered a correspondence between selected principles of revelation and the laws of human history and personality. Liberalism validated its vision of Christianity by appealing to its universality, not its biblical particularity. This appeal constituted its apologetics toward the "modern world."

Given the theological climate in which King was trained, his own analytic step back from the text frequently led him to the authority of psychology. The Baptist tradition in which he was brought up had dwelt upon the motives and gradations of the individual's response to God, to the point that evangelism had become a science of conversion. By mid-twentieth century, psychology had become the secular successor of pietism and had established itself as an objective science of human behavior. The laws of psychology were premised upon an idealized essence of humanity that can be known apart from the authority of biblical revelation. This is the credo of liberalism, and, insofar as King was a participant in that theological subculture, it is discernible in many of his sermons.

In his sermon "Interruptions," as in many others, he begins no fewer than five sentences with variations on "Psychologists tell us," and proceeds to regale Ebenezer with a superficial explanation of the "psychosomatic" causes of illness. In the same sermon he cites the importance of achieving an integrated personality as opposed to those who have fallen prey to the "success complex" because "they have not balanced their personality with a failure complex." Two weeks later at Ebenezer he is doubting Freud's obsession with sex and leaning toward Adler's emphasis on power. In "Mastering Our Fears," based on 1 John 4:18, "Perfect love casteth out fear," he quotes "the psychiatrists" on "phobia-phobia," Karen Horney on schizophrenia, and Freud on the distinction between normal and abnormal fears, and attributes certain fears to a "basis in a childhood experience" now stored in the subconscious. He barely alludes to his text's Christ-centered message as a viable alternative to Tillich's therapeutic "courage to be."

On another Sunday he explains to Ebenezer the perennial dilemma of human identity: "And then we turn to depth psychology. And we try to say that really our problem is a result of inner conflicts, inhibitions, or to use Freudian terms, a battle between the id and superego." In "Training Your Child in Love," he impresses upon parents the importance of teaching self-esteem to their children, which is pastorally helpful to his congregation, but in "Interruptions" explains wife abuse as the expression of a subconscious need, which is not so helpful. Any sort of repression was taboo to King's generation, and he frequently counsels his parishioners not to "repress" their disappointments. Those who criticize excessive behavior in others are often repressing their own frustrated desires to do the same thing! Even daydreaming can be bad for your health, he warns, because it can lead to schizophrenia. He frequently quotes the Robert Schuller of his day, Rabbi Joshua Liebman, whose *Peace of Mind* was a bestselling brew of psychological and religious truisms.

Were it not for his special themes and the power with which he conveyed them, King's infatuation with psychology would be indistinguishable from that of his liberal contemporaries. Never mind that the "modern" that often prefaces "psychology" in the sermons of this period usually means about fifty years out of date. The black preacher would argue that if the Bible enrolls the *whole* world, no knowledge is too secular, advanced, or obsolete for inclusion.

King's black religion was inclusive, and his liberalism was apologetic. Psychological jargon served both purposes, but especially the latter, for it provided King a nonsectarian idiom for appealing to those outside the walls of Ebenezer. Like the literary "greats" whose quotations certified the Movement as Western mainstream, King's psychological authorities lent an aura of universality and even scientific inevitability to his biblical claims. The members of Ebenezer didn't particularly *want* a universal ground of authority other than the Bible, and didn't believe in the inevitability of anything but divine retribution, but the white audiences to whom these sermons were eventually preached were moved and impressed by these values. Not that King was feigning a white worldview; the Boston graduate unreservedly accepted the authority of psychology along with the personalist carton it came in and the apologetic operating instructions. Furthermore, psychology appears to have exerted a therapeutic influence on King himself. Like the New Testament Gospels that occasionally try to explain *why* people refused to believe in Jesus, the psychological theories helped King rationalize human behavior and especially the hate and rejection provoked by his own ministry of love. Implicit in his psychologizing was the belief that political repression has its roots in psychological repression or, at the least, mirrors it. How else could one explain the irrationality of white racism? Although he did not publicly pursue this theory, King knew that the anxieties he aroused in white America operated on a level far deeper than politics or economics.

III

An interpretive master code guided King's reading of Scripture and enabled him to escape the moralisms of liberal homiletics. Despite the influence of psychology on his thinking, King ultimately let the *church* guide his interpretation of Scripture. When the Bible is read as the church's book, the interpreter always works within the boundaries of the church's convictions about God. The interpreter and the Christian community already share a history of God's intervention in their lives. They have a tacit understanding of God's purpose for the church and the world. They enjoy general agreement on questions of morality. Together, they read the Scripture in such a way that will confirm and strengthen their mutual tradition.

King's code contains a narrative and a precept. The narrative is the story of liberation, the precept is the command to love. In a 1964 speech to a Baptist assembly he briefly summarizes the two parts of this master code:

> There are two aspects of the world which we must never forget. One is that this is God's world, and He is active in the forces of history and the affairs of men. The second is that Jesus Christ gave his life for the redemption of this world and as his followers we are called to give our lives continuing the reconciling work of Christ in *this* world.

The prototype of liberation is found in the biblical narrative of the Exodus. In keying on the Exodus, King was making a necessary but highly unoriginal choice. Whether free or slave, every group that came to America expressed its aspirations in the language and imagery of a particular facet of the Exodus. The Puritans focused on the journey and the occupation of the land. In worship, the slaves ecstatically crossed and recrossed the Red Sea and prayed for liberation. Boston Puritan Thomas Prince said in a sermon of 1730, "[N]ever was there any people on earth so parallel in their general history to that of the ancient Israelites as this of New England. . . . [O]ne wou'd be ready to think the greater part of the Old Testament were written about *us*, or that we, tho' in a lower degree, were the particular *antitypes* of that primitive people." The Puritans were the beneficiaries of God's Exodus, but no sooner had they occupied the Promised Land than they became the scourge of a new generation of "Canaanites" who, when *their* opportunity arose, turned the same Exodus narrative against their oppressors.

One former slave recalled the complexities of the Exodus typology: "The preachers would exhort us that we was the chillun of Israel in the wilderness and the Lawd done sent us to take this land of milk and honey. But how us gwine take land what was already took?" For slave or free, the way to peoplehood in America lay in becoming a biblical *antitype*. In the generations after the Puritans, Richard Allen, Denmark Vesey, David Walker, Nat Turner, Harriet Tubman, and Marcus Garvey all drew on the Exodus as well as other biblical motifs to plead the cause of their people or to prosecute it with militant actions. Allen characterized God as the "first pleader of the cause of slaves" but tempered the Exodus with the counsel of love. Vesey's insurrection was fueled by a more militant reading of the Exodus, prompting slaveowners to accuse him of "attempting to pervert the sacred words of God into crimes of the blackest hue." Martin Luther King's treatment of the same narrative, though nuanced for his own day, followed both Allen's and Vesey's lines of interpretation.

Historian James H. Smylie comments, "It is remarkable that King as a biblical interpreter alluded so infrequently in his formal writings to the Exodus narrative." This is a reasonably accurate assessment of King's published writings but does not acknowledge his delivered sermons and addresses, many of which either presuppose an Exodus framework or make allusions to the narrative. For example, in "A Christian Movement in a Revolutionary Age," he explicitly links Moses' Exodus to the Civil Rights Movement, as he does also in "The Birth of a New Nation," "Answer to a Perplexing Question," "Facing the Challenge of New Age," "Desirability of Being Maladjusted" (delivered in a synagogue), "The Meaning of Hope," and his finale, "I See the Promised Land."

Smylie provides an extensive analysis of the published version of

"The Death of Evil upon the Seashore," the main concept of which King owed to Phillips Brooks. But the very title of Brooks's sermon, "The Egyptians Dead upon the Seashore," jarred King's typology and violated his sense of magnanimity toward the vanquished. King quickly abandoned Brooks's sermon because it left too many "white brothers" dead on the shore. Smylie does not allude to King's more important interpretations of the Exodus in "The Birth of a New Nation," a less polished but more original work that displays the type-antitype scheme King used throughout his career. The prophecy with which he concluded "The Birth" in 1957 he repeated almost verbatim the night before he was assassinated. In April 1957 he had said, "Moses might not get to see Canaan." In April 1968, with his destiny bearing down on him, he said, "I may not get there with you."

King's method of retelling the Exodus narrative owes more to his slave forebears than to Phillips Brooks, though his debt is obscured by his own liberal vocabulary and his silence with regard to the great company of black preachers that preceded him. Like the slave preachers, King rarely tells the biblical story in a manner that is separable from the suffering of his own people and other oppressed groups. Nor does he relate the story of his own people's enslavement without using the imagery of the Exodus narrative. In keeping with this world's *enrollment* in the world of the Bible, King allows the two stories to enrich and illumine one another. In this practice he follows the example of the spirituals as well as the slave sermons. When he tells his enriched version, the story is more than a means of illustrating truths about God or the human condition. The accumulated force of the stories told by the slave preachers was to *render an agent* of deliverance, that is, not to *illustrate* God but to *create* God and to make him present to the people. Sometimes the agent was the God of the Exodus, sometimes a composite of Moses and Jesus. In King's storytelling, especially when he paints the cruelty of the Middle Passage and the pathos of kidnapped children, the agent he renders is not God but the corporate identity of the black race. What emerges is an agent of deliverance who is larger than King or any single messiah, a bowed but awakening black human who is made of the clay of history and the fire of the Spirit. King renders this identity for the sole purpose of preparing his people to take the necessary steps to continue the saga.

King's master story is the saga of the enslavement of black people in Africa, their exodus-in-reverse to an alien land, the continued humiliation they endure at the hands of American pharoahs, and now the stirrings of freedom at work among them. This story has unfolded in a universe governed by God, who is made angry by the suffering of his people and whose will it is that they should ultimately triumph. This master story provides the basic plot and atmosphere for most of King's sermons. His own moralisms or reliance on psychological authorities do not finally obscure the master story. Even when the Exodus is not explicitly mentioned, the framework of the plot remains. The weariness with captivity, the sense of *passage*, the controlled rage, and the hope of deliverance that is joyfully anticipated in the sermon's climax—all give to his sermons the structure suggested by the story of the Exodus.

King's use of the Exodus plays both a sustaining and a reforming role commensurate with the message of the Sustainers and Reformers who preceded him. His interpretation of the Exodus promises the deliverance of oppressed peoples, *and* it cuts against the security of those who think they are the custodians of its imagery. King does not employ the simple law-gospel formula of biblical interpretation, hurling texts of judgment against the oppresssors and selecting passages of hope with which to sustain the oppressed. His use of Scripture is more complex, more akin to what the popular black tradition refers to as "signifying," for he reads the Exodus and all his favorite texts with *two* voices. One is the voice of liberation, which promises the ultimate victory of his people; the other is the voice of prophetic critique, which discerns in the triumph of God the defeat of God's enemies. He does not choose two *words* with which to make these distinctions, but two *tones* of one Word. Those with ears to hear, let them hear.

The narrative principle of King's biblical interpretation, then, is the structure of hope implicit in the Exodus: "We Shall Overcome." The preceptive principle is the dominant theme found *within* the larger framework. It teaches *how* we shall overcome. We shall overcome through love. The story of the black Exodus has a nonnegotiable proviso: the deliverance of African Americans and other oppressed peoples will occur only if the vehicle of deliverance, the Christian Civil Rights Movement, is perfectly aligned with the loving nature of Christ's ministry and the redemptive quality of his death. The awakening black human must be formed in love.

The second key opens every text to an exposition of the command to love. The revelation of Jesus' love provides the instrument of redemption. We are "breaking aloose from an evil Egypt," he cries in "The Birth of a New Nation," but "let's be sure that our hands are clean in this struggle." "If there is any one word that describes the ethic of Jesus Christ," he said ten years later to Ebenezer, "it is the word love." He had learned from Benjamin Mays that what Christ really wants to know in Matthew 25 (one of King's favorite passages) is "How did you treat the man farthest down—the naked, the starving, the sick, the stranger, the man in jail?" "You love everybody because God loves them. That is the meaning of 'Christian love.' Somehow you come to the point of loving every man. It's a spontaneous overflowing love that seeks nothing in return. This is what Jesus meant when he talked about love." God is love, and every-thing God touches is infused with love. Even the universe is "friendly" because God made it so. God touched Jesus in a unique way and made him a witness to divine love. Preachers and believers meet that witness in the figure of Jesus, whose teaching and example fill the pages of the New Testament. Therefore love is not the topic of selected texts; it is in *every* text. Sometimes it is hidden beneath the marginalia of an ancient culture, like the exotic data in the Book of Revelation or the opacity of a parable, but the preacher who is schooled in the first principles of theol-ogy, as well as the rudiments of allegory, will know how to find love in the text and make it plain to all.

"Love" has several functional equivalents both in the black tradition and in King's message. Bishop Daniel Payne proclaimed "God our Father; Christ our Redeemer; Man our Brother" as the official motto of the African Methodist Episcopal Church in 1896. The formula hailed back to Richard Allen and had been used by black preachers throughout the century. It expressed the biblical basis for the black church's hope of reconciliation in America. King also inherited a version of the motto from white liber-alism, though the liberals had streamlined it by omitting the reference to Christ. In one of his student outlines King wrote, "The Fatherhood of God and the Brotherhood of man is the starting point of the Christian ethic." He often recited the liberal version as a homiletic formula, but because it failed to convey the subtlety of his argument on love, he turned to other resources.

For a brief period in his career he exegeted love as Gandhian *satyagraha*, but, as an early biographer noted, "the graveyard of Negro leadership was by 1956 replete with the bones of men who had attempted to establish an American passive resistance movement based on Gandhian methodology." Gandhi was foreign to King's own religious heritage and positively unintelligible to his American audience. By 1960 references to Gandhi disappeared from King's sermons and speeches.

The most pervasive functional equivalent to love in King's speeches is "nonviolence," by which he does not mean passive resistance or mere noncooperation with evil but aggressive, even coercive tactics whose goal is "brotherhood" and whose method of symbolic and physical confron-tation always eschews physical violence. King's more spiritualized oppo-

nents argued that such love doesn't belong in the midst of nasty con-
frontations and shouldn't be used as a tool of social policy. It should rather
be reserved for private relationships or the "Beloved Community" at the
end. In answer to his critics on the right, King responded that the only
love Christianity knows is that which was incarnated in Jesus Christ.
Contra Niebuhrian realism, there is no realm that is off-limits to love.
Reformation theologians such as Luther reserved love for spiritual salvation but attempted to regulate civil society with the law or notions of justice. King's religious tradition ignored such distinctions. In bringing love
into the fray, King rejected the old law-gospel method of interpreting the
Bible (and the world) and reasserted the pervasive influence of Jesus in
a secular society.

What was the role of love in King's preaching? Henry Mitchell says
that every great religion has a performance response, something the congregation can not only *think* or *feel* but *do* as a result of hearing the Word.
Whatever actions King elicits from his hearers, at their center is love or
its functional equivalent. Love is each sermon's performance response,
the reply every hearer can make to the preacher's call. King's mature
preaching presented his hearers with a unique possibility for a focused
response. Because every sermon was an expression of God's solidarity
with the Movement, there was always something its hearers could do,
hope, or suffer in harmony with God's nature and Jesus' teaching. Any
text that has not yielded this fruit has been misread. Any religious address or messianic appeal that does not turn on this principle or call for
this response is not to be trusted.

If the Bible is a repository of preaching texts, the passages King did
not choose are of great significance. The history of black protest in America
made several other interpretive keys available to him, but these traditions
do not appear in his sermons. One is the apocalyptic tradition brandished
by radical and incendiary prophets such as David Walker and Nat Turner,
both of whom drew on the Book of Revelation not as a source of moral
truths but as a prophecy of doom. The apocalyptic tradition envisions a
society torn asunder into two warring kingdoms, one of which must
annihilate the other. It relies on a system of signs to predict the final
cataclysm. Walker, for instance, on the basis of the Apocalypse's absolute standard of purity, predicted that the Almighty "will tear up the very
face of the earth!!!" Turner associated the evils of slavery with the "loos-
ing of the Serpent" in the Book of Revelation and later accepted the divine commission to kill the Serpent. True apocalypticism is a far cry from
the politic theology of "The Three Dimensions of a Complete Life." Al-

though the mature King did not shrink from warnings of doom, when
he conjured an apocalytpic vision it was usually one of cosmic and social reconciliation.

A second, more pervasive tradition ignored by King is the black reinterpretation of the so-called Curse of Ham and the apotheosis of Ethiopia. White chaplains to slaveholders (and later segregationists) had long
"explained" the skin color and subjugation of Negroes as the result of a
divine punishment. Creative black interpreters, however, showed how
the negative aspect of Genesis 9 had been superseded in the Christian
gospel. Black nationalists such as Martin Delany, Edward W. Blyden,
David Walker, Henry Highland Garnet, and J. W. Hood, among many
others, combined the reinterpretation of the "curse" with the prophecy
of Psalms 68:21, "Princes will come out of Egypt; Ethiopia shall soon
stretch out her hands unto God." They produced a powerful affirmation
of the historic black peoples of Africa. Far from being the object of a curse,
Ethiopia was the source of nobility for the church and world. Gayraud
Wilmore has documented the ubiquity of this interpretive key in nine-
teenth- and early-twentieth-century black America, saying, "[I]t is impossible to say how many sermons were preached from this text during the
nineteenth century. . . ." Yet one looks in vain for allusions to the theme
or the psalm verse in King's corpus of interpretation. In the one (and
only) sermon in which he refers to the racist interpretation of the Curse
of Ham, he does not refute it with the promise of Ethiopia. Although in
late career he began mentioning Africa and substituting the names of black
artists and thinkers for the "greats" of his earlier sermons, by that time
his principles of interpretation had been fixed according to the nation's
need for reconciliation. The grandeur of the black peoples of Ethiopia
seemed more appropriate to the rhetoric of the black nationalists.

King bypassed other Bible stories popular with black preachers, not
because he was unaware of his own homiletical heritage but because such
traditions did not serve his purposes. The story of the selling of Joseph
into slavery (Genesis 7), to cite a widely used example, inexorably leads
to an interpretation of the Negro as *sojourner* in America, a theme that
King repeatedly rejected as counterproductive to the cause of full *citizen-
ship* in America. The American Negro does not live in *diaspora*. (Hence
the double entendre in the title of his formative sermon "The Birth of a
New Nation".) Ironically, in death King has been portrayed as a new
Joseph: "Behold, this dreamer cometh. . . . [L]et us slay him . . . and
we shall see what will become of his dreams" (Genesis 37:19-20).

Aside from love, none of the themes of his great black predecessors
was prominent in the white sources King read in graduate school. One

could argue that he simply parroted what he read and allowed his interpretive program to be shaped by alien forces. The *Best Sermons* tradition
did not concern itself with the apocalyptic rending of society, the mobil-
ity of the African peoples, or the sojourn of aliens in a foreign land. The
only form of oppression it regularly combated was psychological. *Best
Sermons'* interpretation of the Exodus was *spiritual* at best.

In using the Bible as he did, however, King did not sever his ties
with the black tradition of protest that preceded him, especially the line
represented by Richard Allen, Daniel Coker, and his own father and mentors. He did not have loyalties to an intellectual agenda of interpretation
as such, but he did believe that his reading of Scripture best served the
midcentury interests of his people and the nation at large. Deliverance
and love were needed by everyone in America. Although he preached
liberation, he did not adopt the univocal voice characteristic of contemporary liberation theologians. In fact, in his choice of the Exodus and
the love of Jesus, he was attempting to recover what he correctly understood to be the center of the Old and New Testament for use in the public arena. The Exodus was *the* salvation event in the Old Testament, and
love was at the heart of Jesus' life, teaching, and death. So confident was
he of this center that he not only introduced the Bible into the political
mainstream but attempted to refocus African-American interpretation on
what he took to be the heart of the Bible.

IV

The New Testament Gospels were not written to make "points" about
Jesus of Nazareth, but by means of the accumulated stories themselves
the Gospels render a figure, an agent of redemption. The Christian church
read, sang, and enacted the biblical words, and continues to do so, thereby
re-creating for itself a living figure from ancient materials. Such representation is characteristic of traditions as varied as Orthodox mysticism, Roman Catholic sacramentalism, Protestant typological interpretation of the Bible, and Pentecostal and black ecstasy. Without the
community's performance of the Word, the chasm between the Book and
contemporary communities of faith is unbridgeable. The Bible remains a
dead letter. The African-American church, in particular, knows no "meaning" that resides in the text awaiting explanation. Meaning is *disclosed* in
the communal enactment of the text, the way the meaning of *King Lear*
or a Beethoven symphony is revealed in its performance.

In their experience of Scripture, the preacher and community together vivify the message and characters of the Bible. No one interpreter can make this happen; performance is the achievement of the community. The preacher holds up the mirror of the Bible to the people, and the people respond by recognizing themselves in it.

One of the most vivid examples of the Bible as mirror occurred at the conclusion of the Montgomery Bus Boycott. The night after the Supreme Court ruling ended bus segregation, eight thousand people came to worship at two churches in Montgomery. At the first service, Lutheran pastor Robert Graetz, the only white clergyman in town who joined the Montgomery Improvement Association, was appointed to read the Epistle lesson, 1 Corinthians 13. Before he rose to read, Ralph Abernathy whispered to him, "Read it like you've never read it before. Put everything into it." When Graetz came to the words, "When I was a child, I spoke as a child, I understood as a child, I thought as a child: but when I became a man, I put away childish things," the congregation burst into spontaneous applause and cheers. The corporate black human had recognized its own coming-of-age in the mirror of the Bible's words. Later, a newspaper reporter asked Abernathy, "Isn't that a little peculiar, applauding the Scripture?" "We are a peculiar people." Abernathy replied.

What literary critic Stanley Fish calls an "interpretive community" describes the black church in general and King's ministry of interpretation in particular. When King cries out, "How long . . . not long," or "You reap what you sow," these are not lessons from the past. Even these brief formulas exercise a typological force, for in them the voice of the Psalmist or Paul not only expresses the convictions of their ancient communities but joins with the cries of the slaves and the segregated peoples of King's day. "You reap what you sow" is not a truth that is once again being explained to a new generation. It is a single sustained cry that, because it emerges from God's own voice in the Bible, continues to echo over the generations.

In all the audiotapes of King's sermons in black congregations, a single voice or sometimes a chorus of voices intones, "Make it plain." When King begins to uncover the problems of contemporary society in light of the Word of God, the voice gives its encouraging interpretive advice: "Make it plain." "Tell our story. Make the connection." In some Pentecostal and Holiness churches a reader "lines" the Scripture and the preacher interprets with commentary. Sometimes the deacons and deaconesses create an elaborate antiphonal effect in answering their preacher. In churches where the preacher chants the sermon, the congregation sometimes anticipates the preacher by humming a tune just before he or she begins. King's congregation helped him make it plain by completing his sentences for him, especially his recitations of Scripture, by echoing his words, and by responding with encouragement and joy. When King begins his "Three Dimensions" sermon about John's vision of the heavenly city, he affects an exaggerated ponderousness that allows his hearers to render the scene with him. He identifies "a man named John on the island of . . ." Out of the void that follows, one hears voices solemnly intoning "*Pat*-mos."

In the interpretive community, certain passages encode an instant meaning that is initially signaled by the audience's vocal response. Because of the black church's heritage of oppression, King or any black preacher will set off a sure response with an allusion to Galatians 6:7: "Be not deceived. God is not mocked. For whatsoever a man soweth, that he shall also reap." Rarely will the preacher complete the sentence without audible "help" from the brothers and sisters. One investigator of rural black preaching in Macon County, Georgia, makes an observation that is equally true of King. When the preacher cries, "How long . . . (as in Psalms 13:1: "How long wilt thou forget me, O Lord, forever?"), it is like touching a match to gasoline.

The sermon and the Bible are dynamically related. The sermon does not reflect on the Word of God but the sermon *is* the Word of God in action. King's first inclination, born of many Sundays at Ebenezer, was to recognize the Word of God wherever people are being moved to verbal response to the preached message. He later came to understand call and response as but a dress rehearsal for the community's *social performance* of the Word of God. The Word must move people to new expressions of discipleship. He altered his definition of preaching by subtly shifting the criterion for its authenticity from eloquence to its potential for enactment. If the sermon is promoting the kinds of liberation and love that God-in-Jesus has been known to sponsor in the Bible, and if it is opposing the kinds of injustice that God has always hated, then the sermon is the Word of God, and its preacher is a genuine prophet.

Finally, in the early and mid-1960s the Word and the Movement assumed an organic relationship not unlike that described in the Book of Acts, where the Word of God is said to have embodied its own community and momentum. What the Book of Acts claims for the earliest Christian preaching could be said of King and his colleagues in Alabama, Mississippi, and Georgia: "So mightily grew the word of God and prevailed." The action in the streets and King's preaching became expressions of one another. Historians may "freeze" the Civil Rights Movement and isolate individual rhetorical performances that appear to be turning points or significant moments, but it is the *ceaseless* activity of biblical interpretation and preaching carried out by King and his colleagues that sustained the Movement as a whole and invested it with transcendent meaning. Under his leadership the quest for equality and justice became a Word-of-God movement.

9

The Ebenezer Gospel

NOT counting his days of student apprenticeship, Martin Luther King, Jr. served Ebenezer Baptist Church as its "co-pastor" for eight years and three months, just under one hundred months in all. He did not preach at Ebenezer every Sunday, but he spoke there often enough to establish what Paul (or any preacher) would have called "my gospel," an evolving, sometimes volatile, interpretation of God's will for Ebenezer and the world. The Ebenezer gospel is what the preacher King had to say to his people over the course of his one-hundred-month pastorate; it is his "message." It is important to gather the fragments of this gospel into a coherent whole because he carried a modified version of it into world history, thus making knowledge of the Ebenezer gospel essential to an understanding of his public message—his quest for justice, yearning for redemption, insistence on nonviolence, embrace of suffering, prophetic rage, and all else that emerged from his Sundays in Atlanta. King's Ebenezer sermons differ from his mass-meeting speeches and his civil addresses, but they are not inconsistent with them; they are the religious subtext for his sermon to the nation. The "first draft" of all that he said, achieved, and suffered from Montgomery to Memphis is present in the audiotape recordings and crudely typed transcripts of his Ebenezer gospel.

I

The bent of King's gospel follows the contours of the Christian story of redemption. It begins with the human condition, which is nothing other than the experience of one of life's many perplexities. The perplexity suggests a larger problem, the problem yields a sin, the sin opens onto a social concern (always related to race or war), the concern invites a generalization about "man" or "life," and the stage is set for the next phase of the message. Each of the following sentences represents a statement of the human predicament from a variety of sermons ranging from 1960 to 1968:

> Probably no admonition of Jesus has been more difficult to follow than the command to "love your enemies." . . .
>
> [Like James and John] we will discover that we too have those same basic desires for recognition, for importance, that same desire for attention. . . .
>
> Life is full of annoying interruptions. Things go smoothly for many days, and many years, often, and then our lives are interrupted. . . .
>
> I guess one of the great agonies of life is that we are constantly trying to finish that which is unfinishable.
>
> One of the great problems of life is that of dealing creatively with disappointment. Very few of us live to see our fondest hopes fulfilled.
>
> There are some things that are as basic and structural in history [as sunrise and sunset] and if we don't know these things, we are in danger of destroying ourselves. . . .
>
> [O]ne of the great tragedies of life is the fact that, after a while, most people lose the capacity to say "thank you."

Any attempt to "fix" the problem with our own resources only makes it worse. We invariably exaggerate or become overly reliant upon one of the possible solutions to the problem, which leads to another sin. In King's vocabulary "sin" is a disordered or one-sided attempt to solve a problem by taking it into our own hands and denying its true nature. Sin entails the denial of the transcendent character of our problems, which means that we fail to see them in the context of God's power to rectify them. In "Pride versus Humility" he concludes, "[T]herefore the sin can't be pardoned because you've lost the capacity of realizing the necessity of being pardoned," which, he adds pointedly, is the white southern Christian's problem. Life's interruptions, to cite a frequently recurring theme, may evoke the reactions of despair and fatalism, or they may cause us to lash out in anger toward others. In either case, it does not occur to us to transform these adversities into opportunities for service.

Our frustrations produce in us an internal civil war. We know what is good, but we always fail to do it. Instead of dealing creatively with the many forces that assail us, we nurture personal and social resentment. In our own community, those who are robbed of their civil rights practice other, more banal, forms of robbery on one another: "The barber-shop, the beauty shop, the card table, the telephone—[are] all centers and instruments of robbery."

What we do to one another is both a microcosm and a consequence of what has been done to us. The little robberies, petty hatreds, and daily insecurities in the Negro community reflect a larger evil: the greater robbery of our very selves by means of slavery and segregation, the gross hatred of one race for another, and the racist policies that produce only shame and self-hate in those upon whom they are perpetrated. The Negro finds himself living in a "triple ghetto" of poverty, race, and misery, forced to pay exorbitant rent for substandard housing, compelled to attend inadequate schools, plagued by low wages or unemployment. By the thousands, he says, Negro men face the embarrassment of not being able to be men because they can't support their children. "Out of embarrassment and out of frustration, they often turn to alcoholism or dope to try to run from it and escape its tragedies." There has been "no greater robbery" than the robbery of the stability of the Negro family. And what of those who escape the triple ghetto? Their response is often the mirror image of the poorer Negro's shame:

> [S]o often in the Negro middle class [as E. Franklin Frazier has said] you find a brother who is ashamed to identify with his relatives. He doesn't like to read poetry that's written by Negro poets, and Negro art doesn't have any meaning to him, and he doesn't like listening to Negro spirituals, because you see this reminds him of the fact that he has a slave and an African heritage. He doesn't want to be identified with that. So he goes and tries to identify with all of the values of the white middle class, and he's rejected by the white middle class, so he's left out there in the middle with no cultural roots. And he ends up hating himself and he tries to compensate for this through conspicuous comsumption—Cadillac cars, fine houses, foreign mink coats. This is this brother's problem.

Even the universal sin of ingratitude assumes a special modality among the Negro middle class. Not only have Negroes been robbed of their cultural memory but those who have escaped the "triple ghetto" have forgotten the contributions of those who made their freedom possible. They enjoy the desegregated facilities and the better jobs, and take it for granted that Thurgood Marshall sits on the Supreme Court, but "[t]hey've forgotten the fact that some people have had to die in order for us to get these things." The black middle class enjoys a perverse partnership with white moderates, both of whom want the good life at the expense of the many who are left behind in "the stench of back waters."

The evil we see in ourselves and in society around us does not merely *remind* us of certain Christian doctrines. Our life in America enacts these doctrines, proves them, with painful empirical certainty. The entire process of humiliation by which Negroes are ensnared in evil and perpetrate it upon themselves King called "original sin." Original sin is not a bad habit but a pervasive "habit structure" that can be shattered only by God. Black shame and self-hatred, no less than white racism, are not merely psychologically unhealthy; they efface the more original condition of dignity, the image of God, in which all people were created. King comforts Ebenezer, "So if you are worried about your somebodyness, don't worry any longer because God fixed it a long time ago. He said, 'I'm making *all* my children in my image, and I will declare that every child of mine has dignity and every child of mine has worth.'" Race supremacy not only contradicts the Declaration of Independence but makes a mockery of the doctrine of creation in the image of God and violates God's continuing will for his creatures. In his 1966 sermon entitled "Who Are We?" King says, "Despite a man's tendency to live on low and degrading planes, something reminds him that he is not made for that." "This is my *Father's* world," he had heard his own father sing many times, to which King adds defiantly, "And God has not yet decided to turn this world over to George Wallace. Somewhere I read, 'The earth is the Lord's and the fullness thereof.'" Like race supremacy, American militarism is not only morally wrong; it is a modern exercise of idolatry, which is nothing less than the denial of the lordship of God above princes and nations. When Negroes and poor people languish in the captivity of poverty or serve as cannon fodder in senseless wars, they are not simply economically or politically disadvantaged but bound in the chains of sin that God means to break.

In the Ebenezer gospel, what the preacher introduces as a perplexity or a universal problem he skillfully correlates with Christian doctrine, thereby exposing its sinful dimension and its potential for redemption. Only after the preacher has elevated his topic to one of universal proportions does he draw it through the ashes of Negro life and history. In each sermon, that which presents itself as a problem to be solved or an interesting "truth" for disquisition, ends as an impassioned appeal for redemption. With Paul, the preacher cries, "Who will deliver us from this body of death?"

II

The answer rests on the character and history of God. In the civil addresses "God" sometimes appears as Jefferson's God, a shadowy guarantor of liberal values, but in the Ebenezer and black-church sermons God assumes his biblical identity as the creator of the world, the liberator of Israel, and the Father of Jesus Christ. It is inconceivable to the preacher that a God who created all people in his own image could have created some "more equal" than others or would sanction the systematic exploitation of one race by another. The sermons do not engage the twisted hermeneutics of the white racist preachers, like the Texas Baptist who proclaimed God "the original segregationist," nor do they attempt to prove racial equality on the basis of the doctrine of creation or the character of God, but in the Bible's own spirit of defiance, the sermons assume the created equality of all.

It goes without arguing that God wills to liberate Negroes from captivity in America because that is the kind of God we have. That is his demonstrated character. We have already explored King's reliance on the Exodus and shown how its dynamic as a "setting free" and a "coming through" informs so many of his sermons. Unlike the static and inactive God of deism, who appears in the civil addresses but rarely in the sermons, the God of Ebenezer is relentless in his forward motion. In a 1963 sermon King declares, "Whenever God speaks, he says, 'Go forward!'" Although the preacher sometimes invests history with a forward tilt or a moral bent, his Ebenezer gospel makes it clear that it is the God of Israel, not inevitable laws, who is the motor of deliverance. With the character of God's relentlessness King establishes one of several links between the deity of the Old and New Testaments, for the father of Jesus "is a seeking God" who never quits searching for the lost. A parent who loses a child doesn't say, "Oh well, I've got four more," but he or she continues to seek and to love the prodigal, even the child in prison.

The most prominent characteristic of the New Testament's revelation of God, however, is his love. Whatever hopeful laws are discernible in the universe are due to the Creator's loving character. Like the pulsating life of the universe, "God's love is unceasing and eternal. Love is not a single act of God but an abiding part of his nature." The preacher observes, "You cannot say that man 'is' love but only 'man loves or men love,' but they also hate. Only 'God' and 'love' are joined by 'is.'"

One could say that Jesus is a "chip off the old block," in that God's love had to produce a Savior like Jesus, and Jesus' character witnesses to the full extent of God's love. We understand the nature of God, the preacher explains in a 1967 sermon, by turning to Christ: "And so this morning I know that God is love, because Christ is love. . . . I know that God is just because Jesus Christ is just. And I know that God is a *merciful* God, full of *grace* and glory, because Jesus Christ is merciful."

Although King had been trained in liberal, personalistic Christology, when he addressed his own congregation he preached as evangelical a message—as "sweet" a Jesus—as could be found in any black congregation. When it came to Jesus, Negro Christianity cared as little about the liberal "laws" of Jesus' personhood as it did about the orthodox formulas for his divinity. Instead of dwelling on the metaphysical proportions of Jesus' divinity and humanity, by which orthodoxy thought to safeguard the uniqueness of Christ, the Negro church simply showered its Lord with glory and praise. Now, long after the Greek formulations have lost their significance to many Christians, the hallelujahs and Yes Lords continue to ring out in the African-American church. The effusions King heard from his father on the Lily of the Valley and the Bright Morning Star he repeated many times in the Ebenezer pulpit. In a 1967 sermon he merges his voice with a couple of gospel songs in praise of the Lord:

> Thank you Lord, thank you Jesus,
> for you brought me from a mighty,
>   mighty long way. . . .
> How I got over. And I just want to thank
> you Lord for bringing me over.

In telling what this Jesus could do for the believer—what church doctrine calls "soteriology"—King's rendering attained its evangelical peak. It was another matter altogether when he spoke of the transformation of society, for then he evoked the Pillar of Fire of the Exodus God and the angry God of the Prophets. The Jesus of Ebenezer was the Savior of troubled individuals. There is nothing in King of the old Greek theology of Jesus as the head, the vanguard, of the human race, nor of Nat Turner's Jesus the Liberator, who came to cast fire upon the earth. Jesus' role in the Movement was to be the supernatural partner in Negro suffering, to model a nonviolent response toward racist oppression, and to console the preacher in his moments of discouragement. This is not to say that Jesus played a marginal role at Ebenezer. He is everywhere in King's sermons, for true to their Baptist roots, the Ebenezer sermons portray a Jesus who alone is able to rescue the individual from sin and transport him to eternal life.

> I know he is a-ble to raise us up
>   when we are down,

his Daddy sang.

> Are you torn within
> And somehow beaten from without?
>   Cry, "God is able" . . .
> Are you giving up on your journey?
>   Why? God is able!

the son proclaimed.

Whenever we are threatened with the sins of personal disinte-
gration, such as sexual immorality, dope, or despair, we find rescue in
the Savior. Jesus doesn't merely forgive sins but he has the power to
change sinners, to do for us what we are unable to do for ourselves. The
sermons are quiet on the substitutionary Atonement of Jesus on the
cross, and they say little about the necessity of the Negro's unmerited
suffering—much less than his public utterances. It may be that face-to-
face with his segregated brothers and sisters he could not speak these
words. In the later sermons, King makes it clear that if anyone will bear
the sins of all and die for them, it will be the preacher. Although the sermons make little of Jesus' vicarious Atonement, they do not express reliance on human goodness to effect change. The intricacies of the Atonement were never a concern to the traditional Negro pulpit, but God's
exclusive power to transform lives was its stock and trade. The uneducated Negro preacher reminded his rural Macon County flock, "He make
an ugly person look pretty well. Can't he make you *pretty*, so to speak?"
The sophisticated Atlantan draws on the same tradition in the following
set piece:

> "You may be a dope addict,
>   But I can deal with dope addicts . . .
> You may be an alcoholic,
>   but I can deal with alcoholics . . .
> You may be engaging in sexual promiscuity,
>   but I can make you a faithful husband," says Christ.
> "Come unto me, just as you are,
>   and accept your acceptance."
>   . . . I know Christ:
> He can change the lying man
>   into a truth-telling man.
> He can change a dishonest woman
>   into an honest woman.
> He can change a prostitute
>   into a woman of honor.

Again, a 1963 sermon promises, "'If any man is in Christ,' says Paul, 'he
is a new creation. The old has passed away, the new has come.'" What
does this mean for the Ebenezer Christian?

> The things I used to do,
>   I don't do them now.
> The places I used to go,
>   I don't go there now.

> The thoughts I used to think,
> I don't think them now,
> because God through Christ
> has put his hands on me.

"I know Christ," the preacher repeatedly testifies. "And I tell you this morning that one of the great glories of the gospel is that Christ has transformed so many nameless prodigals." "I recommend him to you this morning, because I know from my own experience that He can make a way out of no way."

King's celebrations of God constituted the climax of his sermons, which is perhaps the element most characteristic of the African-American sermon. This is the gospel, the distinctive genius of Christian speech, and the very thing that is unfailingly eliminated from his printed sermons and that he himself apparently could not *say* to predominantly white congregations. He could not say it because the nature of his gospel, which was the Ebenezer gospel, required the shared articulation of a people's suffering and hope, and that he could not find at Saint John the Divine or the National Cathedral. What remains in his printed and white-church sermons is the residue of psychology and borrowed moralisms for which the preachers of his generation were well approved.

III

The message of Jesus Christ demands a response of the hearer's whole life. Although Ebenezer was a verbally responsive congregation, its co-pastor did not measure the effectiveness of his message by the emotions it aroused on Sunday morning. Churches that were satisfied with getting people "happy" in worship but had no social mission he frequently called "entertainment centers." King's Baptist tradition put great store in the congregation's response to the spoken word, but King himself never "begged" for a verbal response or even subtly suggested it. What he had in mind was a far more nuanced spiritual and behavioral change.

That response begins with repentance and ends in commitment to others, particularly to the struggle for racial freedom. Although he ocasionally called Ebenezer to repentance, King usually reserved his most radical religious language, his appeals for repentance or his exhortations to be "born again," to the political sphere. To be sure, Ebenezer must be "transformed by the renewing of its mind," he said, but the evil from which it turns is not as savage or pervasive as white racism.

Whatever change that occurs will take place in the believer's partnership with Christ. In personal matters as well as the Civil Rights Movement, reliance on our own activism is as big an error as supine resignation. God works through us, and his agency includes our hard work and commitment. "Do we want peace in this world?" King asked. "Man can not do it by himself. And God is not going to do it by himself. But let us cooperate with him. . . ." In his early years, the preacher tended to warn against human self-sufficiency; in his middle and later years, he became increasingly infuriated with those who used God's all-sufficiency as an excuse not to march or commit themselves to the struggle.

At Ebenezer, King modeled a spirituality that abolished the traditional distinctions between belief and action. Much traditional theology, including his own Baptist heritage, isolated "faith" and "love" as successive moments in the life of the believer. In King's vocabulary the two are inseparably joined. There can be no discussion of faith as the intellectual or spiritual preparation for love, or of love as the inevitable response to faith. Faith assumes two modalities at the same time: trust and love.

Faith has an "in-spite-of" quality that always contradicts prevailing conditions. When Job's wife counsels him to despair of God's favor, Job colloquially responds, "Honey, I'm sorry, but my faith is deeper than that." The preacher confesses to being a terrible swimmer because he couldn't relax in the water—until he learned the "dead man's float." God will carry us if we lie back and ride his wave to freedom. Almost canceling the metaphor, he adds, "We can't do it alone. God will not do it alone. But let's go all out on it and protest a little bit and He will change this thing and make America a better nation."

Floating is something like accepting. Using Paul Tillich's formula for faith, "accept your acceptance by God," to which Ebenezer was enormously responsive, King patiently explained the difference between grace and works. He reminds Ebenezer (as if they needed reminding) of the many organizations and clubs with closed memberships. But God's acceptance is not like getting into a fraternity "where they will paddle you a bit to get you in to see if you can pass the test." Not at all, for God specializes in hopeless cases and failed projects. "No matter what your marks were in school, come on just as you are." Only the preacher King could fuse Paul Tillich and the pathos of an evangelical hymn: "Faith is accepting the fact that you are already accepted. And this is why we sing 'Just As I Am'. Christ will take you, if you will just have faith and accept your acceptance. No matter who you are. . . . That's all it takes."

Floating and acceptance finally lead to freedom from fear. The preacher knows that some in his own congregation have refrained from joining the Movement because of fear. He can address their fear because he has experienced it himself. He has also experienced the supernatural relief that possesses one who has been set free from this most elemental bondage. On the basis of his call-experience in the kitchen, the preacher declares himself to have been forever liberated from fear, but he cannot help but notice how fear continues to rob the Movement of white and black support. He exhorts his hearers to stand up against unjust laws, the way the three men in the fiery furnace defied Nebuchadnezzar, the way Julian Bond has "dared to speak his mind" in Georgia. In his sermon on Julian Bond he recounts a poignant meeting with a Duke Divinity School student, a white man, who had been convinced by King's speech but sorrowfully confessed, "[I]f I even talked about brotherhood from my pulpit, they would kick me out."

The definitive form of faith is love. Perfect love casts out fear. The ubiquity of love in King's sermons has already been remarked upon. In the heat of the Movement, theological ethicist James Sellers attacked King's doctrine of love as a theological mistake, and soon after, but on different grounds, the black nationalists derided it as a political betrayal. The main point of Sellers's criticism was that traditionally (by which he meant "in Niebuhr") love is the goal of life, and justice is the means to the goal. To "sloganize love as a technique" places love in the midst of conflict, coercion, and all manner of morally ambiguous situations where it cannot possibly remain true to its own character as disinterested *agape*. To carry out a protest "in love," nonviolently, requires that each person involved in the protest be possesed by that love, which necessitates an absurdly idealistic view of ordinary people. King's belief in the power of love, said Sellers, depends on his belief in the goodness of humanity, both black and white. A more realistic and theologically responsible alternative would be to follow Calvin's notion of "forced and extorted righteousness" and admit that God's righteousness in society takes other forms beside love. Sellers's article, which appeared in *Theology Today* in 1962, complemented the more savage political critique of love to which King was subjected for the rest of his life.

For all their repetitiveness and disorganization, the Ebenezer sermons reveal what a solid theologian and pastor King was. Although he did not have the leisure to publish an academic theology of the Civil Rights Movement, the sermons show that he well understood the grounds on which he was repudiating the position represented by his critics. Sellers's argument never quite touched earth and, perhaps because of the abstract tendencies of most formal theology, it never got around to confronting the sociopolitical alternatives to King's teaching on love.

Against Sellers' criticisms, King in fact did *not* exhort his members to love on the basis of the laws of the universe or the goodness of humanity but on the basis of the revealed nature of God. God created the universe and continues to preserve it out of love. If there is any doubt about the character of God's love, one need only look to the cross of Jesus for clarification, for the cross is "God's photograph." In the Incarnation, God saw fit to plunge Jesus into a politically ambiguous situation, subjected him to suffering, and finally took his life. Jesus' ministry and crucifixion have become the paradigm for the Christian's life in society. "I have long since learned that being a follower of Jesus Christ means taking up the cross," he said in 1967. We are conditioned to expect success and personal fulfillment; our preachers like to preach "nice little soothing sermons on how to relax and how to be happy" or "go ye into all the world and keep your blood pressure down and I will make you a well-adjusted personality." But "[m]y Bible tells me that Good Friday comes before Easter," and the cross is not a piece of jewelry you wear but something you die on."

In this 1967 sermon the preacher is not merely musing on his own fate but pastorally attempting to explicate the meaning of the cross for daily life. The cross may take the form of the "death of the budget of your church," the loss of a job for the sake of principle, or the loneliness of a life without friends. To restrict the cross to the realm of personal spirituality or to the end of history, after all the nasty battles for justice have been fought, makes a mockery of the Incarnation. It also ignores the *teaching* of Jesus on nonviolence in the Sermon on the Mount and restricts it to personal relations. In fact, it was Reinhold Niebuhr who remarked, "We may be able to put God back in nature by a little serious thought, but we cannot put God back in society without much cross-bearing." Immersed in a conflict whose meaning was not yet historically defined, King was trying to do precisely what Niebuhr and the Lutherans were unable to accomplish: to provide a Christological foundation for civil society from which the love of Jesus had been banished by theological rationalizations.

King's agenda did not blind him to the uses of love as a political tactic, nor did it dull his appreciation of the law. No African American worked harder for the passage of civil rights *laws* in the 1960s, and no one labored under fewer illusions about life in a society ruled by love than Martin Luther King. Scores of times he repeated the maxim, "The law can't make a man love me, but it can restrain him from lynching me." Six weeks before he died, struggling with a case of laryngitis, he explained

how laws help avert paternalism by putting all citizens on an equal footing. But this was a sermon on the Parable of the Good Samaritan, and
even though King had preached this same text to Ebenezer many times
before, on this Sunday he found something new to say. Two years earlier
he had focused on the *robbery* of the black man's dignity; today he said
("hoping that my voice will hold up") that he would discuss the difference between law and love. Of course, stronger laws are necessary to
combat housing discrimination and acts of violence against blacks. Even
though "you can't legislate morality" the law can change people's habits
and "pretty soon they adjust." "So I['m] not underestimating the enforce-
able obligations," he explained. "They are real, and they are necessary."

But the point of the parable and of King's own mission is that if "genuine integration, in terms of genuine person-to-person relations, genuine
inter-group relations, [and] mutual acceptance" is to become a reality, it
will occur when people are "obedient to the *un*-enforceable." Integration
"really will not come through the law" until the white person accepts the
black person as a brother or sister. Just as the Samaritan's behavior surpassed the demands of the law, so must the white race be prepared to
exceed the provisions of civil rights legislation. Only then shall we have
integration—not the tokenism with which too many are satisfied but integration as "true spiritual affinity" between the races. It is important to
remember that in terms of the development of King's thinking, this is an
extremely late sermon, spoken in his life's eleventh hour. Furthermore,
it is an old sermon whose text and structure he could have repeated at
Ebenezer with impunity, but he chose instead to reshape it and fill it with
material reflective of his final thinking on love.

King not only preached love but modeled it for his congregation.
Sellers criticized King's doctrine for the unreasonable demands it placed
on its practitioners—they must be transformed by the love of Jesus in
the very practice of it—which was precisely King's spiritual aim and the
aim of any serious expositor of God's love. The practice of love is self-
validating. Although tactics have their place in social change, ultimately
the Christian does what is right not merely to avoid hell or to gain entrance into heaven. "You must love, ultimately, because it's lovely to love.
You must be just," he chides Ebenezer, "because it's right to be just."

The pastor knew that his own congregation as well as his national
followers were having a hard time with his doctrine of nonviolent love.
Even his father had proclaimed from Ebenezer's pulpit, "I'm not ready
for the dogs to be turned loose on me. The dogs and somebody else gonna
get *hit* if they turn 'em loose on me." The son responded with pastoral

direction for his people. He reminds them that those we find most hateful and irritating need our love and can be healed by it. Therefore, don't respond to racists by cursing them or calling them "crackers" when they call you "niggers." When people gossip on you, try to think the best about them. When your kids aggravate you, don't beat them all the time but show them love. He goes so far as to say that even the Blackstone Rangers (Chicago's most notorious gang) can be changed (and have been, he claims) by a little love and attention.

Whether Ebenezer was convinced by its preacher's testimonies to the power of love remains uncertain, but his congregation could not have been indifferent to its preacher's own demonstration of selfless love. When, for example, he tells of the Duke Divinity student who walked away from his convictions, the preacher does not withhold compassion from him. The young man is not a preacher's typical bad example but a tragedy. As prowar sentiments grew throughout the nation, King's radical love became more prophetic and personalized. Finally, in a 1967 sermon he burst out with, "I ain't goin' to study war no more. And you know what? I don't care who doesn't like what I say about it." Likewise, as the pressures for violence increased among segments of the black community, King's insistence on love grew with passionate intensity. At the close of a pedestrian discourse to Ebenezer entitled "Levels of Love," the preacher suddenly reached through the veil that occasionally descends between pulpit and pew and confessed, "I say I love you. I would rather die than hate you."

No one in the Ebenezer congregation would have been persuaded to love by a single sermon, any more than the preacher King had discovered love in any one moment of his life. The movement from repentance to commitment occurs over time, and King repeatedly stresses that it is a movement, not an instantaneous change. In King's vocabulary, salvation does not await the believer at the end of the journey but is the journey itself. Both preaching and salvation are cumulative activities. Early and late in his Ebenezer pastorate King presented salvation as a long and uncertain road: "In this earthly life we never get there totally. It's always a process and never an achievement. But you know what *really* being good is? It's being on the right road." In this respect, as in others, Martin Luther King adhered to the theology of his namesake, Martin Luther, who said,

> This life is not righteousness but growth in righteousness, not health but healing, not being but becoming, not rest but exercise. We are not yet what we shall be, but we are growing toward it. The process is not yet finished, but it is going on. This is not the end, but it is the road.

IV

But what, if anything, lay at the road's end? True to his focus on the way, King saved up little for the end of the struggle. While he appreciated the sustaining role played by the Negro preachers who preceded him, he discarded their vivid portrayals of the afterlife in favor of his own predictions of social transformation. He often said, "It's wonderful to talk about the New Jerusalem. One day we will have to start talking about the New Atlanta. . . . [A]ny preacher who isn't concerned about this isn't preaching the true gospel." His favorite sermon, "The Three Dimensions of a Complete Life," was a psychological and moral interpretation of the Heavenly City's dimensions given in the Apocalypse. Although he occasionally spoke of death as "the door that leads us into life eternal" or of those who have died (the three little girls in Birmingham) as "home today with God," he customarily rendered the future with the imagery of a transfigured society.

In his early sermons he borrowed the idealist Josiah Royce's phrase "the Beloved Community" to evoke the period of social harmony and universal brotherhood that would follow the current social struggle. The Beloved Community corresponds roughly to Karl Marx's postdialectical "Kingdom of Freedom," characterized by the full exercise of individual talents within the community of the whole. What neither thinker provided was a political or economic plan for the transition from class struggle to social coherence. As John Cartwright has observed, King placed less emphasis on a specific blueprint for the future than he did on the qualities that persons must acquire in order to hasten the coming of the new community—in other words, he was a preacher and not an economist.

As King's social idealism was succeeded by more realistic appraisals of human evil, references to the Beloved Community gradually disappeared from his sermons, their place taken by the theological symbol "the Kingdom of God." On the basis of King's published writings and utterances, scholars have debated the question of the kingdom's this-worldly as opposed to other-worldly character, but they have not commented on the radical conversion implicit in the shift from the humanism of the "Community" to the theology of the "Kingdom." The former carries overtones of utopian idealism; the latter acknowledges God's claim upon all human achievements.

In the Ebenezer sermons the symbol "Kingdom of God" represents God's hidden but dynamic role in the transfiguration of earthly relationships, cities, and institutions. In an early published sermon King sketches a few historical possibilities for the manifestation of the kingdom: "And though the Kingdom of God may remain *not yet* as a universal reality in history, in the present it may exist in such isolated forms as in judgment, in personal devotion, and in some group life. 'The Kingdom of God is in the midst of you.'" The image of the kingdom in this passage, like that of the early chapters of the New Testament Gospels, is one of a quietly germinating presence. In his later sermons the kingdom constitutes a rupture in the "old order"; it subjects current laws and institutions to the judgment of God, a picture that corresponds to the violence of the later chapters in the Gospels. According to King's gospel, the last judgment of God is being realized ahead of time in the turmoil of contemporary events. When governors and state legislatures make policies that contradict the law of God, they are arrogating to themselves an authority that belongs only to God. But "God is still in control. And God has not yet turned over this universe to Lester Maddox and Lurleen Wallace." Four years earlier he had said at the funeral for the three childen, "God is the Supreme Court beyond which there is no appeal." In his later sermons, the preacher realizes that it is not sufficient to witness to these false authorities, but the evil they represent must be defeated before "we begin in our little way to build God's kingdom right here."

Nowhere is the present reality of the kingdom given greater force than in his 1966 sermon on "the acceptable year of the Lord." On the basis of Isaiah's prophecy, Jesus strode into his hometown synagogue and announced that his ministry was the fulfillment of God's promise to Isaiah. The Bible's prophetic and messianic utterances are often nested in concrete historical circumstances, such as the return of exiles or the beginning of Jesus' ministry, but they quickly transcend local occurrences and reveal God's universal lordship. In the Bible, what originates as a prediction of David's kingship is eventually celebrated as God's rule over all nations; Jesus' prediction of local persecutions (as in Mark 13) by the end of the chapter has escalated to catastrophes of cosmic significance.

King's sermonic performance of "The Acceptable Year" follows the biblical pattern, for he begins by locating the kingdom in the midst of his congregation's experience of history: "You know the acceptable year of the Lord is the year that is acceptable to God because it fulfills the demands of his kingdom. Some people reading this passage feel that it's talking about some period beyond history, but I say to you this morning that the acceptable year of God can be *this* year." He continues with a set piece whose rhetorical form is as explosively eschatological as its content:

> The acceptable year of the Lord is any year
>   when men decide to do right. . . .

> The acceptable year of the Lord is any year
>> when men will stop throwing away the precious
>> lives that God has given them in riotous living.
> The acceptable year of the Lord is that year
>> when people in Alabama will stop killing civil
>> rights workers. . . .

The piece continues with a survey of contemporary history and international relations until the preacher suddenly removes his blinders and envisions the whole world beneath the lordship of God. He does this by reciting a series of Bible passages in an ascending and ever-widening order of universality concluding at the highest emotional peak:

> The acceptable year of the Lord is that year
>> when every knee shall bow and every tongue
>> confess the name of *Jesus*, and everywhere
>> men will cry, "Hallelujah! Hallelujah!"
> The kingdom of this world has become the kingdom
>> of our Lord and of his Christ, and he shall
>> reign forever and ever. "Hallelujah! Hallelujah!"
>>> [tumult]

If the world is really governed by God, and if that governance will one day be made plain, then the most appropriate Christian stance in the world is one of hope. A war rages between hope and despair in the preacher's own heart and in the lives of discouraged people everywhere. The vision of God's imminent triumph lends resources to hope. Hope does not come to the Christian as easily as optimism, for it is surrounded by the evidences of its foolishness. Yet, to those with eyes to see, "what is hoped for is in some sense already present" in the witness of those who love peace and are willing to work for it. Like faith and love, hope belongs to the Christian's journey through life. In its communal character (King is here speaking at a church anniversary service), hope transcends individual desire: "[Y]ou can never hope for something you don't hope for somebody else, and for many others." Finally, if you want to cultivate the spirit of hope in yourself, look to those, like the slaves of old, or our forebears in this church, or the handicapped, or the terminally ill who refuse to give up, and be guided by their example.

V

Here and there one encounters individual examples and guides in the art of hope. The Baptist cannot be expected to laud the saints, and he doesn't, but he does identify the church of Jesus Christ as the earthly agency of redemption for the world. The suffering it experiences and the love it practices represent God's will for all people. But humanity is so dazzled by its newfound autonomy and technological prowess that it must look to the church to rediscover its true character as God's creation. Although God has written the divine laws of suffering, love, and equality into nature and the human mind and, most definitively, into the face of Jesus Christ, modern humanity nevertheless claims independence from all such laws. The bravado of the human spirit has taken modern civilization to the limits of its finitude, an achievement frequently celebrated by King in his sermons. "Look at what we've done!" he exclaims in one of his poetic set pieces on "The Greatness of Man." When the human spirit coalesces in purpose with the divine will, the preacher King pronounces a blessing upon it, but when it inevitably diverges from the divine, then it is the church's responsibility to demonstrate a better way.

King's preaching rendered an agent of deliverance identified in Chapter 8 as the "awakening black human," a corporate figure who would experience the new birth of redemption and lead others to the same. The Ebenezer gospel provides additional definition to that figure: it is the church—not the whole church but the African-American church. There was a time when King was able to speak of the church as an undifferentiated body of Christians and even to lament the existence of a white and black church. There was a time when he viewed the Civil Rights Movement, especially the Selma campaign, as a movement representing the whole church and as the catalyst for an ecumenical awakening in America. However, when the white churches had an opportunity to prove themselves, they failed.

He frequently told how he had expected the support of the white ministers, if not the congregations, in the Montgomery struggle, only to have his hopes dashed. The cause of freedom for God's people has been repeatedly stymied by pious men of God and their congregations. He frequently complained to Ebenezer of the outrageous hypocrisy of the white congregations—such as the Baptist church that sent money to Africa but fired its pastor for allowing a black man to sing in the choir; such as the churches that emphasize the aesthetics of the liturgy above the demands of the prophets; such as the Dutch Reformed Church in South Africa that sanctions apartheid; such as the churches that crusade to keep the county dry but remain "wet" when it comes to prejudice; such as the good southern Christians who participate in bombings and lynchings.

[D]o you know that [the] folk [who] are lynching them are often big deacons in Baptist churches and stewards in Methodist churches, feeling that by killing and murdering and lynching another human being they were doing the will of the almighty God? *The most vicious oppressors of the Negro today are probably in church.* [Governor] Ross Barnette teaches Sunday School in a Methodist church in Mississippi. Mr. Wallace of Alabama taught Sunday School for years.

The "white preachers" are no better than their lay members. "[T]hey can look at racial injustice and never open their mouths against it." King could only conclude that "we are no better than strangers" to those Christians "even though we sing the same hymns in worship of the same God."

Nor did he spare the complacency of his own African Baptist church. Even among the membership of Ebenezer there were influential voices raised against his activities, but they were few and ineffectual. Ebenezer's parent-church organization, however, the National Baptist Convention, led by the autocratic Joseph H. Jackson, did not support King's demonstrations or violations of the law. King's dream of harnessing the Civil Rights Movement to the enormous train of the National Baptist Convention was dashed at the denomination's convention in 1960 when delegates rejected the King-backed challenge to Jackson's presidency and stripped King of his official position in the organization. From that time onward, he stepped up his criticism of the other-worldly, individualistic nature of the black Baptist churches that had spurned him.

One type of black church "burns up" in mindless emotionalism while another "freezes up" in bourgeois dignity. The latter group does not understand the American paradox, that the first step to becoming fully American is to celebrate one's own distinctiveness. Its members are "ashamed that their ancestral home was Africa." Its preacher "preaches a nice little essay on Sunday. . . . And they don't sing Negro spirituals and gospel songs because that reminds them of their heritage. So they are, you know, they are busy trying to be ashamed that they are black." King was concerned that many black congregations were not providing measurable support to the Movement, but even more concerned that they were not living up to their unique messianic assignment in America. They had been refined in the furnaces of slavery and segregation for a purpose, and now the proving of that purpose had begun. How would they respond? "This," he cried out to Ebenezer with evangelical and historical urgency, "is your great opportunity."

In the later sermons King's criticisms of the Negro church receded into the background, as another model of the black church assumed greater definition. In his earlier period of accommodation, King practiced a rhetoric of balance in which his assessment of the church consistently mirrored black with white failures—for example, black otherworldliness appears as a reflection of white secularity, and so forth. But the moral bankruptcy of the white churches and the downright evil of many of their leaders upset the balance both in King's rhetoric and his theology, and the black church finally emerges in all its splendid *contrast* to the churches around it. The black church serves as an agent of redemption because it continues to live as a "colony of heaven" in an alien world. Unlike the white church, it continues to enact the Christian story and has not rationalized away the harder demands of the prophets and Jesus. It still understands that the church is called to heal the brokenhearted, preach deliverance to the captives, and to announce the acceptable year of the Lord. Because of the evil it has caused or sanctioned, the white church has defaulted on its task and is no longer worthy to bear the name "church." King was willing to acknowledge a spiritual *ecclesia*, a germ of authenticity, in the white church, but little more.

Instead of understanding the black church as the mirror-image of the white, King interpreted its mission as an extension of the early church, which defined itself *over against* its pagan environment. The early church, King reminded his hearers in 1967, refused to fight wars, engaged in civil disobedience, accepted persecution, and knew how to suffer. Like Martin Luther, who made suffering one of the "marks" of the church, King asserted that the greatest strength of the black church is its willingness to suffer on behalf of the entire nation in order to bring about reconciliation. "I am grateful to God," King wrote in 1967, "that, through the influence of the Negro church, the way of nonviolence became an integral part of our struggle." What is also clear from his comments on the church is how fully King broke with civil religion and eventually with his rhetorical strategy of identification. By differentiating the black from the white church and identifying the latter as the oppressor-church, he effectively eliminated the illusion of commonality between all Christians and demolished the theological facade of the white churches.

In the crucible of the Movement and weekly preaching, theology retained its importance for King, as he continued to believe in the power of ideas to shape social behavior, but theology assumed a sense of immediacy—and danger—that it had not possessed at Crozer or Boston. The safe interval between theory and practice had disappeared. In his sermons to Ebenezer about suffering, commitment, and the witness of believers, King was redefining the contemporary church. In his emerging theology, the black church would no longer appear as a footnote to mainline Protestantism, for that god had failed. Instead, the black-church experience would become the new criterion of faithfulness against which all theology and practice must be measured.

VI

The final element in the Ebenezer sermons and one, like the gospel climax, that never appears in King's printed or white-church sermons, is the invitation. In many black and evangelical churches the sermon concludes with an exhortation to accept the preacher's offer of salvation. This is followed by a final appeal or invitation for the hearers to come to the altar as an expression of their conversion to Christ. The conclusions of King's sermons follow this two-step process. His concluding words usually take the form of an offer whose theme and imagery are closely related to the main topic of the sermon. By the sermon's end, however, the complications and ambiguities of life in a racist society are set aside, as the preacher offers nothing less saluatory or more complex than a personal relationship with Jesus. As one listens to King speak such words Sunday after Sunday, one senses that these conclusions are not merely preacher-formulas for him but represent expressions of his own settled peace with his vocation. He appears to derive immense satisfaction from them, perhaps because they are "pure" gospel, pure grace, which, when it is verbally applied to discouraged people, is something like anointing a wound with healing oils. For example, in his 1966 sermon on the seeking God entitled "The God of the Lost," the preacher concludes with the following offer:

> And that is it this morning, my friends. Somebody here is lost. Somebody here is away from the fold this morning. God is searching. God is still seeking. And the beauty of our gospel, the beauty of our religion, is that when you are found, heaven rejoices. . . . This is it. God is still seeking and searching for someone here.
>
> > "Softly and tenderly, Jesus is calling,
> > Calling for you and for me."
>
> There is a voice saying, "Come home, come home.
> Ye who are weary, come home. You've been out
> of the fold a long time. It's terrible to be
> lost out there! Come on back home."

In the 1963 sermon in which he compared faith to the "dead-man's float" he concludes on the same theme: "This is a magnificent time. . . . Just throw your all out on the divine. He will hold you up. You can't sink. You don't need to struggle any longer. God will hold you up." Then with the choir singing softly in the background, he implores his hearers, "Wherever you are, will you come this morning? Wherever you are."

Out in Macon County in the 1940s, William Pipes recorded the country preacher's concluding invitation, "Won't He take care of you, chillun? Be consolation for you? My, my Lord! I done tried! I know He *do* take care of you, preserve your soul, and den what I like bout You, carry you home. Here rain wouldn't fall on you. Beautiful land. Gwine to sing and open de doors of the church." King's Ebenezer maintained the traditional "We open the doors of the church now," and invited worshippers to come forward to join the church by baptism (in the evening service), by profession of faith, or for "watch care" (for sojourning students carrying a letter of recommendation). But the general tenor of the invitations indicates that their main purpose was to encourage repentance and conversion in the seeker. "This is the time to make your decision," King cries. As a worshipper comes forward, he exalts, "God, be thank[ed] for this young man. Is there another who will come this morning? Is there another?" On a sweltering Sunday in July 1963, with the organ playing softly in the backgorund, he intoned, "We open the doors of the church now. Someone here this morning needs to make a decision for Christ. . . . Who this morning, wherever you are, whether you're in the main sanctuary, or whether you're downstairs in the first unit of the church, whether you're in the balcony, come now and take that step." The choir begins singing, *Where He Leads Me,* and King, speaking above the choir continues, "Who this morning will take that step?" until he too begins singing in makeshift bass harmony with the choir. As people file forward, he is pleased to report, "There's another this morning, there's another. 'Where he leads me . . .'" and the preacher is humming, then singing once again.

The services at Ebenezer often ended on a chaotic note when the invitation would degenerate into general announcements. The earnest invitation of the above sermon is quickly followed by the business of the day: "We're urging each of you to be back tonight for the Lord's Supper and for baptisms. Since we have some 25 or 30 candidates for our baptism, it will take a little longer tonight than usual, and we are asking the candidates to be here at 6:30. . . ." On another occasion, when his father was to be away, King reminded the congregation that he would be available for emergencies and "on hand to take care of things next Sunday." On other mornings his father would pop up at the very end to comment on something his son had said in the sermon or to bullhorn one final announcement, as on the Sunday when he announced that his emergency discretionary fund had run low. "I want fifteen cents from everybody," he shouted. "Right quick!"

If the conclusions show anything about Martin Luther King, Jr., they demonstrate the appropriateness of his role as a Christian civil rights leader whose political stance and rhetorical performance were deeply informed by the church's gospel. These invitations, as well as the Ebenezer gospel as a whole, reveal the springs of his religious appeal to America: to repent of its racist and militarist sin; to honor the provisions of its covenant with the divine Being; to discipline its life according to the demands of brotherhood; to embrace the higher Power who makes new life possible; and to anticipate with joy the physical and spiritual transfiguration of the nation. The invitations in particular, with their emphasis on decision, had their counterparts in the religiously charged atmosphere of the mass meetings in Negro churches throughout the nation. There King translated the personal and religious altar calls of Ebenezer into corporate and political appeals for racial solidarity.

10

Bearing "The Gospel of Freedom": The Mass Meeting

SHORTLY before King embarked on the Albany campaign, he dispatched two of his closest aides to confer with members of the evangelist Billy Graham's staff. Even though the two preachers were worlds apart on political questions, King had long known and admired Graham, trusted his goodwill, and had once played a small part in one of the evangelist's crusades. Now King's people wanted to know how their man could win the approval of the American public and how the SCLC could run its operations as smoothly and successfully as Graham's. Graham's recommendations were carried back to King but never implemented. The bottom line was that King was spreading himself too thin. Graham's team spent thousands of hours preparing for a few, carefully orchestrated appearances each year; King was dashing in and out of every little church and hamlet that wanted him, answering and inspiring Macedonian calls wherever he went. His model of evangelism seemed designed eventually to exhaust the evangelist.

It is significant that Martin Luther King turned, however briefly, to Billy Graham, for with that gesture King was acknowledging the possiblity of translating an evangelistic tradition the two men shared into a method of social transformation. Evangelism and politics had never been strangers to one another in American life. Their entanglement dated back to the hegemony of the Puritan pulpit over every social institution in New England. In the nineteenth century, evangelism and politics separated into two streams: revivalism and social reformism. A few reformist preachers before King, like the fiery abolitionist Theodore Weld or the Social Gospeler Rauschenbusch, had dreamed of fueling social improvement with evangelical power, but each attempt seemed fated to choose between political agitation and personal religion. Billy Graham had made his choice; now King wanted to make his. After the success of the Montgomery Movement, he was poised to effect a critical transfer from the Ebenezer Gospel to the Gospel of Freedom. Henceforth everything he had learned in the black sanctuary would be modified for reception in the larger culture and loosed upon the injustices of American society. The vehicle of King's social gospel would be the mass meeting.

I

The mass meeting was born in Montgomery with the December 5, 1955, rally at the Holt Street Baptist Church. From the beginning the meetings served an indivisibly sacred and civic agenda. At Holt the throng listened to Bible readings, sang *Onward Christian Soldiers* and *Leaning on the Everlasting Arms*, and took courage from a sermonic speech by the young Dr. King. Once the Movement got under way in Montgomery, meetings were held every Monday evening at two churches on the east side of town and every Thursday at two churches on the west side. "You had to get there by 4," Deacon R. D. Nesbett remembers, "to get a seat for a 7 o'clock service." The meetings began with songs and hymns and moved quickly to reports and requests for help. People signed up to drive cars, provide food, or to serve in other capacities. Then came testimonies from the "martyrs," as the assembly named them—those who had been arrested or who had suffered in a special way. The Lutheran pastor Robert Graetz recalls that the meetings were highly emotional social and religious occasions, with each having the character of a "victory celebration" not unlike the early church's celebration of its suffering and martyrdoms. Their religious character was heightened by the music. Coretta Scott King remembers that the Montgomery meetings featured a worshipful atmosphere and the hymns and spirituals of the church, such as *What a Fellowship, What a Joy Divine* and *Lord, I Want to Be a Christian*. The meetings helped create "the feeling that something could be done about the situation, that we could change it." The presence of the ministers of Montgomery, who led the assemblies in prayer and exhortation, along with the religious music and the symbolism of the sanctuary itself, confirmed the traditional partnership of the Negro church with its people's social aspirations.

Preachers like King usually moved from meeting to meeting, making a "grand entrance" just before their time to speak. They made their case for freedom on the basis of the Constitution and the Bible, illustrating their lessons and exhortations with stories about people everyone knew and incidents from the week's events. Dexter member Zelia Evans would not characterize the ministers' speeches as sermons, "but sometimes they [King and the preachers] *made* them sermons" by exhorting from the Bible and evoking a religious response—by preaching. Indeed, it was in these meetings that King proved his extraordinary power as an orator. One journalist said his delivery was "like a narrative poem." She thought his voice had such depth of tenderness that it could "charm your heart right out of your body."

As the Movement spread throughout the South so did the meetings. John Lewis observed that they were attended by sharecroppers, the poor, and common folk because the meetings were held in church, and the people felt at home in church. Only in the sanctuary could one experience the religious expression of all that is wrong with this world *and* the hope for dramatic change. These meetings were *church*, and, for some who had grown disillusioned with Christian otherworldliness, they were better than church.

As in Montgomery, the mass meetings held throughout the South also served to solidify a sense of community among the participants. The meetings provided a continuous social commentary on fast-breaking events, a forum for information and tactical planning, a school for correction and instruction in nonviolence, a place of praise and encouragement, but, most of all, a way of keeping together. In their atmosphere of fellowship, ordinary people got a chance to speak up and express their dreams for a better life. At a meeting in Albany a plain man stood up to make testimony—not about an abstract notion of freedom but about the kind of life he longed for in his hometown: "It is a funny thing," he said. ". . . As much hell as we've caught here in Albany, I still love it. It's home. I love to fish, to pick the magnolias, to pick blueberries. I love peanuts. . . . I want to stay here. I want to raise my children here. I want to live a decent, law-abiding, self-respecting and dignified life here."

The mass meetings were not media events. Their poignancy was never translated into public relations or Nielsen ratings. Had they gained access to mass media, the meetings could not have functioned as the secret life of African Americans in the South. The meetings represented the ripening of a people's social and spiritual preparation for freedom. If the slogan in pre-Revolution France was "The bread is rising," in Montgomery, Selma, Albany, Greenwood, and countless other southern towns, it might have been a lyric from one of the most popular meeting songs, "I see freedom in the air." The paradox of the mass meetings was that they *had* to belong to African Americans in order to minister to African Americans, but they also possessed the squandered potential for influencing the whole society. One night Jackie Robinson told the packed Sixth Avenue Church in Birmingham, "The inspiration in this church tonight should be shown throughout the world."

Reporter Pat Watters, who more than any other professional observer of the Movement recognized the genius of the mass meetings, expressed his frustration at their elusiveness to historians:

> I sit and lament anew that the movement did not reach southern whites, lament the southern proscriptions that made it impossible for whites to enter such churches, hear such eloquence, feel the southernness of those meetings, and lament as much the forces, the compulsions of American culture that prevented any serious attempt by the media (television being surely the most appropriate) to present what was said and felt by the Negro people in those meetings. Back then, even then, I understood enough to say that if ever they would just put one mass meeting on television, for however long it might take, it would all be over.

What record we have of the meetings we owe to reporters like Watters and the FBI and the local police departments. Though they did not intend to document moving celebrations of the gospel and human freedom, the police surveillance files and bugged recordings of the proceedings provide the most thorough account of what went on. Testimony to one of the most vibrant sociopolitical movements in recent American history is stored in yellowed files and reel-to-reel tapes in the basement of the Birmingham Public Library, ironically, among the collected papers of Eugene "Bull" Connor. The mass media may have been indifferent to these events, but the police were not.

The rich portrait of the meetings found in the files of the Birmingham Police Department is accompanied by the commentary of the hapless officers assigned to attend them. One can read their reports and imagine the dispatches first-century Roman agents must have relayed to their supervisors concerning the strange practices of a new and dangerous religion. These are "Dear Sir" letters, the "Sir" being Jamie Moore, chief of police. "Dear Sir," the officer wrote, "Tonight was their victory night. The church was filled to capacity and later about 1,500 school children came in with their flags." He makes no comment on the Movement's tendency to combine early Christian themes like "victory" with (to the white citizenry) maddening displays of patriotism.

"Dear Sir," another detective wrote, "Martin Luther King made his grand entrance at approximately 8:15 to a standing ovation" at the First Baptist Church of Ensley. The detective then adds an image—a thin slice of atmospherics—to which only an eyewitness has access: "Negroes were sitting in the windows." On the worship practices of the group, one detective wrote, "Dear Sir":

> The more he [the preacher] talked, the more he screamed and hollered. He kept saying that God would save Bull, God save Wallace, God save Khruschev [sic]. He screamed, "What do you want me to do, Lord?" over and over again. . . . He kept hollering and toward the last said that he would like to take Bull in his arms and take him to God and save him. . . .

He adds, "It was very cold and there was sleet and ice frozen over when we came from the church."

At the cavernous Sixth Avenue Baptist Church, the policeman described the atmosphere in the church: "At this time he [James Bevel] led the church in singing, and the Negroes got all worked up while singing, stomping their feet and waving their arms and screaming. There were about 300 people standing and marching [in the church]. The entire audience was between 1,800 and 2,000." A week later, with the meetings still at full throttle, the same detective reported how a visiting rabbi asked everyone in the audience to put an arm around the person sitting to either side. "Of course Officer Watkins and myself were sitting between two negroes and they really gave us the treatment."

The officers' own moral thickness is nowhere clearer than in the brief summary for April 9, 1963. At the 16th Street Baptist Church the great Al Hibler sang and King had just performed what would later become his most famous set piece. The officer reported sarcastically, "I had a dream tonight."

The police tapes capture the good humor of the Movement, though that, too, appears to have been lost on the detectives. The comedian Dick Gregory visited one of the meetings, but the officer on duty reported little appreciation for "his dry jokes." When one of the most feared judges in Birmingham died suddenly, two days after Rev. Fred Shuttlesworth had predicted that *if* he died he would probably get 180 days in hell and 100 dollars plus costs, one of the leaders asked for prayer for the judge's family "because someone must have loved him." A couple of nights earlier Rev. Gardner suggested that the people pray with their eyes wide open, for "if you pray with your eyes shut you might be crucified." Several times the meetings elected "Bull" Connor an honorary member of the Civil Rights Movement for all his tactics had done to further the cause.

Even King, who usually hid his gifts for mimicry and funny stories from the public, skillfully used humor to defuse the crowd's nervousness at the prospect of facing Connor's dirty tricks. About the police dogs, he joked, "And dogs—well, I'll tell you, when I was growing up I was dog-bitten for *nothing*, so I don't mind being bitten by a dog for standing up for freedom." Of Connor's armored vehicle, which King derisively called a "white tank," he said, "[T]hey can get his white tank, and our black faces will stand up before the white tank." Of the fire hoses, he reminded his audiences that many of them were Baptists, and Baptists know what *deep* water is all about. The humor does not appear to have gotten out of hand or become an end in itself. In a Christian context, laughter in the face of danger, like praise in a time of sorrow, represents trust in God's authority over earthly rulers.

From the beginning, observers and participants noticed that the mass meetings resembled evangelistic church services. Like the typical week-long revival, these assemblies tested the endurance of the faithful; the mass meetings in Birmingham, for example, began in January and ended in May 1963, and at one stretch ran an astonishing fifty consecutive nights. All the preaching and exhorting was borne along by the lively and some-times haunting music of the Movement. One evening a soprano led the assembly in *He'll Give You the Victory*. Her voice soaring above the other singers, their chorus filled Brown's Chapel with a sound as plaintive as the whistling wind. As the song descended into applause and amens, one of the ministers can be heard to joke, "The doors of the church are open." Also like revivals, the meetings closed with an appeal to members of the audience to make a decision to participate in the Movement and to signal their commitment by coming forward.

The early meetings of the Alabama Christian Movement for Human Rights featured complete sermons on biblical texts by various local min-isters. King "preached" too, but one hears only fragments of his famous sermons, like "A Knock at Midnight," in these tumultuous meetings. Later the sermon was relegated to the forty-five-minute period of preparatory worship or eliminated entirely in favor of speeches and sermonlike ad-dresses. Fervent prayer sessions, however, did continue throughout the mass meetings. Like the praise session held before church in some Baptist and Pentecostal traditions, the prayers were a time of spiritual intimacy preceding the meeting proper. As the reporter Watters noticed, the prayers were every bit as eloquent as the oratory. Before a march in Albany one of the preachers prayed:

> We feel much akin
> To those who went out
> Two by two

> In the days of old.
> We will march around
> Those jail house walls
>   That symbolize segregation.
> We will walk around them
> Like unto Joshua
> Until the walls
> Come tumblin' down
> Take care of us
> Take care of the policemen
> Take care of Chief Pritchett
> Take care of the mayor
> And the city council.
> We pray that as they see
> A powerful and peaceful people
> their hearts will be moved.
> Consecrate, dear God
> This whole community.

The marchers then began singing and moving, "We are marching to Zion / Beautiful, beautiful Zion . . ." before his words ended.

Despite the nimbus of "church" that accompanied the meetings, King and his associates intentionally modified the traditional Sunday-morning gospel in order to accommodate the equally traditional African-American quest for political freedom. This was first apparent in the music. Although the Movement produced some of its own musical expressions of redemptive suffering, such as Sam Block's hauntingly beautiful *Freedom Is a Constant Dying*, usually the words of spirituals and gospel songs were altered by a less creative process and made to refer to the freedom movement.

> Woke up this mornin'
> With my mind stayed on *Jesus*

became

> Woke up this mornin'
> With my mind stayed on *freedom*.

> Over my head I see *Jesus* in the air

was changed to

> Over my head I see *freedom* in the air.

> When I'm in Trouble, Lord, Walk with me

became

> Down in the Jailhouse, Lord, Walk with me.

The ecstatic repetition of the name of Jesus, frequently heard in Pentecostal churches, gave way to frenzied repetitions of "freedom" in the mass meetings. The traditional Baptist refrain "When we all get to heaven" was changed to "When we all know justice." In the midst of the Albany campaign marchers sang,

> This little light of mine
> I'm gonna let it shine . . .
> All on Chief Pritchett
> I'm gonna let it shine.

In his book on Birmingham, King also noted that the final appeal for soldiers in the "nonviolent army" had its counterpart in the Sunday morning altar call. In words that echoed his own invitations at Ebenezer, he cried out at Brown's Chapel, "This will be Selma's great opportunity." With many other Ebenezer formulas he exhorted them to step forward, first for training in nonviolent techniques, then to march.

The meetings continued for months after King's death in preparation for the Poor People's Campaign, but the modifications to the gospel became more aggressive in tone and their rhetoric incorporated blatant suggestions of violence and revenge. *We Shall Overcome* (some day) gave way to *We Shall Overrun* and to new victory songs like Len Chandler's *Move on Over or We'll Move on Over You*. The singing, which had been indispensable to the spirit of King's SCLC, came to symbolize for more militant groups like SNCC, CORE, and other adherents of Black Power, a sellout to false harmony and religious sentimentalism. "Whoever heard of a revolution where they lock arms" and sing? Malcom X asked. "You don't do that in a revolution. You don't do any singing, you're too busy swinging." When activist Sammy Younge was murdered in January 1966, James Foreman remembers, "people were just filling the streets, and they weren't singing no freedom songs. They were mad." As a matter of symbolic principle, the music of the Civil Rights Movement eventually fell silent. Reporter Watters chronicled not only the growing silence but the change in the religious atmosphere of the meetings. He vividly remembers a meeting led by Ralph Abernathy after King's assassination at which participants responded with Amens and Preach! but in mockery of black church worship.

In their "classic" form, the mass meetings through 1965 were the expression of a social movement in America whose Christian character was not limited to rhetorical gestures. Many of King's former aides now make a simple distinction between their leader's approach and that of competing organizations: "Martin was a Christian." Under King's leadership, the Movement used Judeo-Christian symbols and methods to effect change at the grassroots level. The meetings and all that followed them were governed by the two master keys that had unlocked the Bible for King: Exodus from bondage and love of enemy. The rhetoric of Exodus was everywhere in those years, and it lent to the Movement its most evocative and enduring symbol, the march. At the close of the march from Selma to Montgomery, King stood on the steps of the capital and cried out ecstatically,

> Be jubilant my feet.
> Our God is marching on!

The theology and joy of marching seemed to come second nature to King's organization, but to the militant groups, which were more intent on seizing power than standing up for righteousness, the march appeared as harmless as a Sunday school parade and as nonsensical as loving one's opponents.

In the Montgomery, Albany, Birmingham, and Selma meetings the preachers blessed their persecutors (police and local politicians), prayed for those who despitefully used them, and incessantly exhorted the people to love their enemies. At Saint James Baptist in Birmingham, the police recorded King's pleas that his hearers not only boycott the after-Easter sales in white department stores but do so with purity of heart: "For even ole' Bull is a child of God. We love ole' Bull." The preachers prayed for the sheriffs and their kin when they were sick, and on the rare occasions when the authorities came to the church, the marchers treated them with respect. In Birmingham, Rev. Gardner even commended the undercover detectives for their perseverance in attending so many meetings in a row. When James Bevel lashed out at "white trash" and gestured toward the detectives, or when Abernathy ridiculed Sheriff Jim Clark and his bumbling deputies, King and others invariably took the high road of love and understanding. At one meeting the police recorded Andy Young telling the audience,

> The Police don't know how to handle the situation governed by love and the power of God. We really can't blame the white man in some ways, because from the time he was a child he has been brought up to think we are animals. Remember this, for some day you may be on the police force or riding in the same car with these officers.

A week later King was telling the same crowd at the Sixth Avenue Baptist Church, "There may be more blood to flow on the streets of Birmingham before we get our freedom, but let that blood be our blood instead of the blood of our white brother." When protesters or marchers engaged in violence, King's organization held itself accountable.

The SCLC was one of the few activist agencies in recent memory that took repentance seriously. Because racism represented not only a structure of evil in society but "sin," it required curtailment or change, to be sure, but also "repentance," which mirrors inner transformation. The young man who had professed, "I'm a Baptist preacher, and that means I'm in the heart changing business," not only called on recalcitrant cities like Birmingham and Selma to repent but as late as Memphis 1968 declared a day of repentance for his own Movement because of the flurry of rock throwing and window breaking kicked up by some demonstrators.

After Selma, however, the language of redemptive suffering and love of the "white brother" all but disappeared from King's oratory. The evangelicalism of the Birmingham and Selma meetings was replaced by angry analyses of racism and militarism in America. Love of enemy, which was the most prominent theological theme of the Montgomery, Albany, Birmingham, and Selma campaigns, was succeeded by general appeals for sacrifice and tactical proposals based on black and white demographics. Nonviolence remained the moral staple of King's Movement, but it was a nonviolence impelled by an African-American version of Realpolitik. In a late speech in Yazoo City, Mississippi, King told a crowd,

> I would be misleading you if I made you feel that we could win a violent campaign. It's impractical even to think about it. The minute we started, we will end up getting many people killed unnecessarily. Now I'm ready to die myself. . . . [But] when I die I'm going to die for something, and at that moment, I guess, it will be necessary . . . [but] we can't win violently. We have neither the instruments nor the techniques at our disposal, . . . but we have another method, and I've seen it, and they can't stop it.

II

If one of the distinguishing characteristics of the 1955–1965 mass meetings was their Christian character, a second was their stunning oratory. Although the theological terms of King's appeal changed in his final years, its rhetorical power was not diminished. He remained the designated preacher in the Movement's Church of Freedom. Although not the deep thinker he was portrayed to be, King's ability to dramatize ideas made him the most formidable orator in America. If the occasion demanded that the crowd be *moved*—either to action or restraint—there was no one in the world more capable of delivering the desired effect than Martin Luther King. His rhetorical genius stymied opponents on all fronts and won critical battles from Holt Street in Montgomery to Mason Temple in Memphis. He was not only a mass-*media* phenomenon whose voice was nationally recognized but also a mass-*meeting* phenomenon whose voice was required to activate small-town marches and voter registrations. The people came in crushing numbers to hear him and to touch him. There was no occasion so large that it could overwhelm his voice, and no church so small that he would not get down with the country folk and preach the gospel in it. One of the preachers of the Movement, C. T. Vivian, said, "You see, we all knew that we could not do the things needed until Martin spoke. That's [why] we gathered people together—for Martin to speak! It was after crowds would hear Martin that we knew that they were ready." And after the march, it was often King's voice that interpreted the accomplishments of the group or dressed its wounds.

The Civil Rights Movement spoke several "languages" at the podium. Among them was the speech to educate, usually given at major conventions or college commencements; the inspirational address (or speech of encouragement), often given at friendly rallies around the country; and the civic address, whose focus on civil religion was dictated by the place and occasion of the speech. No genre of address, however, came so close to fusing the word with the souls of its hearers as the mass-meeting speech. It stirred ingredients of education, encouragement, and civil religion in the primal cauldron of the African-American sermon, producing an effect without parallel in twentieth-century America. Born in preaching and uttered from the watch fires of a hundred African-American sanctuaries, the mass-meeting speech created a mystic bond between the speaker's word and its enactment by those who heard it. Ask some of the preachers who participated in the meetings what these speeches were about, and their eyes mist over as they say, "They were the same, the same."

Scores of preachers besides King gave these speeches, and many went on to prominence in the church or politics. Outside observers of SCLC and other civil rights organizations were often struck by the sheer quantity of black preachers involved in their day-to-day operations. "SCLC is not an organization," one of its officials told a reporter, "it's a church." The one who presided over this "church" was habitually surrounded by clergy or closely associated with preachers. The cohort included Ralph Abernathy, Andrew Young, Wyatt Tee Walker, C. T. Vivian, Jesse Jackson,

Bernard Lee (who was ordained later), Walter Fauntroy, James Lawson, Charles Sherrod, Bernard Lafayette, John Lewis, Jim Bevel ("the foremost preacher" in the SNCC), and many others. If these preachers were the princes of the Movement, Martin was the preacher king. Leaders in cities outside Atlanta tended to be pastors of significant black churches as well: Fred Shuttlesworth in Birmingham, Kelly Miller Smith in Nashville, Thomas Kilgore in Harlem, C. L. Franklin in Detroit, and C. K. Steele in Tallahassee are representative. Black ministers anchored the leadership of nearly every local movement, and their sanctuaries served as bases of operations. As preachers, they were all partakers of black-church eloquence, each with his special rhetorical gifts and trademark. Bevel was fiery, Vivian urgent, Lewis sincere, Lawson intellectual, and none of them, including King, could match Abernathy's ability for charming an audience with folk humor. If, in the words of C. T. Vivian, they often found themselves "preaching the people into the streets," the preachers also knew when to turn down the heat in order to avoid a self-defeating and bloody riot. In this respect, King's preachers functioned analogously to the old-time Negro preachers whose apocalyptic visions ignited hope and kept it alive but who also knew how to contain the fires they had set. The overwhelming presence of clergy at all levels of the Movement—in the planning councils, behind every pulpit, on the streets, in the jails—was a visible reminder of the historic importance of the church and its preachers in the long tradition of black protest in America. When the dust had settled over the Selma campaign, a white southern congressman lamented, "The preachers did us in."

None of the preachers had the philosophical temperament to elevate or idealize the moment as King did, but many had rhetorical equipment that was in no way inferior to his. The nobility of King's effect was possible only by contrast to those who preceded and followed him on the platform. Before him came gifted speakers who were not shy about using invective, sarcasm, vilification, mimicry, and slapstick in order to excite the crowd. And those who followed him to the microphone provided the less-exciting organizational details for the prospective march or voter registration.

One of the most effective preachers on the stump was the prophetic James Bevel. His mass-meeting speeches and sermons, including his impassioned appeal for nonviolence after the murder of Jimmie Lee Jackson, have been preserved only in the memory of witnesses because, by his own admission, he was living for the moment and the Movement "without a sense of history." In those years, he recalls, the preachers wanted only to push the gospel to the limits of its social implications. On April 18, 1963, the police reported that Bevel worked the Sixteenth Street Baptist Church into a "frenzie" with a sermon on Mary and Joseph in Bethlehem. Just as Joseph could not secure a room for his pregnant wife, so Bevel and his pregnant wife could not get a motel room in Alabama. The man who turned them away was a deacon in the Baptist Church. "Some Negroes," he went on,

> think that Bull Connor can give life, but only God can give life; and I know God is looking after me. Doesn't the stupid white man know that Martin Luther King is a disciple of God? . . . You can put us in jail, but you can't stop us. When the Holy Ghost gets to a man, nothing can stop him. The Negro has been sitting here dead for three hundred years. It is time he got up and walked.

At the close of the same meeting, Wyatt Tee Walker jumped up and announced, "I am looking for two dozen Negroes who are willing to die for freedom!"

The orator King owed more than anyone has acknowledged to preachers like Bevel, Vivian, Walker, and, most of all, to his opening act, Ralph Abernathy. As a speaker Abernathy was everything King was not: homespun, down-to-earth, a master of mimicry, uproariously funny. No one had mastered the Lowndes County vernacular as Abernathy had, and no one could warm and charm a crowd as Abernathy could. Although often rumored to be jealous of King, he had no qualms about introducing his friend as "one of the greatest men that God has ever breathed life into" with "a Ph.D. from Boston University." If King was the Boston-trained orator, a.k.a. "de Lawd," Abernathy was the Marengo County folk-preacher.

Abernathy's stories were masterpieces of local dialect and characterization. One evening in Selma, the police recorded Abernathy's story about a hired hand who decided to call his white boss by his first name.

> The bossman said, "Now Charlie, what's wrong with you? Have you lost your mind or somethin'?" And he said, "No, I haven't lost *my* mind. . . ." "Well, I *know* you've lost your mind. You've worked for me for forty years, and you've been faithful. Let me know if you've been sick, so I can take you to the doctor." [Abernathy mimics the boss's phony, solicitous style of speech.] And he said, "No, I'm not sick, but you know the Freedom Riders—they come through town, and they have told us, *John* [more laughter at the thought of calling a white man by his first name], *John*, I'm just as good as you . . . I want you to tell *Ann* [more laughter] that from now on there's not going to be any *Mr.* John and *Miss* Ann. In fact, it's not going to be Miss *anything*. It's not gonna even be *Miss*-issippi. It's just gonna be plain ole *Sippi*." And somebody must get the word

downtown to Sheriff Clark that Negroes are not afraid of him. He can come out and wave his stick all he want to. He can push us and he can shove us, but we are not afraid. We're gonna stand our ground.

The same evening Abernathy began to address the little recording device on the pulpit that relayed the speeches to a police mobile unit outside. To the crowd's delight, Abernathy called the device a "doohicky." "This is the little doohicky that takes the message downtown," he explained,

> and I want you, doohicky, to tell 'em: [crowd chants, *doohicky! doohicky!*] you go places we can't go, and will you tell the good white folks of Selma, Alabama, that we are not afraid and that we aren't gonna let nobody turn us around. Now, doohicky, I want you to tell 'em that we were down there today, but we'll be back tomorrow. And you tell them, doohicky, that we have checked with our lawyers, and we know that we have a right to stand in *our* court house and that they may as well get ready for more than one hundred down there. . . .
>
> Now, we want you to get this other message over, doohicky [at every mention of the word, the crowd squeals with delight], today we had to march to town two abreast, twenty feet apart. Now, doohicky, this is not right, for, for, this is not Russia. If this was Russia, doohicky, we could understand. But this is the land of the free and the home of the brave, and we have the right, doohicky, to walk with whomever we want to walk with. And we are sorry to disappoint you, but we ain't gonna walk only two tomorrow, and we're not gonna stand any twenty feet apart. Doohicky, you tell 'em that we're gonna walk *together*.
>
> Now, now, little doohicky, I hope that these few words will find you well. [Now to the crowd] . . . I want *you* to talk to the doohicky. You see, they got it out, they got a rumor out that only a few Negroes want to be free, and we gonna *all* talk to this doohicky tonight. . . .

The Doohicky speech exposed the absurdity of the authorities' attempts to control the Movement. Decades later it continues to bring pleasure to researchers in the basement of the Birmingham Library, for without the Police Department's doohicky we would know far less about the texture of the mass meetings.

Always with empathy, humor, and a powerful rhythmic partnership with his audience, Abernathy prepared the way for King. One winter evening in Selma he electrified his hearers with a defiant blast at the plight of domestics who take the *early* bus:

> Are you through with the 'early'? [*Yeah!*] Are you going back on the *early* in the mornin'? [*No! No!*] Are you going to the back door in the morn-

ing? [*No! No!*] And the fourth and final thing I want to know, Are you gonna take the town over tomorr'a? [*Yeah!*] Go on, children, go on, go on, children, march in the streets of Selma. Stand in the face of Sheriff Clark. . . . I looked in Mr. Clark's eyes, and I could see in the back of their minds that they were saying, "Boys, we're like tenants here. It's all over now." [huge response] Just tell me one time: what do you want? [*Freedom!*] What do you want? [*Freedom!*] What do you want? [*Freedom!*] When do you want it? [*Now!*]

As the meeting begins to break up, one can hear the leaders of the assembly shouting over the din, "This meetin's not over yet!"

III

By the time it was King's turn to speak to the mass-meeting audiences, Abernathy and the musicians would have prepared the church to receive its new Moses. Entering to choruses of

> I'm so glad . . .
> Integration is on its way
> Singing glory hallelujah
> I'm so glad!

the smiling King would make his way down the center aisle, now greeted by shouts of "Martin Luther King says 'freedom,' Martin King says 'freedom'!" The music and shouting would finally subside, and the symbol of his audience's aspirations would take his place behind the pulpit. Most of King's mass-meeting speeches had a dual purpose. He visited the meetings in order to inspire commitment to the local movement and to insure that the commitment remained nonviolent, which meant that his speeches were always marked by the tension between exuberance and restraint.

In a mass-meeting speech, King typically (1) justified his presence in a city not his hometown; (2) proclaimed the world-historic significance of the present moment; (3) rehearsed the terrible conditions under which the people have lived; (4) vilified and warned the local authorities who continue to oppress the people; (5) reaffirmed the "somebodyness" of those in his audience, often on the basis of biblical revelation; (6) reviewed the accomplishments of the Civil Rights Movement; (7) called for a specific act of commitment; (8) insisted upon and lectured on the nonviolent character of the Movement; and (9) promised a new and better day for blacks and whites in America. Perhaps the speeches were "the same," as his former associates insist, but the chemistry in each church was different, and taken as a whole, the speeches created a voice for an awakened body of protest.

On January 14, 1965, King strode into Brown's Chapel to address the Negroes of Selma. The victory of Birmingham lay in the past; "Bloody Sunday" and the Pettus Bridge fiasco lay two months in the future. The quality of the audiotape is so clear that the historical moment seems alive once again. The following account is in a collection of surveillance recordings made in 1965 by representatives of the Dallas County, Alabama, Sheriff's Department. Early in the speech King *justifies his presence and criticizes his detractors*:

> Of course there are some, including the mayor, who would consider some of us outside agitators [*That's all right*, the crowd is quick to respond]. Maybe I ought to tell you why I'm here [*Preach it, now: go ahead and tell us!*] I'm here because I was *invited* here [*That's right . . . glad you're here!*]. And I have organizational connections here. The Dallas County Voters' League is an affiliate of the Southern Christian Leadership Conference, so I *belong* [a glissando: *be-law-ong*] here, applause]. I am here because injustice is here [*That's right*, more applause]. I cannot stand idly by when any community finds itself caught in the shackles of man's inhumanity to man because injustice anywhere is a threat to justice everywhere [*Yeah!* applause]. Just as the eighth-century prophets left their little towns and villages captured by "thus saith the Lord," just as the apostle Paul left his little village of Tarsus, and decided to "hotfoot" it all around the Greco-Roman world to carry the gospel of Jesus Christ, *I am compelled to carry the gospel of freedom* everywhere men are oppressed [applause, *All right!*]. . . . So I'm in Selma because [crowd roars back, *You belong here!*].

King impresses on his audience the *historic significance* of what they are about to do when he says,

> Some years ago a great [3-second pause] writer from France said, "There is nothing more powerful in all the world [*yeah?*] than an idea whose time has come." [*Yeah. O.K. It's here! It's here!*] My friends, I believe people all over the world realize now that the idea whose time has come today is the idea of freedom [*Yes, yes. Come on, come on!*]. I am absolutely convinced [long pause] that the Negro people of Selma, Alabama, have been captured by this idea whose time has come [*Yeah, all right*]. I can tell by the way you sing [*oh*]. I can see it on your faces [*Make it plain*].

Because the Selma campaign will focus on voters' rights, he *rehearses the voting injustices* the people have suffered. But first he attacks the root of the problem, which is segregation:

We know that segregation is nothing but a new form of slavery covered up with certain niceties of complexity [huge applause and shouts]. Segregation is the cancer in the body politic which must be removed before . . . [his voice is drowned by a tumult]. . . . Alabama has not been fair to her citizens of color [*All right, yes*]. Alabama, Alabama—has issued such complicated literacy tests that a person with a Ph.D. or a law degree . . . could not pass [*laughter*]. Alabama has given all kinds of stumbling blocks and conniving methods to keep the Negro from becoming a registered voter . . . [lost in applause and shouting]. But we are determined to register and we are determined to vote [*Yessir*].

This is the *specific proposal*, King announces:

There is a plan for Selma. We are going to make Monday "Freedom Monday" [applause]. We are not merely to engage in high-sounding words [*No sir*]. We are not merely to talk about freedom [*Tell it, doctor*]. We are here to do something about it. . . . Monday will be a day of massive testing and challenge. [On Monday we are going to] test the public accommodations [*All right, we're with you, you know*]. Every restaurant, every theater [*All right!* louder], every hotel on the highway and every motel of the community [*Yessir*] will stand under the scrutiny. . . . On Monday we are going to say to Selma in no uncertain terms, "We are through with segregation, now, henceforth, and forevermore" [huge uproar of applause and laughter]. On Monday we're going to get some jobs, like being policemen and other things "just like white people" [applause and shouting]. [Finally] we will challenge the registration policies by a massive march on the court house [*All right!* great applause].

And what is the deepest legitimating basis of these bold actions? The *intrinsic worth* of the people of Selma.

I know we have been abused and we've been scorned [murmur of approval]. We've known the long night of injustice [*yes*]. Two hundred and forty-four *years* we knew slavery. Another hundred years we've known segregation. And so often we feel that we don't count [*All right*]. We lose our sense of dignity and a sense of self-respect [*Yessir*]. I come to tell you tonight in Selma, You may not have a lot of money [*No*], you may not have degrees surrounding your name; you may not know all the intricacies of the English language; [*No*] you may not have your grammar right [*Right*] but I want you to know that you are as good as any Ph.D. in English [shouts and applause]. I come to Selma to say to you tonight that, oh, [*Speak! speak!*] that you are God's children and therefore you are *some*-body [tumult].

The speaker *reviews the triumphs of the Civil Rights Movement* and thereby places Selma into an immediate historical context: So much has happened in the South, he says.

Don't let anybody fool you [*No*]. The South is better today than it was five years ago. Don't let anybody fool you about that [somewhat restrained applause]. . . . The South is better today because Medger Evans lived in Mississippi and died in Mississippi [*My Lord! Tell it, tell it*]. It's better because three young civil right workers who had only one desire and that was to help work with people as they tried to register and vote died in the soil of Mississippi. The South is better today [soft murmurs of agreement] because back in 1955 Rosa Parks lived in Montgomery, Alabama. Like Martin Luther of old who said, "Here I stand, I can do no other, so help me God," she said in substance, "Here I sit, and I can do no other, so help me God" [great applause and shouting, *It's better! It's better!*].

With such a legacy of courage, Selma can *look forward to a better day*.

If we will march by the hundreds, our success will enter the conscience of this nation [*My Lord!*] and after a while the forces in power in Alabama will have to say, "You can't stop a people like this" [*No!*]. . . . We will appeal to the conscience of Selma. "This is Selma's opportunity to repent!" [*It's time, yes!*] This is Selma's opportunity to say to the nation, "We've gone down the wrong path [*All right*] like the Prodigal of old. We've strayed to the far country of brutality. . . . Monday can be a day when Selma will "come to herself" and move back to the Father's house.

The preacher concludes lyrically,

> Somehow, we can say, America,
> For your security,
> We have sailed the bloody seas of two world wars,
> For your security, America
> [*The truth, the truth!*]
> Our sons died in the trenches of France,
> In the foxholes of Germany,
> In the islands of Japan,
> And now we are saying, America,
> We just want to be free.
> [*Doctor! Doctor!*]

We know we cannot be our full selves unless we have freedom. . . . We will sing again as we did earlier,

> "Before I'll be a slave
> I'll be buried in my grave [uproar rising]
> And go home to my Father
> And be saved." [pandemonium]

The tape breaks off before King can fully explain the nonviolent character of his plan for Selma.

A few of the standard elements of King's mass-meeting oratory are missing from the above speech. He was not always above the vilification of the local authorities. He never tired of announcing that they had turned "Ole Bull" into a steer and put him out to pasture, and he regularly referred to Sheriff Jim Clark and his "posse" as "bad news." Nor did he shy away from leaning on the preachers who weren't involved. In Birmingham the police heard him say, "If your preacher isn't with it, you put a fire under him and tell him to get with this drive!"

In some speeches King occasionally went into great detail in his teaching of the principles of nonviolence. On April 3, 1963, the police overheard his four steps in nonviolent action: (1) collect the facts—understand the evil you are battling against; (2) negotiate with your opponent—negotiation calls for "self-purification," by which he meant the recognition of one's own anger, hatred, and other "mistakes"; (3) direct action creates "enough tension" to draw attention to the conflict in order to "get to the conscience of the white man" (here Connor was forewarned of this strategy but nevertheless reacted clumsily); and (4) nonviolent action refuses to give in to more militant calls for violence or to white pleas for moderation. The time for action is always "now."

On many occasions King went to extraordinary lengths to teach the evangelical principles behind nonviolence. With thirty-five hundred people, including scores of children, in Birmingham jails, King came to a packed church, one of five he addressed that evening, where he calmly delivered the theological rationale for the campaign. After setting the situation in historical perspective—"Never in the history of this nation have so many people been arrested for the cause of freedom and human dignity"—he justified the use of children in the demonstrations on the basis of Jesus' visit to the temple. "I must be about my Father's business," he said to the appreciative sighs of the audience. In the main part of the speech he taught his familiar distinction between *eros* and *agape*. The crowd seems less interested in *agape*-love for the white man than in King's soulful explication of *eros*: You love your lover, he explains, because there is something about your lover that *moves* [*mooooves*, wave of response] you. It may be the way [the response is growing; King can be heard to say in exasperation, "shoot!"]—"that's *eros*." It may be the way your lover *talks* or *walks* [again, the crowd voices its pleasure]. Love is greater than "like."

> God's love was poured out on the cross of Christ for unlikable people, people who don't *move* us, whose ways are not our ways. With this kind of love, we, that is, all in this room, will transform those who have persecuted us all these years and create a new Birmingham. But it won't happen without a sacrificial effort on our part. We have to keep push-

ing economically and marching for it. So keep this movement rolling.
. . . If you can't fly, run. If you can't run, walk. If you can't walk, crawl.
But by all means, keep moving.

Today, it is difficult to imagine how anyone could address a situation of comparable volatility with such rationality and patience. Yet King
made a career of doing just that. Two years later in Selma, King was giving
the same speeches, making the same seminary distinctions, arguing with
the same passion for the nonviolent love of Jesus. One evening at Brown's
Chapel it was clear from the moment he rose to speak that he wasn't there
to inspire or excite his audience but to teach. After a discussion of voter
statistics and average per capita incomes in Alabama and Mississippi, he
abruptly turns to the causes of violence in the South. He theorizes that
the white South harbors a deep sense of guilt for its racism, "a haunting,
agonizing guilt." But like the alcoholic who feels guilty when he sobers
up and therefore turns to drink, the white southerner is trying to drown
his guilt in the racist guilt-provoking behavior. (The crowd responds with
some whistles, as if to indicate that this is a tough line to follow. This is
no Abernathy rouser.) So, he continues, when you are confronted with
the white man's violent behavior, remember where it comes from. One
can meet anger this deep only with an even deeper love. He then pon-
derously and for the thousandth time explains the difference between
philia, eros, and *agape,* this time mixing in an allusion to Plato's dialogues.
(The crowd has grown somber and respectful. This is *definitely* not an
Abernathy pep talk.) *Agape,* King continues, transcends affection; you love
everyone because *God* loves everyone, because "Christ *died* for the segregationist as well as the integrationist."

The building in which King delivered this speech, Brown's Chapel,
is located just off an unpaved street beside a single-story housing project
in one of the dreariest neighborhoods in Selma. The setting alone, to say
nothing of the weight of history, economics, and the burden of racism in
America, seems to disqualify such theology as an impossible dream. Add
to this the many militant criticisms of King's approach (such as that of
Malcolm X, who showed up during the Selma campaign and announced,
"I don't believe in *any* kind of nonviolence"), and one can only question
the feasibility of *agape* as a social force. Not only its conclusions, but even
the mounting of King's argument seem impossible today; but then the
argument and its conclusions were not *possible* three decades ago either.
Prophetic speech always thrusts itself into impossible situations. The language of Christian love drew whistles and raised eyebrows in Brown's
chapel too, especially with the jails of Selma bursting with Negroes.

IV

King's dream of redemption is the legacy for which he is best known and fondly remembered. But the texture of his prophetic speech—notably, its obsession with victory through sacrificial love—underwent a marked change in the last three years of his life. Only a year and a half later in Chicago, one hears defiance, desperation, and anger not evident in Birmingham and Selma. In the course of planning his assault on one hundred segregationist real estate offices in the Chicago campaign, he lashes out at Christians who piously say, "Lord, Lord," but refuse to do God's will. In an uncharacteristic fit of pique, he ridicules "sincere stupidity," adding, "You take sincerity and put it in a small, closed mind, [and] you have the most dangerous force in the world." Gone are references to God's love for the white brother, replaced by angry denunciations of the "terrible cancer" in Chicago, for which King and his associates are "the social physicians."

King's Chicago oratory proved the ancient distinction between Cicero and Demosthenes. According to proverb, when Cicero finished speaking the people said, "How well he spoke," but when Demosthenes finished speaking, the people said, "Let us march." In one of his most eloquent and impassioned mass-meeting speeches, entitled "Why I Must March," King tells his black audience,

> We don't have much education.
> We don't have much political power.
> We do have some:
> We have bodies, we have souls.
> We will be able to change this city
> And make it an open city.
> They were saying to us yesterday,
> "Don't march."
> I listened to 'em, and I said to myself,
> "I'm marching."
>
> I say to Chicago, "You want us
> to stop marching?
> Make justice a reality.
> If you want a moratorium on
> demonstrations,
> Put a moratorium on injustice.
> If you want us to end our moves into
> communities, open these communities.
> We'll be glad to stop marching."

I don't mind saying to Chicago or anybody,
"I'm tired of marching, tired of marching
   for something that should have been mine at first."
I don't mind saying this to you this night . . .
I'm tired of the tensions surrounding our days.

I don't mind saying to you tonight
That I'm tired of living every day
Under the threat of death.
I have no martyr complex.
I want to live as long as anybody
   in this building tonight,
And sometimes I begin to doubt
Whether I'm gonna make it through.
I must confess I'm tired.

Yes, I'm tired of going to jail;
I'm tired of all of the surging murmur
   of life's restless sea,
But I'll tell anybody,
I'm willing to stop marching.
I don't march 'cause I like it;
I march because I must
And because I'm a man
And because I'm a child of God.

The power in this speech and others like it depends on its quality as testimony. Its cadences are not rehearsed or artistic. It appears to be free of
memorized set pieces imported from other speeches. "Why I Must March"
captures the spontaneous, impassioned defiance not only of a weary
people but of one weary man who represents the people. At Holt Street
in 1955 the young preacher had moved the people by intuiting *their* rage;
in Chicago he spoke for himself.

His Chicago oratory also reflects what C. T. Vivian described in 1968
as the Movement's long odyssey through an inherited world of false assumptions. At the beginning, King and his associates imagined a nation
with an enormous moral capacity that had been weakened by regional
conflict, irrational fears, and the political system of segregation. The
Movement assumed that if it could communicate the real issues to the
nation, its reservoir of morality was such that America could and would
sustain the necessary changes. Through the noble example of their black
brothers and sisters, white Americans would face up to the truth and solve
the nation's racial and economic problems. Such a model of communi-

cation could have arisen only in a Christian culture, for the model combined elements of soul winning or evangelism with Rauschenbusch's moral-influence theory for the stated purpose of redeeming the soul of the nation. By word and deed, blacks would win over key white leaders in every community, one by one, who in turn would convince other whites of the righteousness of the cause. Blacks would freely admit how they had contributed to their own dilemma, and, ultimately, the power that blacks and whites hold in common—their religion and democratic principles—would achieve the goal of social redemption.

But along the way, Vivian continues, blacks discovered that the nation does not have a soul or the moral capacity to renew itself. America does not suffer from a political flaw or the results of a historical accident. Its racism is a cancer so systemic that it is immune to even the most eloquent and heroic communication of the issues. No matter how rationally the political and economic considerations are presented, they all bow to a system "that pivots on our very flesh," its blackness, Vivian argues. What could be more foolish than to expect a system so conceived and so dedicated to solve the problems of African Americans? Sometime after Selma, King began to abandon the assumptions about morality that had guided him thorugh repeated crises and that had underlain his hundreds of mass-meeting speeches. In Chicago he came face to face with the truth about racism as he had never done in the South. He did not stop speaking or preaching, but his oratory assumed the character of the defiant witness who spits the truth into the face of his interrogators, uttering it at any cost. Hence his speech is not entitled "Let Us March" or "Won't You March with Us," but "Why I Must March."

Against the backdrop of Chicago, Memphis, and the denouement of the Poor People's Campaign, the mass meetings of 1955-1965, for all their turbulence and conflict, represent a period of social and theological homeostasis, when the preacher King could still dare to coordinate his vision for a new society with the promises in the Word of God. Thus the move from the Ebenezer Gospel to the Gospel of Freedom of the mass meetings was as natural as opening the Bible, lining a hymn, or issuing an altar call. At Ebenezer he preached explicitly from Bible *texts*, in the meetings implicitly from Bible *themes*, but no matter where, he preached the universals of love, justice, and dignity. In both settings he stood in sanctuaries surrounded by the symbols of Christianity and before audiences that were truly "congregation" to him. In both settings King treated the congregation as the conscience or guardian of the community, and his audiences responded by playing the chorus to his lead. In either setting the preacher was never removed from the inspirational hymns of the church. In the meetings they enjoyed "freedomized" versions of the same spirituals and gospel songs they sang at Ebenezer on Sunday mornings. Both the Ebenezer Gospel and the Gospel of Freedom demanded a response from the congregation. At Ebenezer the altar call summoned worshippers to a closer walk with Jesus. In the mass meetings the preaching evoked the more focused and courageous response of signing up to march or attempting to vote, decisions that many of the Christians who came forward doubtless understood as just another form of following Jesus.

The relationship of the Ebenezer Gospel to the Gospel of Freedom is revealed not so much in the content of King's speeches as in the atmosphere of their reception. The mass meetings *felt* like church to the people of Birmingham and Selma because their common theological tradition had taught them not to distinguish between salvation in the church and freedom in the city. Just as God is one, so also God's gift of salvation is indivisible. Jesus / is like Moses / is like Sojourner Truth / is like Martin King because our God's saving activity manifests itself in a long and steady stream of witnesses punctuated by timely messianic leaders. The celebratory atmosphere of the mass meetings resembled the call-and-response pattern at Ebenezer because there is only one worship of the one true God. St. Paul proclaimed One Lord, One Faith, and One Baptism; the black church continued, One Spirit, One Freedom, and One Beat. Great music is great music. It didn't really matter *where* you heard King preach; it was all one sermon anyway. The mass meetings felt like church because in the truest sense of African-American theology they *were* church. In addition to promoting the political freedom that was at hand, they conserved what had always been central to life in Africa and in the new world. With their celebration of salvation as freedom, the mass meetings reinforced the most traditional of African-American values, the unity of religion and life.

Epilogue

THE first sentence of the *New York Times* editorial on April 7, 1968, began "Martin Luther King was a preacher, a man from Georgia . . ." The editorial was, of course, an epitaph, as is any book on King written in these days. Dead for more than a quarter century, King and his achievement demand a summing up, and several recent biographies have done just that. But no appraisal will be completely accurate that does not begin where the *Times*'s began. No portrait of King that neglects his ministerial identity and commitments will do justice to the true character of his achievement. What might be assumed of any Baptist preacher may legitimately be said of King: that he discovered his identity and calling in the church, fashioned his world in the image of the Bible, trusted the power of the spoken Word, endeavored to practice Christian love at all times, and couldn't shake the preacher's chronic infatuation with conversion. When such unexceptional observations are applied to King's public career, they illumine its fundamentally Christian character. Reckonings of Martin Luther King, Jr. will be made by church historians and theologians as well as political scientists.

What is not so easily determined is the full measure of his contribution. No epitaph can be comprehensive, especially that of a preacher. Preachers live by the open-endedness of the Word of God, which, like a metaphor or a promise, possesses meanings and levels of fulfillment far beyond the preacher's ken. Whenever he was asked about the gains that had come about as a result of his activity, King answered with a metaphor rather than hard data. He would smile his tired smile and say that the black man had straightened his back, and that you can't ride a man whose back isn't bent. Although the Civil Rights Act and the Voting Rights

Act were monuments to his efforts in his own lifetime, he usually evaluated his achievement in terms of the spirit and promise for the future, as if he could see children of other cultures learning to pronounce his name.

In the Motse Maria High School in the Lebowa homeland of the Transvaal, the American volunteer teacher introduced her class of fourteen-year-olds to the life of Martin Luther King, Jr. They had never heard of him. Several days later she gave them a copy of his final speech, "I See the Promised Land," and played the audiotape. The youngsters listened as if hypnotized; some moved their lips with King as they followed along. At the words, "I've been to the mountaintop," one little girl leaped from her chair and danced before the class. "Oh, Martin," another girl exclaimed in lilting Sotho accent, "your words moved down from my ears and touched my heart and soul."

In the United States appraisals of King's effect are less poetic. When his former associates like Jesse Jackson, Ralph Abernathy, C. T. Vivian, James Bevel, and Wyatt Tee Walker, all Baptist preachers, were asked for this book about the Movement's contributions, they replied with varying socioeconomic analyses of the situation of the African American, ranging from pride in the rising middle class to despair over drugs and violence. The view from Wyatt Tee Walker's church, a remodeled Loew's Theater in Harlem, is bleak. "It's worse here," Walker said. "Across the board, it's worse now than when I first went to work for Dr. King." Yet all the veterans shared an interpretation of their leader's achievement that allows for the intangible gains of the spirit: dignity, pride, and the kinship of all peoples. They also recognized that the preacher's appeal to the spirit led to material benefits for millions of blacks and poor whites in America. His altar call to the nation resulted in unprecedented civil rights legislation in the areas of access to public facilities, housing, voting, and jobs.

King's organization set the standard for communication and protest by which succeeding social movements have and will be measured. Its power to inspire similar forms of protest in Europe, Asia, and Africa is a tribute to the universal qualities King awakened in the SCLC and his supporters. The students sang *We Shall Overcome* in Tiananmen Square. More than one preacher announced, "I have a dream," in Poland, Hungary, and East Germany. And in South Africa one little girl danced.

In America, King announced that oppressed people can seize the rights that are due them without losing their soul in the process or destroying those who abuse them. He made his case so eloquently that for a transfixing moment in American history many agreed with him. Since his simple claim flew in the face of reason and history, he had no choice but to argue it on the basis of faith. When he preached civil religion, his message relied on the moral logic of the nation's foundational documents. When he preached from the New Testament, his sermons reflected the irony and precariousness of the Christian's existence in a hostile world. One doesn't sit down at a segregated lunch counter or face fire hoses and dogs because such an action makes sense or is guaranteed by the Constitution, but only because there is a greater logic that resolves the contrarities of African-American existence on a higher level. Walter Fauntroy said that when Rosa Parks sat down in the front of the bus, "she was making a statement as to whether or not God could be trusted."

Long after King himself began to doubt the goodness of the "white brother" and the tainted principles of civil religion, his expression of hope in the kinship of the races endures, as the Sermon on the Mount endures, as a mark to aim at in a sinfully divided society. The more pessimistic he grew with regard to humanity, the more optimistic he became about God. Even in the darkest period of his own discouragement, he continued to say to African Americans, "Go ahead! *God* can be trusted." The power of God still shapes the behavior of individuals and institutions.

After King's death, his old mentor Pius Barbour said in a sermon, "Mike was a great believer in this, the attitude of Jesus: He believed spiritual power could down any power. Can it?"

It is a measure of the preacher's abiding influence in our lives that we still ask the question and want to answer, Yes.

Notes

Abbreviations

Barbour Collection	Typed and holograph sermons, sermon notes, outlines, and audio recordings of the Reverend J. Pius Barbour, Philadelphia, Pennsylvania.
Connor Collection	The Papers of Eugene "Bull" Connor, containing inter-office police reports and verbatim accounts of mass meetings held in Birmingham. Birmingham Public Library Archives, Birmingham, Alabama.
JAH	*Journal of American History* 78: 1 (June 1991).
MLK, Atlanta	Martin Luther King, Jr. Center for Nonviolent Social Change, Archives, Atlanta, Georgia.
Birmingham	Birmingham Public Library Archives, Birmingham, Alabama.
MLK, Boston	Special Collections of the Mugar Memorial Library, Boston University, Boston, Massachusetts.
MLK, Duke	Duke University Divinity School, Media Center, Durham, North Carolina.
MLK, Howard	Howard University Divinity School Library, Tape Recording Collection, Washington, D.C.
MLK, Union	Reigner Recording Library, Union Theological Seminary, Richmond, Virginia.
Moorland-Spingarn	Ralph J. Bunche Oral History Collection, Moorland-Spingarn Research Center, Howard University, Washington, D.C.

Shortened Titles

Afro-American	Milton C. Sernett, ed., *Afro-American Religious History, A Documentary Witness.*

Afrocentric	Molefi Kete Asante, *The Afrocentric Idea*.
"Autobiography"	Martin Luther King, Jr., "An Autobiography of Religious Development."
Bearing	David Garrow, *Bearing the Cross: Martin Luther King, Jr. and the Southern Christian Leadership Conference*.
Christianity	Walter Rauschenbusch, *Christianity and the Social Crisis*.
"Letter"	Martin Luther King, Jr., "Letter from Birmingham City Jail."
Life Experience	Richard Allen, *The Life Experience and Gospel Labors of the Rt. Rev. Richard Allen*.
Measure	Martin Luther King, Jr., *The Measure of a Man*.
Souls	W. E. B. Du Bois, *The Souls of Black Folk*.
Strength	Martin Luther King, Jr., *Strength to Love*.
Testament	Martin Luther King, Jr., *A Testament of Hope*, ed. James Melvin Washington.
Where?	Martin Luther King, Jr., *Where Do We Go from Here: Chaos or Community?*

Prologue

Page

3. "a Baptist preacher": "The UnChristian Christian," *Ebony* 20:10 (August 1965): 77.

4. "editorial assistance": David J. Garrow, *Bearing the Cross: Martin Luther King and the Southern Christian Leadership Conference* (New York: Morrow, 1986), pp. 105, 280, 312, 544-45, and index, "Stanley Levison."

4. King's published sermons: *Strength to Love* (Philadelphia: Fortress Press, 1981 [1963]); *The Measure of a Man* (Philadelphia: Fortress Press, 1988 [1959]).

5. a product of the black church: See James Cone, "Martin Luther King, Jr., Black Theology—Black Church," *Theology Today* 40:4 (January 1984); Taylor Branch, *Parting the Waters: America in the King Years 1954-63* (New York: Simon & Shuster, 1988), pp. 1-68; David L. Lewis, *King: A Critical Biography* (New York: Praeger, 1970), pp. 3-6; Stephen B. Oates, *Let the Trumpet Sound: The Life of Martin Luther King, Jr.* (New York: New American Library, 1982), pp. 1-7.

5. "Pilgrimage": In *Stride Toward Freedom: The Montgomery Story* (New York: Harper & Brothers, 1958), pp. 90-107.

Chapter 1: Surrounded

Page

15. "witnesses": Hebrews 12:1.

15. autobiography: "An Autobiography of Religious Development," *The Papers of Martin Luther King, Jr.*, vol. 1, *Called to Serve January, 1929-June 1951*, ed. Clayborne Carson (senior ed.) (Berkeley: University of California Press, 1992),

pp. 360-61. See the same document in holograph, n.d. (Fall semester 1950) in the Martin Luther King, Jr. collection in the Special Collections Section of Mugar Library of Boston University (hereafter cited as MLK Boston).

16. Ebenezer's architecture: Clifford Geertz has shown the causal relationship between *cosmos* and *ethos* in *The Interpretation of Cultures* (New York: Basic Books, 1973). Worldview yields a way of living in the world (pp. 89-90).

16. the chain of authority: This is an old philosophical and religious image. Geneva Smitherman sketches a version of the chain in *Talkin and Testifyin* (Boston: Houghton Mifflin, 1977), pp. 109-10. The importance of both social and architectural structure has been pointed out by Melvin D. Williams, *Community in a Black Pentecostal Church: An Anthropological Study* (Pittsburgh: University of Pittsburgh Press, 1974), pp. 109-72, though one must be careful not to equate the small and impoverished Pentecostal congregation he analyzes with the much more sophisticated Ebenezer Baptist Church. On the chain of command in an African-American congregation, see Charles V. Hamilton, *The Black Preacher in America* (New York: Morrow, 1972), p. 117.

16-17. world-building and worship: See Peter L. Berger, *The Sacred Canopy: Elements of a Sociological Theory of Religion* (Garden City N.Y.: Doubleday, 1967), pp. 17-26. Also Williams's chapter "The Physical Setting," pp. 143-56. On the importance of drama: Hugh Dalziel Duncan, *Communication and Social Order* (New York: Bedminster Press, 1962), p. 113. Duncan draws on Kenneth Burke's dramatistic theory outlined in *A Grammar of Motives* (Berkeley: University of California Press, 1945, 1969), pp. 3-9.

16. King in pulpit: King often entered the sanctuary at the "meditation" point just before the sermon. The "late" arrival traditionally enhances the importance of the sermon. Interview, Lillian Lewis. (An interview is hereafter abbreviated Intr. Unless otherwise credited, all interviews were conducted by Richard Lischer. See "Interview Records" in Bibliography for details.)

17. "Afro-Baptist sacred cosmos": The phrase and the rich concept belongs to Mechal Sobel, *Trabelin' On: The Slave Journey to an Afro-Baptist Faith* (Princeton: Princeton University Press, 1979), pp. 139-80. See also James M. Washington, *Frustrated Fellowship: The Black Baptist Quest for Social Power* (Macon: Mercer University Press, 1986), pp. 4-16.

18. "Lord, I'm Coming Home": Audiotape of the service containing the sermon "What a Mother Should Tell Her Child," May 12, 1963, Martin Luther King, Jr. Center for Nonviolent Social Change Archives, Atlanta, Georgia (hereafter cited as MLK, Atlanta).

18. "nurtured in its bosom": "Letter from Birmingham City Jail," *A Testament of Hope: The Essential Writings of Martin Luther King, Jr.*, ed. James Melvin Washington (San Francisco: Harper & Row, 1986), p. 298. Toward the conclusion of "Ingratitude," June 18, 1967, MLK, Atlanta, King confesses, "And I'm just thankful to be alive. I'm just thankful to be able to be in church this morning." On the importance of "place" to preachers, see Phillips Brooks, *Lectures on Preaching* (Grand Rapids: Baker Book House, 1969 [1877]), p. 190.

18-19. "Sweet Auburn": Alexa Henderson and Eugene Walker, "Sweet Auburn: The Thriving Hub of Black Atlanta, 1900-1960" (Denver: U.S. Department of the Interior/National Park Service, n.d.). L. D. Reddick, *Crusader Without Violence: A Biography of Martin Luther King, Jr.* (New York: Harper & Brothers, 1959), pp. 30-41. Stephen Birmingham, *Certain People* (Boston: Little, Brown, 1977), pp. 216-32.

19. "lovely relationships": "Autobiography," MLK, Boston; *Papers*, 1:360.

19. suicide attempts: Reddick, pp. 60-61; Oates, pp. 6-7, 10-11.

19. "the Veil": W. E. B. Du Bois, *The Souls of Black Folk* in *Three Negro Classics* (New York: Avon Books, 1965 [1903]), p. 214.

19. recognition scenes: "Autobiography," MLK, Boston; *Papers*, 1:362. James Farmer, *Lay Bare the Heart: An Autobiography of the Civil Rights Movement* (New York: Arbor House, 1985), p. 37. For Marcus Garvey's similar experience, see Milton C. Sernett, *Afro-American Religious History: A Documentary Witness* (Durham: Duke University Press, 1985), p. 380.

20. "Baptist Band": *Songs of Zion* (Nashville: Abingdon Press, 1981), #163. See Milton Sernett, "Black Religion and American Evangelicalism: White Protestants, Plantation Missions, and the Independent Negro Church" (Ph.D. diss., University of Delaware, 1972), p. 185.

20. irony of white Baptist church: "A Knock at Midnight," June 25, 1967, in Lucy A. M. Keele, "A Burkeian Analysis of the Rhetorical Strategies of Dr. Martin Luther King, Jr., 1955-1968" (Ph.D. diss., University of Oregon, 1972), Appendix 2, pp. 311-12.

20. "common master and Savior": "The Drum Major Instinct," *Testament*, p. 263.

20-21. on the Negro church; E. Franklin Frazier, *The Negro Church in America*, and C. Eric Lincoln, *The Black Church since Frazier* (New York: Schocken Books, 1974 [1963], p. 48. Du Bois, *Souls*, p. 340. William E. Montgomery, "Negro Churches in the South, 1865-1915" (Ph.D. diss., University of Texas, 1975), p. 171.

20-21. Ebenezer's membership: Reddick, p. 84.

21. "mixed type": St. Clair Drake and Horace R. Cayton, *Black Metropolis*, vol. 2, rev. ed. (New York: Harcourt, Brace & World, 1945, 1962), pp. 673-74.

21. "The Frenzy": Du Bois, *Souls*, p. 338.

21. "rivers of Africa": An allusion to the controversial question of the continuity between Africa and American Christianity. In his few references to the question, King appears to have assumed a continuity in form but not theological content. The two most important books on the subject are by E. Franklin Frazier, *The Negro Church in* America, and Melville J. Herskovits, *The Myth of the Negro Past* (New York: Harper & Brothers, 1941). King's later sermons contain brief but noncommittal allusions to these works. See, for example, his satiric comments on black congregations who want to deny their African past in "A Knock at Midnight," 1967 or 1968, Duke University Divinity School Media Center, audiotape (hereafter cited as MLK, Duke). Frazier argued that the conditions of slavery in the United States made the survival of African culture impossible. He

discerned virtually no continuity between African religion and the Negro church. Herskovits argued the other side, citing "patterns of motor behavior" and, specifically, the continuity between the African river cults and Baptist baptism by immersion to make his point (see p. 232). Homiletician Henry Mitchell has followed this line of argument by tracing Christian doctrines and practices to their putative African sources. For a theological extension of Herskovits's arguments, see Henry H. Mitchell, *Black Belief: Folk Beliefs of Blacks in America and West Africa* (New York: Harper & Row, 1975), pp. 58-94. The best discussion of the debate is in Albert J. Raboteau, *Slave Religion: The "Invisible Institution" in the Antebellum South* (New York: Oxford University Press, 1978), pp. 55-75. For further comments on continuity, see C. Eric Lincoln and Lawrence H. Mamiya, *The Black Church in the African American Experience* (Durham: Duke University Press, 1990), p. 347, and Wilson Jeremiah Moses, *Black Messiahs and Uncle Toms: Social and Literary Manipulations of a Religious Myth* (University Park: Pennsylvania State University Press, 1982), pp. 17-29. On the nature of African religion, see John S. Mbiti, *An Introduction to African Religion* (London: Heinemann, 1975), on God, pp. 10-64.

21. "upon me": "Guidelines for a Constructive Church," June 5, 1966, MLK, Atlanta.

21. the ring; See Raboteau, pp. 70-71; also Eileen Southern, "The Religious Occasion," in C. Eric Lincoln, ed., *The Black Experience in Religion* (Garden City, N.Y.: Anchor Press, 1974), pp. 57-63. On the "Pythian madness," Du Bois, *Souls*, p. 338. On communication as the "achievement of the group": Smitherman, p. 109.

22. the ring and Daddy King: Intr. Gordon Midgette, November 28, 1988.

22. "hearts and souls": "A Knock at Midnight," 1967 or 1968, MLK, Duke, audiotape.

22. African *nommo*: Janheinz Jahn, *Muntu: An Outline of Neo-African Culture*, trans. Marjorie Grene (London: Faber & Faber, 1958, 1961), pp. 124-27, 132-33; and Molefi Kete Asante, *The Afrocentric Idea* (Philadelphia: Temple University Press, 1987), "African Foundations of *Nommo*," pp. 59-80.

22-23. Ebenezer history: Henderson and Walker, pp. 17, 24. Reddick, pp. 82-86. See *Ebenezer Baptist Church: The Centennial Celebration, 1886-1986* (Atlanta, 1986), p. 1. See *Papers*, 1, Introd., p. 13. Also, Intr. Joseph L. Roberts, Jr. "The Lord has helped us": 1 Samuel 7:12. Rev. A. D. Williams: Henderson and Walker, p. 24. Oates, pp. 4-5.

23. "thousand in number": Du Bois, *Souls*, p. 338. See Frederick L. Downing, *To See the Promised Land: The Faith Pilgrimage of Martin Luther King, Jr.* (Macon: Mercer University Press, 1986), p. 47.

23-24. Michael King, Sr.: Martin Luther King, Sr. (with Clayton Riley), *Daddy King* (New York: Morrow, 1980). Coretta Scott King, *My Life with Martin Luther King, Jr.* (New York: Holt, Rinehart & Winston, 1969), pp. 75-80. "You can do it": For the story of King Sr.'s new financial system and his relationship to Ebenezer, see Branch, pp. 40-44.

24. Social Gospel and black church: See Ralph E. Luker, *The Social Gospel in Black and White: American Radical Reform, 1885–1912* (Chapel Hill: University of North Carolina Press, 1991), esp. chap. 7, "Urban Mission," pp. 159–90.

25. name change: Branch, p. 44. Reddick tells the story differently. He says that Michael Sr. changed his name when he received his passport *before* his trip abroad (pp. 50–51). On the rebaptism, Intr. Gordon Midgette, based on the reminiscences of 103-year-old William Frank Paschal, who claimed to have rebaptized King Jr. at the Beulah Baptist Church in Atlanta.

25. PK: Farmer, p. 34. On the preacher and his wife, p. 33.

26. "missing link in life": "Autobiography," MLK, Boston; *Papers*, 1:360.

26. "More Like Jesus": Oates, p. 8. "Master M. L. King": Recorded in Clarence M. Wagner, *Profiles of Black Georgia Baptists* (Atlanta: Bennett Brothers, 1980), p. 94.

26. letter to his father: *Papers*, 1:103.

26–27. Martin's conversion: "Autobiography," MLK, Boston; *Papers*, 1:361.

27. fundamentalism: "Autobiography," MLK, Boston; *Papers*, 1:361. On Daddy King's leniency on dancing, Intr. Minnie Showers.

27. doubts and "real father": "Autobiography," MLK, Boston; *Papers*, 1:360–61.

27. "to serve humanity": "Autobiography," MLK, Boston; *Papers*, 1:363.

28. King's first sermon: Branch, pp. 65–66, and Lewis Baldwin, *There Is a Balm in Gilead: The Cultural Roots of Martin Luther King, Jr.* (Minneapolis: Fortress Press, 1991), p. 281.

28. Willis Williams: *Papers*, 1, Introd., p. 4.

28. "great-grandson of a Baptist preacher": *Ebony* 20:10 (August 1965): 77. Cf. his self-introduction to Mount Pisgah Missionary Baptist Church in Chicago in "Thou Fool," August 27, 1967, MLK, Atlanta.

28. "Black Fathers": Henry H. Mitchell, *Black Preaching* (Philadelphia: Lippincott, 1970), pp. 52ff. For a historical survey of African preachers in the United States, see ibid., pp. 52–94; W. E. B. Du Bois, ed., *The Negro Church* (Atlanta: Atlanta University Press, 1903), pp. 22–37; Carter G. Woodson, *The History of the Negro Church*, 3d ed. (Washington, D.C.: Associated Publishers, 1972 [1921], pp. 34–86, 146–63; Charles Emerson Boddie, *God's 'Bad Boys'* (Valley Forge: Judson Press, 1972): Henry J. Young, *Major Black Religious Leaders, 1755–1940* (Nashville: Abingdon Press, 1977).

28. pioneering women preachers: William L. Andrews, ed., *Sisters of the Spirit: Three Black Women's Autobiographies of the Nineteenth Century* (Bloomington: Indiana University Press, 1986).

28–29. Sustainers and Reformers: The terminology used here roughly parallels the "pastoral" and the "prophetic strands" identified by Peter J. Paris in "The Bible and the Black Churches" in Ernest Sandeen, ed., *The Bible and Social Reform* (Philadelphia and Chico, Calif.: Fortress Press and Scholars Press, 1982), pp. 136–44. Paris also identifies two additional types, pp. 144–51. Compare Benjamin E. Mays's "compensatory" and "constructive" types of black religion in *The Negro's God* (Boston: Chapman & Grimes, 1938), pp. 14-15, and Thomas J. S. Mikelson, "The Negro's God in the Theology of Martin Luther King, Jr.: Social Commentary and Theological Discourse" (Th.D. diss., Harvard University, 1988), pp. 62-72.

29. "interim strategy . . . otherworldliness": Gayraud S. Wilmore, *Black Religion and Black Radicalism: An Interpretation of the Religious History of Afro-American People*. 2d ed. rev. (Maryknoll, N.Y.: Orbis Books, 1983), p. 51. Cf. James Cone, *God of the Oppressed* (New York: Seabury Press, 1975), p. 61.

29-30. otherworldly religion: Benjamin E. Mays and Joseph W. Nicholson, *The Negro's Church* (New York: Negro Universities Press, 1933), p. 17. Cf. Drake and Cayton, 2:619.

30. "too dangerous": Coretta Scott King, p. 31.

30. sermon as "temporary escape": Gary T. Marx, "Religion: Opiate or Inspiration of Civil Rights Militancy Among Negroes," in *The Making of Black America*, vol. 2 (New York: Atheneum, 1969), p. 366. According to Marx's data, Episcopalians took a more militant stance than Baptists in the Civil Rights Movement (p. 367). King's Baptist-based Movement, however, contradicts the author's thesis: "As civil rights concern increases, religiosity decreases" (p. 372). Marx's sociological method does not do justice to the tradition of protest in the Christian religion and appears oblivious to the formative presence of the African-American church in the Movement. For Lincoln and Mamiya's critique of Marx, see pp. 221-27.

30. "dream": Reported in William H. Pipes, *Say Amen, Brother! Old-Time Negro Preaching: A Study in Frustration* (New York: William-Frederick Press, 1951), p. 118. Pipes's work is a pioneering study of Negro folk preaching in Macon County, Georgia, which the author carried out by means of field observation and recording. The book contains a wealth of information on rural Negro preaching, including transcriptions of several sermons, and a brief history of the Negro pulpit in America. Pipes recognizes in the Negro sermon the influences of African religion and the second Great Awakening. He understands its purpose to be escape from the cruelties of this life. He may be correct in construing Negro religion as an escape valve for social discontent, but he does not appreciate the constructive uses of eschatology in the life of a downtrodden people (cf. pp. 67-71). Nor does he accept the validity of black preaching as a folk art. He apologizes, "The reader might well wonder if preaching like that mentioned above and the sermons which have been recorded in this book are indeed preaching. . . . This work assumes no obligation to prove that Negro preaching is genuine preaching . . ." (p. 2).

31. "bring new courage": "What a Christian Should Believe about Himself," *Papers*, 1:281. On "you are not niggers," see "Is the Universe Friendly?" December 12, 1965, MLK, Atlanta.

31. white audiences: "By organizing New England towns around autonomous local churches, and by authorizing ministers—and only ministers—to speak on all occasions of public note, the founders established patterns of community

that would ensure the sermon's place at the center of New England society, and with it, New England's identity as a unique 'people of the Word.'" Harry S. Stout, *The New England Soul: Preaching and Religious Culture in Colonial New England* (New York: Oxford University Press, 1986), p. 13.

31. powerful and the powerless: David S. Reynolds, "From Doctrine to Narrative: The Rise of Pulpit Storytelling in America," *American Quarterly* 32:1 (Spring 1980): 486. See Cone, *God of the Oppressed*, p. 60.

31. "an actor": James Weldon Johnson, *God's Trombones: Seven Negro Sermons in Verse* (New York: Viking Press, 1927), Introd., p. 5.

31. Jasper's sermon: In William E. Hatcher, *John Jasper* (New York: Fleming H. Revell, 1908), p. 164, italics added, spelling altered.

32. "ire of ruling powers": Eugene Genovese, *Roll, Jordan, Roll: The World the Slaves Made* (New York: Vintage Books, 1972), pp. 266, 268.

32. "signifin(g)": Henry Louis Gates, Jr., *The Signifying Monkey: A Theory of Afro-American Literary Criticism* (New York: Oxford University Press, 1988), pp. 46-47.

32. Turner and Vesey: Reddick, p. 14.

33. King's development at Morehouse: *Papers*, 1, Introd., pp. 38-40. See his transcript, which includes courses in his major, sociology, including Contemporary Social Trends in America, Social Anthropology, Social Institutions, and Intercultural Relations, pp. 39-40. The Reverend Samuel Proctor, friend and contemporary of King's and former pastor of Abyssinian Baptist Church in Harlem, is certain that King never had a formal course that would have surveyed nineteenth-century black preachers. King's formal education was "Eurocentric" (Intr.).

33. "condition of a slave": *The Life Experience and Gospel Labors of the Rt. Rev. Richard Allen* (New York: Abingdon Press, 1960 [1833]), pp. 69-70 (punctuation altered). On Allen, see Young, pp. 24-40, and Frederick E. Maser, *Richard Allen* (Lake Junaluska, N.C.: Commission on Archives and History, United Methodist Church, 1976). God the "first pleader": *Life Experience*, p. 72.

34. "Man Our Brother": *Life Experience*, Introd., p. 9.

34. "to cut my flesh": Daniel Coker, *A Dialogue Between a Virginian and an African Minister*, in William Loren Katz, ed., *Negro Protest Pamphlets* (New York: Arno Press and New York Times, 1969 [1810]), pp. 24, 39. For Coker's use of "suffer" and for an insightful literary and theological analysis of Coker's *Dialogue*, see Theophus H. Smith, "The Biblical Shape of Black Experience; An Essay in Philosphical Theology" (Ph.D. diss., Graduate Theological Union, 1987), pp. 188, 187-96.

34. "win you in the process": King, *Stride*, p. 217. King is using almost verbatim E. Stanley Jones's account of Gandhi in *Mahatma Gandhi: An Interpretation* (New York: Abingdon-Cokesbury, 1948).

35. Walker's *Appeal*: "Our Wretchedness in Consequence of the Preachers of Religion" in Sernett, *Afro–American*, pp. 192-93. See Young, pp. 41-51, and Vincent Harding, *There Is a River: The Black Struggle for Freedom in America* (New York: Harcourt Brace Jovanovich, 1981), pp. 81-94.

35. Garnet's speech: In Carter G. Woodson, *Negro Orators and Their Orations* (Washington, D.C.: Associated Publishers, 1925), pp. 150-57. On Garnet's speech and career, see Asante, *Afrocentric*, pp. 139-47; Young, pp. 85-97; Hamilton, pp. 66-68; Harding, pp. 140-53; Ernest G. Bormann, ed., *Forerunners of Black Power* (Englewood Cliffs, N.J.: Prentice-Hall, 1971), pp. 146-54; and Joel Schor, *Henry Highland Garnet* (Westport, Conn.: Greenwood Press, 1977).

35. ". . . Go to Hell": Garrow, *Bearing*, p. 622.

35-36. northern and southern preachers: Moses, pp. 83-84. Hamilton makes the same distinction and the same judgment (p. 43). On God's love, see Mays, *The Negro's God*, pp. 162-88, and Mikelson, p. 71. Congressman John Lewis recalls that growing up in a black Baptist Sunday school he learned that color means nothing to the Holy Trinity (Intr. Ralph J. Bunche Oral History Collection, Moorland-Spingarn Research Center, Howard University (hereafter cited as Moorland-Spingarn).

36. two audiences: Moses, pp. 46, 63.

Chapter 2: Apprenticed to the Word

Page

39. "grown up with": Intr. Donald Smith, MLK, Atlanta.

39. "osmosis": Intr. Wyatt Tee Walker.

39. spoken word: On oral language the most important work has been done by Walter J. Ong, *The Presence of the Word* (New Haven: Yale University Press, 1967) and *Orality and Literacy* (London: Methuen, 1982). On Plato's fear, see Northrop Frye, *The Great Code* (New York: Harcourt Brace Jovanovich, 1982), p. 9. On metaphor, Aristotle, *Poetics*, 22, On the Yoruba, Asante, *Afrocentric*, p. 70.

39. verbal performance: Smitherman, pp. 79, 90-99. Cf. Genovese's comments, p. 432.

40. "with his words": *The Autobiography of Malcolm X* (New York: Ballantine Books, 1964), p. 154.

40. "move the world": Intr. C. T. Vivian. "what people want": Intr. Fred Shuttlesworth, Moorland-Spingarn.

40. Martin and words: Oates, pp. 7-8. Preaching contests: Intr. C. T. Vivian.

40-41. Greek training: Richard Lanham, *The Motives of Eloquence: Literary Rhetoric in the Renaissance* (New Haven: Yale University Press, 1976), pp. 2-3.

41. "pleasure for boys": Augustine, *On Christian Doctrine*, IV, 3.

41-42. Chandler's influence: Bryant and Wallace, Intr. G. L. Chandler (Donald Smith), MLK, Atlanta. "Word gods": Oates, p. 17. Increasing . . . : Branch, p. 363.

42. "unfit to preach": Quoted in Montgomery, p. 248. See Mays and Nicholson's study, *The Negro's Church*, p. 17. 200 seminarians: William Brink and Louis Harris, *The Negro Revolution in America* (New York: Simon & Shuster, 1964), pp. 99, 202. In a recent assessment, Lincoln and Mamiya write, "Unfortunately,

this rate of professional training for the ministry, as measured by a seminary degree, has not improved very much since the time of Mays and Nicholson's study" (p. 129).

42. Mays on Tuesday: "Our Task Here at the College," Moorland-Spingarn, circa 1958. Although his remarks were recorded after King was a student, the tape gives a flavor of Mays's famous Tuesday morning sessions.

43. King on cars: E.g., "The Good Samaritan," August 28, 1966, MLK, Atlanta. See also H. Beecher Hicks, Jr., *Images of the Black Preacher: The Man Nobody Knows* (Valley Forge: Judson Press, 1977), pp. 43-61. Wilmore writes, "The myth that all black preachers drove Cadillacs during the Depression and all black churches had plenty of money belongs to black folklore" (p. 161).

43. interrelatedness: Mays, *Disturbed About Man* (Richmond: John Knox Press, 1969), p. 22. The published sermons are records of Mays's earlier sermons. Cf. Harry Emerson Fosdick, *Riverside Sermons* (New York: Harper, 1958), pp. 251-52. Compare King's "Letter from Birmingham City Jail," *Testament*, p. 290; "What a Mother Should Tell Her Child," May 12, 1963, MLK, Atlanta; "The Man Who Was a Fool," *Strength*, p. 70. Keith D. Miller has traced common quotations in "The Influence of a Liberal Homiletic Tradition on *Strength to Love* by Martin Luther King, Jr." (Ph.D. diss., Texas Christian University 1984; later published as *Voice of Deliverance: The Language of Martin Luther King, Jr. and Its Sources* [New York: Free Press, 1992]), and "Martin Luther King, Jr. Borrows a Revolution: Argument, Audience, and Implications of a Secondhand Universe," *College English* 48:2 (February 1986): 249-65.

44. King's use of Mays: Intr. Lawrence Carter. "Fleecy locks": e.g., "The American Dream," *Testament*, p. 212. Oxenham: cf. Mays, *Disturbed About Man*, pp. 40-41. On longevity: cf. ibid., p. 117, and "I See the Promised Land," *Testament*, p. 286.

44. "Jesus was powerful": Mays, *Disturbed About Man*, pp. 31-32.

45. "pretended to be ignorant": Intr. Donald Smith, MLK, Atlanta.

45. "hard preaching": Lewis, *King*, p. 4.

45. Daddy's domination: Intr. Bernard Lee.

45-46. Daddy's preaching style: Audiotapes, n.d., Ebenezer Baptist Church, Atlanta; Duke Divinity School Media Center; Intrs. Bernard Lee, Murray Branch, J. T. Porter, Gardner Taylor. "No antics": Intr. J. T. Porter. "to develop the church": Intr. Bernard Lee.

46. "played with": "Nonconformist—J. Bond," January 16, 1966, MLK Atlanta.

47. ". . . Father's world": Audiotape, n.d., Ebenezer. For a corroboration of the widespread use of the formulas in King, Sr.'s sermon, such as "a thousand hills and cattle on every one of them," see Williams, p. 119.

47. "Awww, Dad": Intr. Bernard Lee.

47. father's warning: Letter, December 2, 1954, MLK, Boston.

48. "ideal of worship": Henderson and Walker, p. 23. On Borders, see James English, *The Prophet of Wheat Street* (Elgin, Ill.: David C. Cook, 1967).

48. "biggest and best": Intr. Juel Pate Borders. "study Borders": Branch, p. 641. "whole answer": Intr. Juel Pate Borders.

49. Borders's sermons: audiotapes, n.d., Wheat Street Baptist Church; Intr. Juel Pate Borders.

49. "I Am Somebody": For the complete text of the praise-poem, see *Forty-fifth Pastoral Anniversary, 1937-1982, Wheat Street Baptist Church*, p. 28. "Black man": Ibid., p. 26. Garth Baker-Fletcher has devoted a book to the concept: *Somebodyness: Martin Luther King, Jr. and the Theory of Dignity* (Minneapolis: Fortress Press, 1993).

50. Ray's style: Audiotapes, Cornerstone Baptist Church, Brooklyn, N.Y. Intrs. Gardner C. Taylor, Evans Crawford. Cf. Sandy Ray, *Journeying Through a Jungle* (Nashville: Broadman Press, 1979) for representative sermons.

50. "My Lord!": Audiotape, n.d., Duke Divinity School Media Center.

50. "vowels distended": Lewis, *King*, p. 22.

51. Taylor's theology: Gardner C. Taylor, *How Shall They Preach?* (Elgin, Ill.: Progressive Baptist Publishing House, 1977), p. 110. King's model: Intrs. Bernard Lee, Murray Branch, J. T. Porter.

51-52. King's student papers: See *Papers*, 1:225-30, 257-62, 273-79. On King's tendency to exaggerate his learning in social philosophy, see Clayborne Carson, et al., "Martin Luther King, Jr. as Scholar: A Reexamination of His Theological Writings," *Journal of American History* 78:1 (June 1991): 95.

52. King's intellectual development: Kenneth L. Smith and Ira G. Zepp, *Search for the Beloved Community: The Thinking of Martin Luther King, Jr.* (Valley Forge: Judson Press, 1974), and John J. Ansbro, *Martin Luther King, Jr.: The Making of a Mind* (Maryknoll, N.Y.: Orbis Books, 1982).

52. "fundamentalism": "Autobiography," MLK, Boston; *Papers*, 1:361. "Liberalism provided me with an intellectual satisfaction that I had never found in fundamentalism" ("Pilgrimage to Nonviolence," revised version, *Strength*, p. 147).

52. "Pilgrimage": original version in *Stride*, pp. 90-107.

53. Gandhi and King: C. F. Andrews, *Mahatma Gandhi's Ideas*, p. 60, quoted in Smith and Zepp, p. 49. It is puzzling that Smith and Zepp do not apply Andrews's thesis to King's relationship to his own religious tradition.

54. "social hope": Walter Rauschenbusch, *Christianity and the Social Crisis* (New York: Harper & Row, 1964 [1907]), p. 106. "Prophets as moral reformers": For an opposing view of the prophets see Stanley Hauerwas, "The Pastor as Prophet. Ethical Reflections on an Improbable Mission," in Earl Shelp and Ronald Sunderland, eds., *The Pastor as Prophet* (New York: Pilgrim Press, 1985), pp. 42- 43. See Rauschenbusch, *Christianity*, pp. 4-14. Cf. "Pilgrimage," *Stride*, pp. 91-92.

54. liberalism and Jesus: William McGuire King, "The Biblical Base of the Social Gospel," in Sandeen, especially the section, "The Preoccupation with Jesus," pp. 62-65.

54. Jesus and social change: Rauschenbusch, *Christianity*, pp. 60-61, 64.

55. Davis: For an informative summary of Davis's evangelical liberalism, see Smith and Zepp, pp. 21-31.

55. "EXCELLENT": "Autobiography," MLK, Boston.

55. "Preposterous": "How Modern Christians Should Think of Man," *Papers*,

1:275. King is quoting but not citing the well-known Christian ethicist John C. Bennett. See the comments of Ansbro, p. 88.

56. "living religion": King, "A Comparison of the Conceptions of God in the Thinking of Paul Tillich and Henry Nelson Wieman" (Ph.D. diss., Boston University, 1955), p. 244. Already at Morehouse, King had learned from George Kelsey "that behind the legends and myths of the Book were many profound truths which one could not escape" (*Papers*, 1, Introd., pp. 42-43, quoting King's "Autobiography").

56. "friendly and good": Quoted in Kenneth Cauthen, *The Impact of American Religious Liberalism* (New York: Harper & Row, 1962), p. 73. See "Is the Universe Friendly?" December 12, 1965, MLK, Atlanta, for King's adaptation of Fosdick. King once characterized Fosdick as the greatest preacher of the twentieth century (Robert Moats Miller, *Harry Emerson Fosdick: Preacher, Pastor, Prophet* [New York: Oxford University Press, 1985], p. 335).

57. "optimistic about human nature": "How Modern Christians Should Think of Man," n.d., MLK, Boston. Cf. *Papers*, 1:274.

57. "also reap": Galatians 6:7. See Pipes, p. 42, and Richard Lischer, "The Word That Moves: The Preaching of Martin Luther King, Jr.," *Theology Today* 46:2 (July 1989): 177. On the African-American laws of history, see Genovese, p. 246. On black history as theodicy, see Sernett, "Black Religion and American Evangelicalism," p. 327.

57. "eschatological": James H. Cone, *For My People* (Maryknoll, N.Y.: Orbis Books, 1984), p. 150. Cf. Cone, *God of the Oppressed*, pp. 18-20.

57. "rejected by Yale": Branch, p. 81.

57-58. Boston Personalism: For summaries see Smith and Zepp, pp. 99-118; Ansbro, pp. 71-90; Brian M. Kane, "The Influence of Boston Personalism on the Thought of Dr. Martin Luther King, Jr." (Th.M. diss., Boston University, 1985).

58. "earnestly arguing ideas": Intr. William Smith.

58. race-related dissertation: Intr. Evans Crawford.

58. De Wolf's definition and "reaches its peak": Notebooks, holograph, MLK, Boston.

58. "fourth-generation": Kane, p. 25.

59. "could not touch": Richard W. Fox, *Reinhold Niebuhr: A Biography* (San Francisco: Harper & Row, 1985), p. 30.

59. "Jesus saves": Systematic Theology II notes, MLK, Boston; Cf. Cauthen, p. 211.

59. personalist laws: Edgar Sheffield Brightman, *Moral Laws* (New York: Abingdon Press, 1933), pp. 229-49 *passim*. King: "There are certain laws. . . ." ("Discerning the Signs of History," November 15, 1964, MLK, Atlanta). King often spoke of Good Friday and Easter as "laws of life," e.g., "Guidelines for a Constructive Church," June 5, 1966, MLK, Atlanta.

59. "flame of spirit": Cf. Brightman's discussion, pp. 276-84.

59. "everybody somebody": Systematic Theology I notes, MLK, Boston.

59-60. "little me": See "a man in a man" in Clifton H. Johnson, *God Struck Me Dead: Religious Conversion Experiences and Autobiographies of Ex-slaves* (Philadelphia: Pilgrim Press, 1969), pp. 113-14. "My personality": "The Dimensions of a Complete Life," *Measure*, pp. 51, 54, italics added.

60. "de big Massa": Genovese, p. 167.

60. Niebuhr and personality: Fox, pp. 140-41.

60. "confused": Smith and Zepp, p. 91.

60-61. Niebuhr's prophecy: *Moral Man and Immoral Society* (New York: Scribner's, 1952 [1932]), pp. 252-54. Cf. Smith and Zepp, pp. 93-94. King alluded to the prophecy in a March 25, 1967, speech given in Chicago, MLK, Atlanta.

61. critique of Niebuhr: "Reinhold Niebuhr's Ethical Dualism" (Boston University term paper, May 9, 1952), MLK, Atlanta.

61. Niebuhr's influence: That influence continues to be overestimated, e.g., in Branch, who attributes the King sermon "The Answer to a Perplexing Question" to the inspiration of Niebuhr (p. 93), when in fact the sermon derives from a sermon by Phillips Brooks.

61. "superficial optimism": "Pilgrimage," *Stride*, p. 99.

61. "parochial concerns of theology": See Carson et al., p. 95. "understanding of man": "How Modern Christians Should Think of Man," n.d., MLK, Boston. Cf. *Papers*, 1:274.

62. "thesis . . . synthesis": for examples of his many syntheses, see Smith and Zepp, pp. 115-16.

62. King's plagiarism: See "The Student Papers of Martin Luther King, Jr.: A Summary Statement on Research," *JAH*, 23-31.

63. "intellectual jive" and "going through the motions": Garrow, "King's Plagiarism: Imitation, Insecurity, and Transformation," *JAH*, 89-90.

63. "academic falsification": Keith D. Miller, "Martin Luther King, Jr., and the Black Folk Pulpit," *JAH*, 121. Miller compares the way King merged his voice with Professor Brightman's to the methods of the folk preacher who identifies with the "I" of a gospel song. But it is quite a stretch from the genre of Ph.D. thesis to a folk sermon.

63. "highly implausible": Garrow, "King's Plagiarism," *JAH*, 88.

63. Lewis's critique: See David L. Lewis, "Failing to Know Martin Luther King, Jr.," *JAH*, 82.

64. superficiality: Lewis was the first to note King's "derivative" intelligence, in *King*, pp. 44-45. On self-falsification, see Miller, "Martin Luther King, Jr.," *JAH*, 121-23.

64. "dramatization of ideas": Lerone Bennett, Jr., *What Manner of Man: A Biography of Martin Luther King, Jr.* (Chicago: Johnson, 1968), p. 187.

64. Keighton: Keighton's less than laudatory remarks are quoted in Mervyn Warren, "A Rhetorical Study of the Preaching of Doctor Martin Luther King, Jr., Pastor and Pulpit Orator" (Ph.D. diss., Michigan State University, 1966), p. 39.

64. "nobler manhood": Quoted in Batsell B. Baxter, *The Heart of the Yale*

Lectures (New York: Macmillan, 1947), p. 3. For a taste of Keighton's approach to homiletics, see his *The Man Who Would Preach* (New York: Abingdon Press, 1956).

64. homiletics bibliography: holograph, MLK, Boston.

65. sermon forms: See Branch, p. 77. Also W. Blackwood, *The Preparation of Sermons* (New York: Abingdon-Cokesbury Press, 1948), pp. 136-51; William E. Sangster, *The Craft of Sermon Construction* (Philadelphia: Westminster Press, 1951), pp. 88-89; Halford E. Luccock, *In the Minister's Workshop* (New York: Abingdon-Cokesbury Press, 1944), pp. 134-47. On the "Hegelian" sermon, see Blackwood, pp. 148-49. King writes in one of his sermons, "Jesus recognized the need for blending opposites" (*Strength*, p. 9).

65. "altruism": "On Being a Good Neighbor", *Strength*, pp. 26-35. "Perplexing Question", *Strength*, pp. 127-37. Compare Phillips Brooks, *Sermons Preached in Country Churches* (New York: Dutton, 1883), pp. 179ff.

66. "fire for form": Robert Frost, "What Fifty Said."

66. *Best Sermons*: The annual series, published by Harper & Brothers, served as an almanac for preachers. Wallace Hamilton's most important book (for King) was *Horns and Halos in Human Nature* (Westwood, N.J.: Fleming H. Revell, 1954). Also important were George A. Buttrick, *The Parables of Jesus* (Garden City, N.Y.: Doubleday, Doran, 1929), the many sermon volumes of Harry Emerson Fosdick, and others cited in Miller, *Voice of Deliverance*.

66. "Three Dimensions": See Phillips Brooks, "The Symmetry of Life" in *Selected Sermons*, ed. W. Scarlett (New York: Dutton, 1949 [n.d.]), pp. 195-206. King made a rare admission of his dependence on another preacher (Brooks) in an interview with Marvyn Warren, "A Rhetorical Study of the Preaching of Dr. Martin Luther King, Jr.," p. 105. See Coretta Scott King, p. 6. Additional one-page sermon outlines are listed in the Calendar of Documents in *Papers*, 1:454, 463.

66. "throughout his life": The titles of his student sermons preached at Ebenezer are listed in *Papers*, 1:86-90.

67-68. Barbour: Intr. Almanina Barbour. J. Pius Barbour, typed and holograph sermon notes and outlines, includes audiotapes and photographs (Barbour Collection). In October 1948, young King wrote his parents, "Since Barbor [sic] told the members of his church that my family was rich, the girls are running me down. Of course, I don't ever think about them I am to [sic] busy studying. I eat dinner at Barbors [sic] home quite often. He is full of fun, and he has one of the best minds of anybody I have ever met" (*Papers*, 1:161).

67. first Negro graduate: Intr. Kenneth Smith.

68. Barbour mediated between races: Kirk Byron Jones, "The Activism of Interpretation: Black Pastors and Public Life," *Christian Century* 106:26 (September 13-20, 1989): 817-18.

68. liberal sermons: "Rising Above Circumstances," April 2, 1950; "How Can You Get Strength . . . ?" March 27, 1949, Barbour Collection.

68. "my experience with God": Sermon outline, "Never Man Spake Thus," August 28, 1949, Barbour Collection.

68. "poor little ole' me": Intr. Almanina Barbour.

68. "Steps toward certainty": Sermon outline, "The Unknown God," July 29, 1951, Barbour Collection.

69. "moral cleanliness": Sermon outline, "The New Israel," October 21, 1951, Barbour Collection.

69. "warm up to Christ": Audiotape, n.d., Barbour Collection.

69. "make a synthesis": Sermon notes, no title, 1948, Barbour Collection.

69. Aeschylus, Sophocles, Euripides: Sermon outline, "Human Tragedy," November 26, 1950, Barbour Collection.

69. King's doodlings: Theology notebooks, n.d., MLK, Boston.

69. jokes about Barbour: Intr. Frank Young. "Living his life through us": Intr. Samuel Proctor. A July 21, 1955, letter from Barbour to King is perhaps a reflection of Barbour's own frustration with his situation in Chester, Pennsylvania: "I warn you. Don't get stuck there [in Montgomery]. Move on to a big metropolitan center in THE NORTH, or some town as ATLANTA. You will dry rot there. I feel sorry for you with all that learning." Correspondence, MLK, Boston.

69-70. "telephone book": Intr. Evans Crawford.

70. the drill: Intrs. Almanina Barbour, Sam Proctor.

70. Sam Proctor: Intr. Almanina Barbour. Proctor had graduated from Crozer and "Barbour University" a few years earlier. He would later dedicate a book on preaching to Barbour (*Preaching about Crises in the Community* [Philadelphia: Westminster Press, 1988]).

70. grades: Barbour sermon notes and outlines, Barbour Collection.

70. "gospel live for me": Intr. Bernard Lee.

Chapter 3: Dexter Avenue and "The Daybreak of Freedom"

Page

72-73. Montgomery: King, *Stride*, pp. 27-28; Oates, pp. 54-59.

73. Emmett Till: In a 1963 Mother's Day sermon King emotionally recalled the tragedy of Emmett Till: "You've seen it [racial injustice] in the crying voice of a little Emmett C. Till screaming from the rushing waters in Mississippi" ("What a Mother Should Tell Her Child," May 12, 1963, MLK, Atlanta).

73. "baffled and demoralized": James Baldwin, "The Highroad to Destiny," in C. Eric Lincoln, ed., *Martin Luther King, Jr: A Profile*, rev. ed. (New York: Hill & Wang, 1984), p. 94.

74-75. Vernon Johns: "current of the years": quoted in Boddie, p. 74; see Branch, pp. 15, 17, 24; Johns, *Human Possibilities: A Vernon Johns Reader*, ed. Samuel Lucius Gandy (Washington, D.C.: Hoffman Press, 1977), p. xv; Zelia S. Evans and J. T. Alexander, *Dexter Avenue Baptist Church, 1877-1977* (1978), p. 64; Intrs. Wyatt Tee Walker, G. Murray Branch, Gardner C. Taylor.

75. "semi-annual visit to the church": Branch, p. 15.

75. "safe to murder Negroes": Evans and Alexander, p. 64. Cf. Branch, p. 24.

76. Friendly Universe: Johns, *Human Possibilities*, p. 73. See "Transfigured Moments," in Joseph Fort Newton, ed., *Best Sermons, 1926* (New York: Harcourt Brace), pp. 333-350. The editor's introduction reads: "Mr. Johns is the first colored preacher to appear in *Best Sermons*, and it is both an honor and a joy to bid him welcome, alike for his race and his genius (p. 334).

76. sit-ins: Farmer, pp. 90-95, 106-10; Garrow, *Bearing*, pp. 26-27. Other forerunners: Garrow, *Bearing*, pp. 11-17.

76. King's predecessors: Evans and Alexander, pp. 16-68.

76. installation service: Service bulletin, MLK, Boston.

77. "without antagonizing his parishioners": Intr. R. D. Nesbitt.

77. "from pulpit to pew": "Recommendations to the Dexter Avenue Baptist Church for the Fiscal Year" September 1, 1954, MLK, Boston. Cf. Evans and Alexander, pp. 71ff.

77. "Brother Pastor": Intr. R. D. Nesbitt.

78. "the figures": Informally related by Patsy Sears.

78. "to commend people": Intr. Zelia Evans.

78. "Recommendations": MLK, Boston, 1954-1955. Cf. Evans and Alexander, pp. 71-139.

78. Deacon Randall: Intr. J. T. Porter.

79. "anything well": "Recommendations," 1956-1957. MLK, Boston. Cf. Evans and Alexander, p. 109.

79. "pamper you nor spare": "Comforting Tidbits for Our Pastor and Club Member," MLK, Boston.

79. "official welcome": Intr. Ralph Bryson.

79. "under fire": Intrs. J. T. Porter, Almanina Barbour.

79. "great problems of life": Reddick, p. 13. General background on Dexter Avenue Baptist Church: Intrs. Zelia Evans, Ralph Bryson, R. D. Nesbitt, Thelma Rice, G. Murray Branch, Mary Lucy Williams.

80. King the liturgist and preacher: Intr. J. T. Porter.

80. "class church": Intr. J. T. Porter; Lincoln, *The Black Church since Frazier*, p. 116; Drake and Cayton, 2:536-39.

80. Dexter worship: worship folders, MLK, Boston. On black worship, see Montgomery, pp. 192-93. "last person who shouted": Intr. G. Murray Branch.

80. "total silence": Intr. J. T. Porter.

80. "bronze Buddhas": Intr. G. Murray Branch.

81. "Schuller type": Intr. J. T. Porter.

81. King's Dexter sermons: Much of our information concerning the sermons King preached at Dexter Avenue derives from the edited sermons in *Strength to Love*; from later versions that he is known to have preached in his first congregation; and from hearsay reports, church bulletins, and other circumstantial evidence. Nevertheless, his trial sermon and other transcripts make it possible for us to piece together an account of his development at Dexter Avenue Baptist Church. Many of his sermons at Dexter were tape recorded by a member of the congregation, Roscoe Williams. To my knowledge they have not been transcribed.

The King estate has not made the tapes available to scholars. (Intrs. Ralph Bryson and Mary Lucy Williams).

81. "3-D in Religion": Intr. Ralph Bryson. Cf. "The Dimensions of a Complete Life," May 31, 1959, Dillard University, MLK, Atlanta, and "The Three Dimensions of a Complete Life," April 9, 1967, MLK, Atlanta.

82. "it embarrasses me": Quoted in Evans and Alexander, p. 69. No date or source is given. The quotation may be from an oral comment of King's. Its sentiments are consistent with his early statements on black worship and preaching.

82. "another side of the candidate": Intrs. Evans Crawford, Sam Proctor.

82. "Recovering Lost Values": typed transcription. The entire sermon will appear in *Papers*, vol. 2. See Peter Holloran, "Recovering Lost Values: Transcribing an African American Sermon," *Documentary Editing*, September 1991, pp. 49-53.

82. "it's wrong": See Holloran, 51.

83. "little gods": See ibid., 49.

84. sermon on Ghana: "The Birth of a New Nation," April 1957, MLK, Atlanta.

85. "Onward Christian Soldiers": King, *Stride*, pp. 60-63; Intr. Robert Graetz.

85. ". . . get all this?": Matthew 13:51.

86. lack of preparation: King, *Stride*, pp. 59-60.

86. slave preachers: See Johnson, *God Struck Me Dead,* pp. 22, 23, 74.

86-88. Boycott address: Holt Street address, December 5, 1955, audiotape and transcript, MLK, Atlanta. The poetic arrangement is mine. Succeeding references are to this speech.

89. "Members . . . amazed": Intr. Ralph Bryson.

Chapter 4: What He Received: Units of Tradition

Page

93. bombings: Branch, pp. 637-39. Intr. Bevel.

93. "I Have a Dream": The young woman was Prathia Hall, now the Reverend Prathia Hall Wynn. The story is Rev. James Bevel's report. Intrs. Almanina Barbour, James Bevel.

93. "*dream* chill'un": Pipes, p. 118.

94. American Dream: "The American Dream," *Testament*, p. 208. See James H. Cone, *Martin and Malcolm and America: A Dream or a Nightmare?* (Maryknoll, N.Y.: Orbis Books, 1991), p. 66.

94. borrowed . . . overheard: The comment is Evans Crawford's.

94. "received": 1 Corinthians 4:7.

94. "*topoi*": Edward P. J. Corbett, *Classical Rhetoric for the Modern Student*, 2d ed. (New York: Oxford University Press: 1971), pp. 45-167. Ong, *Orality and Literacy*, pp. 110-11. On the "proofs," including examples and maxims, see Aristotle, *Rhetoric*, III, 17.

94-95. outlines: See *Strength*, pp. 26-35; "We Would See Jesus," May 5,

1967, MLK, Atlanta; "Is the Universe Friendly?" December 12, 1965, MLK, Atlanta; "What Is Man?" January 12, 1958, Sunday Evening Club, MLK, Atlanta.

96. published versions: "The Dimensions of a Complete Life" in *Measure*, pp. 33–56; it also appears in the early editions of *Strength to Love* but is omitted from later editions.

96–98. Brooks and King: Compare Brooks's "The Symmetry of Life" in *Selected Sermons*, pp. 195–206, and "The Dimensions of a Complete Life," in King's published version in *Measure*, pp. 35–56, esp. pp. 35–36. See also the version preached at Dillard University, May 31, 1959, MLK, Atlanta.

97. Liebman: *Peace of Mind* (New York: Simon & Schuster, 1946), esp. chap. 3, "Love Thyself Properly," pp. 38–58.

98. Buttrick: *The Parables of Jesus*, p. 150.

98–99. "The Death of Evil": Compare King's "The Death of Evil upon the Seashore," *Strength*, pp. 76–85 with Brooks's "The Egyptians Dead upon the Seashore" in *Selected Sermons*, pp. 105–15.

99. "Standing by the Best": August 6, 1967, MLK, Atlanta, and Fosdick in *On Beinq Fit to Live With: Sermons on Postwar Christianity* (New York: Harper & Brothers, 1946), pp. 108–16.

99. "Bad Mess": April 1966, Howard Divinity School Library, Washington, D.C. Cf. Fosdick in *The Hope of the World* (New York: Harper & Brothers, 1933), pp. 117–25.

99. "Great . . . But": 2 Kings 5:1. Compare King's "Great . . . But," July 2, 1967, MLK, Atlanta, with Fosdick's "What Does It Really mean to Be Great?" in *On Being Fit to Live With*, pp. 45–52.

99–100. Drum Major: *Testament*, pp. 259–67. Miller, *Voice*, pp. 3–6.

100. "Three Dimensions": December 11, 1960 version, Unitarian Church, Germantown, Pa.

100–102. Last version: All further references are to "The Three Dimensions of a Complete Life," April 9, 1967, MLK, Atlanta.

102. "his own black experience": Compare his 1958 "What Is Man?" to his 1966 "Who Are We?" MLK, Atlanta.

103. "He's my everything": "Is the Universe Friendly?" December 12, 1965, MLK, Atlanta. For other versions of the same formula, see "Discerning the Signs of History," November 15, 1964; "The Three Dimensions of a Complete Life," April 9, 1967; "What Are Your New Year's Resolutions?" January 7, 1968, MLK, Atlanta. See also Pipes, p. 29; Drake and Cayton, 2:625; Williams, pp. 110–11, for the black tradition's use of this formula; and Jon Michael Spencer, *Sacred Symphony: The Chanted Sermon of the Black Preacher*. Contributions in Afro-American and African Studies, No. 11 (New York: Greenwood Press, 1987), pp. 35–36.

103. "Sons of God shouted for joy": E.g., "A Christmas Sermon on Peace," *Testament*, p. 258, and Taylor, p. 60. Cf. the conclusions to "What a Mother Should Tell Her Child," May 12, 1963, and "Is the Universe Friendly?" December 12, 1965, MLK, Atlanta.

103. King's quotation booklets: See the Calendar of Documents in *Papers*,

1:463. The booklets are held by the King family and have not been released to scholars.

104. "piercing the American sky": February 1-5, 1965, MLK, Atlanta.

104. "swinging lanterns": "But, If Not," November 5, 1967, MLK, Atlanta.

104. "rawhide whip": Howard Thurman, *Deep River* (New York: Harper & Brothers, 1945, 1955), p. 35. Cf. "Shattered Dreams," in *Strength*, p. 92.

104. "Maladjustment": Charles Templeton, "Peace of Mind Is Not Enough" in Alton M. Motter, ed., *Great Preaching Today* (New York: Harper & Brothers, 1955), p. 227. Cf. *Strength*, p. 24, and the later and less-polished "Nonconformist—J. Bond," January 16, 1966, MLK, Atlanta.

105. Dunbar: *Strength*, p. 111.

105-108. "Shattered Dreams": The sermon appears in *Strength*, pp. 86-95. Compare J. Wallace Hamilton, "Shattered Dreams," pp. 25-34. On the many sermons on disappointment, see Miller, "The Influence of a Liberal Homiletic," p. 143. Miller treats the literary relation of Thurman and King in "Influence," pp. 137-41. Cf. Miller, *Voice of Deliverance*, pp. 120-21. Compare King's sermon and Thurman's treatment of slavery in *Deep River*, pp. 36-37, and *Strength*, pp. 87-88.

106. Buttrick: Miller, "Influence of a Liberal Homiletic," pp. 143-44, and *Voice of Deliverance*, pp. 81-85. Cf. Buttrick's "Frustration and Faith" in *Sermons Preached in a University Church* (New York: Abingdon Press, 1959), p. 110. For the *Rubaiyat*, see *Strength*, pp. 88-89.

106-107. "this one solitary personality": King probably got this piece from Charles L. Wallis, *A Treasury of Sermon Illustrations* (New York: Abingdon-Cokesbury Press, 1950), p. 40, which he owned. King offers a close paraphrase of the piece in many of his sermons, including "The Drum Major Instinct" in *Testament*, p. 266, where he offers no citation. Interestingly, King elaborates on the illustration in order to make it self-referencing. He adds the sentences "They called him a rabble-rouser. They called him a troublemaker. They said he was an agitator."

107. "mastery of anxiety": See *Pulpit*, January 1948, April 1949, December 1950, April 1951, May 1952, February 1953, and July 1956. During that period King was not the only preacher to follow Brooks's "The Symmetry of Life." In the July 1954 *Pulpit*, Harold L. Lunger published "A 3-D Faith" that incorporated the familiar dimensions of a complete life. In the April 1950 issue of *Pulpit*, preacher F. Gerald Ensley defended the practice of homiletical borrowing in "On Being Original" (pp. 95-96).

107. "make testimony": The citations that follow are taken from "Unfulfilled Dreams," March 3, 1968, MLK, Atlanta. See Lischer, "The Word That Moves," 181-82.

109. "one's admiration for King": Miller, "Influence of a Liberal Homiletic," p. 195.

110. the nature of man: See *Pulpit*, April 1948, May 1950, January 1955; Hamilton's sermons, "Remember Who You Are" and "Horns and Halos in Human

Nature," pp. 46-67; Fosdick, "The Mystery of Life" in *The Secret of Victorious Living* (New York: Harper & Brothers, 1934), pp. 129-38; Fulton Sheen, "The Psychology of a Frustrated Soul" in *Best Sermons, 1949-1950*, pp. 28-34.

110. ninety-eight cents: "What Is Man?" January 12, 1958, MLK, Atlanta. For the Jeans quotation see *Strength*, p. 71.

110. a "bad mess" in Crete: Fosdick, "Making the Best of a Bad Mess," *Hope of the World*, pp. 117-25. For King's version see "Making the Best . . . ," April 1966, Audiotape, MLK, Howard Divinity School Library. Proctor's sermon is available in audio and videotape from Duke University Chapel.

111. "Three loves": Miller, "Martin Luther King, Jr. Borrows a Revolution," 250-51. On the Prodigal Son, see Miller, "Influence of a Liberal Homiletic," pp. 73-74, 250.

111. "period of apprenticeship": *Strength*, p. 7.

112. white Protestants had "already approved": Miller, "Martin Luther King, Jr. Borrows a Revolution," 256. Miller's theory draws on theologian James H. Cone's assertion that the primary audience for Martin Luther King was white America. See Cone, "Martin Luther King, Jr., Black Theology–Black Church," in which he asserts, "References to the intellectual tradition of Western philosophy and theology were primarily for the benefit of the white public so that King could demonstrate to them that he could think as well or better than any other seminary or university graduate" (p. 411). In a recent book Cone writes, "King used liberal, Protestant theology to articulate the religious dimensions of what he believed about America, because he knew that its language about God was more acceptable to white people than the spirituality of black people" (*Martin and Malcolm*, p. 132). In the area of preaching, African-American scholarship has not dealt with King's use of sources. Many would agree with the accomplished black preacher and former King colleague J. T. Porter: "Seldom, I think, black preaching has a mentor in the white community" (Intr.).

112. "self-making" and "iconic status": Miller, "Composing Martin Luther King, Jr.," *PMLA* 105:1 (January 1990): 70-71, 79.

112. Lincoln and Parker: See Garry Wills, *Lincoln at Gettysburg: The Words That Remade America* (New York: Simon & Shuster, 1992), p. 107.

113. "everybody could understand": Intr. John Gibson, Moorland-Spingarn.

113. *self* and *style*: These are Richard A. Lanham's assertions in *Style: An Anti-Textbook* (New Haven: Yale University Press, 1974), pp. 115-16, 124.

113. both black and American: Du Bois, *Souls*, p. 215. See *Where Do We Go from Here: Chaos or Community?* (New York: Harper & Row, 1967): "The old Hegelian synthesis still offers the best answer to many of life's dilemmas. The American Negro is neither totally African nor totally Western. He is AfroAmerican, a true hybrid, a combination of two cultures" (p. 53).

114. folk preaching: Bruce Rosenberg, *Can These Bones Live? The Art of the American Folk Preacher*, rev. ed. (Urbana: University of Illinois Press, 1970, 1988). Rosenberg's work has helped renew interest in folk preaching in the United States. On the importance of meter, Rosenberg, p. 149; on rhythm, pp. 128-33. Against Rosenberg's emphasis on meter, Gerald L. Davis reasserts the importance of

theme in black folk-preaching in his *I got the Word in me and I can sing it, you know. A Study of the Performed African-American Sermon* (Philadelphia: University of Pennsylvania Press, 1985), p. 54. On the importance of oral formulas in illiterate societies, see the seminal work of Albert B. Lord, *The Singer of Tales* (Cambridge: Harvard University Press, 1960), a study of the process of transmission of Serbo-Croatian folk epics. Walter Pitts hypothesizes on West African retentions in American folk preaching in "West African Poetics in the Black Preaching Style," *American Speech* 64:2 (Summer 1989): 137-49. See the article on the chanted African-American sermon by Jon Michael Spencer in Richard Lischer and William H. Willimon, eds., *Concise Encyclopedia of Preaching* (Louisville: John Knox/Westminster Press, 1995).

114. traditional set pieces: Rosenberg, pp. 37-38; Miller, "Influence," p. 150; Hortense Spillers, "Martin Luther King and the Style of the Black Sermon," *Black Scholar* 3:1 (September 1971): 14. "The Eagle Stirs Her Nest" is performed by the legendary Reverend C. L. Franklin on a cassette, "The Eagle Stirs Her Nest," Paula Records, 1973.

114. "Deck of Cards": Rosenberg, p. 185.

115. "rock in the weary land": Rosenberg, pp. 192-93.

115. "not a safe town": Recorded in Spencer, *Sacred Symphony*, p. 106. The Scripture text is 1 Peter 5:8: "Be sober, be watchful. Your adversary the devil prowls around like a roaring lion, seeking someone to devour."

115-16. "the Good Negro": "Remaining Awake Through a Great Revolution," delivered in Cincinnati, 1964, MLK, Duke, audiotape.

116. "right speech": Eugen Rosenstock-Huessy, *Speech and Reality* (Norwich, Vt.: Argo Books, 1970), pp. 179-80.

116. "repeated a sermon": Intr. Ralph D. Abernathy. See Rosenberg, pp. 39-40.

116. "original and eternal": Rosenstock-Huessy, p. 162, italics deleted.

117. Augustine's advice: *On Christian Doctrine*, IV, 29. "There are some who can speak well but who cannot think of anything to say. If they take something eloquently and wisely written by others, memorize it, and offer it to the people in the person of the author, they do not do wickedly."

117. Church's homiletical tradition: See Yngve Brilioth, *A Brief History of Preaching*, trans. Karl E. Mattson (Philadelphia: Fortress Press, 1965), pp. 67-73.

117. "studied unoriginality": Stout, p. 154.

117. copyright laws: See Miller, "Influence of a Liberal Homiletic," p. 195.

Chapter 5: The Strategies of Style

Page

119. "The *way* he's saying it": Intr. Ralph David Abernathy.

120. "a better nation": "Answer to a Perplexing Question," March 3, 1963, MLK, Atlanta.

120. on style: Kenneth Burke, *Counter-Statement* (Berkeley: University of

California Press, 1968 [1931]), p. 38. See Richard Lanham's discussion of style in *Style,* esp. pp. 44-45.

120. on pleasure and the epic: Eric A. Havelock, *Preface to Plato* (Cambridge: Harvard University Press, 1963), pp. 152-59. Havelock's observations are equally applicable to Africa. King's biographer speaks of the "deeply pleasurable emotions" that attended his speaking (Lewis, *King,* p. 394).

120. "trilogy of durability": "The Meaning of Hope," December 10, 1967, MLK, Atlanta.

120. "crushing domination": See Spillers, 19. The phrase is found in "The Death of Evil upon the Seashore," *Strength,* p. 78.

120-21. "to see Jesus": "We Would See Jesus," May 5, 1967, MLK, Atlanta.

121. teach, delight, move: Augustine, *On Christian Doctrine,* IV, 1.

121. "veins of history": Holt Street address, December 5, 1955, MLK, Atlanta.

121-22. light and darkness: "Remaining Awake Through a Great Revolution," MLK, Duke, audiotape, italics added for emphasis.

122. prayer on Highway 80: March 9, 1965, MLK, Atlanta.

122. mythication: Arthur L. Smith (Molefi Kete Asante), *The Rhetoric of Black Revolution* (Boston: Allyn & Bacon, 1969), p. 34. See also Mary Rose Sloan, "Then My Living Will Not Be in Vain: A Rhetorical Study of Dr. Martin Luther King, Jr., and the Southern Christian Leadership Conference in the Mobilization for Collective Action Toward Nonviolent Means to Integration, 1954-1964" (Ph.D. diss., Ohio State University, 1977), p. 143.

122. on metaphor: On its "strangeness" and as a mark of genius, see Aristotle, *Poetics,* 22; see also *Rhetoric,* III, 10. On its use in homiletics and for a bibliography, see Richard Lischer, "What Language Shall I Borrow? The Role of Metaphor in Proclamation," *Dialog* 26:4 (Fall 1987): 281-286.

122-23. metaphor in black preaching: Pipes, p. 43. "fiddle in my belly": Raboteau, p. 269 (italics omitted). Brother Carper: H. Dean Trulear and Russell E. Richey, "Two Sermons by Brother Carper: 'The Eloquent Negro Preacher,'" *American Baptist Quarterly* 4 (March 1987): 11, spelling altered.

123-24. metonymy: Northrup Frye, *The Great Code: The Bible and Literature* (New York: Harcourt Brace Jovanovich, 1982), pp. 7-8, 27; Burke, *A Grammar of Motives,* p. 506; Corbett, pp. 481-82. On the distinction between metaphor and metonymy, see Bernard Brandon Scott, *Hear Then the Parable* (Minneapolis: Fortress Press, 1989), pp. 29-30.

123. King's metonymies: The technique is everywhere in King, for example, the following sermons at MLK, Atlanta: "Who Are We?"; "Training Your Child in Love"; "The Meaning of Hope"; and in *Testament,* p. 251 and *passim.* He was not immune to the mixed metaphor: ". . . all reality hinges on moral foundations" (The Man Who Was a Fool," MLK, Atlanta, and *Pulpit,* June 1961).

124. on light and dark: Quoted in Keele, p. 246.

124. Lincoln: Wills, p. 37.

125. "audience did not understand": On the categorical imperative, see, for

example, "Levels of Love," May 21, 1967, MLK, Atlanta. After his professor Harold DeWolf heard him preach a sermon in which King alluded to Hegel, the professor mused, "And Hegel is a figure you don't often hear quoted in a sermon" (Intr. Moorland-Spingarn).

125. *what* and *how*: On the fusion of substance and style, see E. L. Epstein, *Language and Style* (London: Methuen, 1978), p. 80.

125. "You ain't no nigger": "Is the Universe Friendly?" December 12, 1965, MLK, Atlanta.

125. "live with rats": Oates, p. 375.

125-26. "look at what we've done": "Great . . . But," July 2, 1967, MLK, Atlanta, audiotape only. See "Remaining Awake Through a Great Revolution," *Testament*," p. 275. Additional quotations on the greatness of humanity and human sin are taken from this sermon.

127. on 1967: Garrow, *Bearing*, pp. 565-68; Oates, pp. 429-30.

128-29. varieties of repetition: Corbett, pp. 471-76. *Alliteration*: "I Have a Dream," *Testament*, p. 219; *Assonance*: "The Meaning of Hope," December 10, 1967, MLK, Atlanta; *Anaphora*: "Our God Is Marching On," *Testament*, p. 230. "How long?" see Pipes, p. 42. *Epistrophe*: "Remaining Awake Through a Great Revolution," 1964, Duke Divinity School, audiotape. "Nothing has been done": from a sermonic version of the same speech, "Remaining . . ." *Testament*, pp. 274-75; *leitmotif*: "The Three Dimensions . . ." April 9, 1967, MLK, Atlanta; on hell: "Lazarus and Dives," March 10, 1963, MLK, Atlanta. The object of King's wrath was a retired airline executive named Carleton Putnam, whose *Race and Reason: A Yankee View* (Washington D.C.: Public Affairs Press, 1961) asked such questions as "Can you name one case in all history in which a white civilization failed to deteriorate after intermarrying with Negroes?" and "Can you name one case in all history of a stable, free civilization that was predominantly, or even substantially, Negro?" (p. 105).

129. copiousness and intensification: See Kenneth Burke, *A Rhetoric of Motives* (Berkeley: University of California Press, 1969 [1950]), p. 69, and Ong, *Orality and Literacy,* pp. 40-41.

130. "have seen a great light": "Why I Am Opposed to the War in Vietnam," April 30, 1967, MLK, Atlanta.

130. parataxis: For a definition with examples from the Bible, see G. B. Caird, *The Language and Imagery of the Bible* (Philadelphia: Westminster Press, 1980), pp. 117-21. For a superb reading of a biblical narrative that features parataxis, see Robert Alter, *The Art of Biblical Narrative* (New York: Basic Books, 1981), pp. 3-22.

130. "shout for joy": "Remaining Awake . . ." *Testament*, p. 278. Cf. "A Christmas Sermon on Peace," *Testament*, p. 258. For another instance of parataxis, "Pride versus Humility," October 9, 1966, MLK, Atlanta, where in the midst of a discussion of marital infidelity, he exclaims, "All we like sheep have gone astray."

130. "sound track": Several scholars have been indispensable to the study of the musicality of King's preaching. Among them, Jon Michael Spencer, both

in *Sacred Symphony*, pp. 1-16, and, more important, in personal conversations; William Turner, "The Musicality of Black Preaching: A Phenomenology," *Journal of Black Sacred Music* 2:1 (Spring 1988): 21-29, passim; Mitchell, *Black Preaching*, pp. 162-89. An incredibly rich source is the chapter "The Language of Black America" in Smitherman, *Talkin and Testifyin*, pp. 101-45.

131. "same beat": Wyatt Tee Walker, quoted in Gerald Davis, epigraph.

131. "hit a lick": Spencer, *Sacred Symphony*, pp. 4-5. On rhythm, see Rosenberg, pp. 71-72.

131-32. stammer: See Herbert Marks, "On Prophetic Stammering," in Regina M. Schwartz, ed. *The Book and the Text: The Bible and Literary Theory* (Oxford: Basil Blackwell, 1990), p. 67. Also, Mitchell, *Black Preaching*, p. 176. In one African-American congregation the preacher is said to have stammered, "I, I, I, would like to go on, but [looking at the clock] the clock tells me it's late." "It's *wrong*! It's *wrong*!" the congregation responded (Intr. Sam Proctor).

132. run-on rhythm: See "Guidelines for a Constructive Church," June 5, 1966, MLK, Atlanta.

132-33. 'whoop' and 'holler': "Some Things We Must Do," December 5, 1957, quoted in Keele, p. 165.

133. "he *preached* this morning": "A Knock at Midnight," 1967 or 1968, MLK, Duke, audiotape.

133. "breathtaking range": Despite the more precise, contrary assessments of musicologists. See Warren, "A Rhetorical Study of the Preaching of Doctor Martin Luther King, Jr.," who rates King's range as "unimpressively limited"—less than an octave (pp. 126-27).

133. "freedom, freedom": "Birmingham Negroes; Plea for Freedom," October 22, 1963, MLK, Atlanta.

133-34. tonal languages: Smitherman, pp. 134-37; Asante, *Afrocentric*, pp. 84-85. Blue note: Jahn, p. 223.

133-34. jazz: Spencer, *Sacred Symphony*, pp. 15-16. See Gerald Davis's discussion of melisma, "the production of several notes around one syllable" (p. 80).

135. "drum major": "The Drum Major Instinct," February 4, 1968, MLK, Atlanta, audiotape; and *Testament*, pp. 259-67.

135-36. call and response: Smitherman, pp. 104-13; Spencer, *Sacred Symphony*, pp. 5-8; Mitchell, *Black Preaching*, pp. 167-68.

136. "God Almighty!": Pat Watters, *Down to Now* (New York: Pantheon Books, 1971), p. 14.

136-38. congregational responses: See Mervyn A. Warren, *Black Preaching: Truth and Soul* (Washington D.C.: University Press of America, 1977), pp. 27-28. Warren offers examples of several styles of black preaching: effect preaching—styled for audience reaction; truth preaching—cerebral, intellectual; ethics style—based on the character of the speaker; methods style—geared to a mixed audience, both intellectually astute and vocally responsive. He places King in this final category (pp. 33-41). Evans Crawford, longtime dean of Rankin Chapel at Howard

University, has worked out a taxonomy of response that ranges from "Help him, Lord" to "Hallelujah! Amen!" (Intr. Evans Crawford).

136. C. L. Franklin: Quoted in Hamilton, p. 29.

137. "the acceptable year of the Lord": "Guidelines for a Constructive Church," June 5, 1966, MLK, Atlanta.

137-38. delayed feedback: Intrs. Evans Crawford and Bernard Lee.

138. "importance as God's children": This is Cone's point in *Speaking the Truth: Ecumenism, Liberation and Black Theology* (Grand Rapids: William B. Eerdmans, 1986), pp. 27, 130.

138. "fiery mad" and "fiery glad": Henry Mitchell, *The Recovery of Preaching* (San Francisco: Harper & Row, 1977), pp. 67–68.

138. Jonathan Edwards: This description is given in Pipes, p. 2. King actually quotes the same passage in a Crozer term paper written in 1950. See *Papers*, p. 346.

139. "landing strip": Intr. Wyatt Tee Walker. King followed the practice of announcing the sermon's theme in the first sentence.

139. climax: The most informed treatment of the sermon's climax is found in Mitchell's discussion of "celebration" in *The Recovery of Preaching*, pp. 54–73, 159–60.

139. "making gravy": Intr. C. T. Vivian.

140. "Amazing Grace": "Great . . . But": June 2, 1967, MLK, Atlanta, audiotape.

140-41. climax and sexual imagery: Pipes suggests that a woman's cry "Tear it up" "might have a sexual implication." He compares the abandon of the sermon's climax (and the effect on the audience) with the climax of sexual intercourse; "the words 'hon' [*honey*] and 'pull 'em down' suggest this interpretation" (p. 195, n).

140. "the sisters": Intr. William Rutherford, Moorland-Spingarn.

141. pleasure and prophecy: See Lewis, *King*, p. 394.

Chapter 6: From Identification to Rage

Page

142. "hurt and baffle them": Baldwin, "The Highroad to Destiny," in Lincoln, *Martin Luther King, Jr.*, p. 97.

142. identification: Burke, *A Rhetoric of Motives*, pp. 55, 208. See also C. H. Perelman and L. Olbrechts-Tyteca, *The New Rhetoric: A Treatise on Argumentation*, trans. John Wilkinson and Purcell Weaver (Notre Dame: University of Notre Dame Press, 1969), who refer to the "continuous adaptation of the speaker to his audience." They quote Gracian, who says that speech is "like a feast, at which the dishes are made to please the guests, and not the cooks." Demosthenes insisted on only one rule: the "adaptation of the speech to the audience, whatever its nature" (pp. 23–25).

143. wheelbase of their cars: "Lazarus and Dives," March 10, 1963, MLK, Atlanta.

143. cutting the preacher's salary: Intr. John Lewis, Moorland-Spingarn.

143. on the black church's commitment: Intr. Gardner C. Taylor.

143. "indescribable capacity for empathy": Quoted in Oates, p. 272.

144. childhood segregation: Oates, pp. 8-10.

144. amusement park: For example, "Letter," *Testament*, pp. 292-93.

144. Marian Anderson: "Dives and Lazarus," March 10, 1963, MLK, Atlanta. At the conclusion of the sermon Daddy King offered some informal commentary on the Marian Anderson story, drawing a parallel to his own satisfaction with his son.

144. brother-in-law: "New Wine in Old Bottles," January 2, 1966, MLK, Atlanta. The comment is on the audiotape but does not appear in the transcript. King used the line in many sermons and speeches. On this particular occasion it is spoken with something near bitterness. In a later sermon he said, ". . . and everybody [who] argues about intermarriage is an unconscious racist." He gives the example of the "liberal lady" who says, "But I must say, Dr. King, that I wouldn't want a Negro to marry my daughter." King's reply: "I wouldn't want my daughter to marry George Wallace" ("Mastering Our Fears," September 10, 1967, MLK, Atlanta).

144. *nigra*: "Levels of Love," MLK, Atlanta, May 21, 1967.

145. "doors are closed": "The Ballot," MLK, Atlanta, July 17, 1962 (mass-meeting speech).

145. "we don't serve niggers": "What a Mother Should Tell Her Child," May 12, 1963, MLK, Atlanta.

145. "focal instance": Robert C. Tannehill uses the term in "The 'Focal Instance' as a Form of New Testament Speech: A Study of Matthew 5:39b-42," *The Journal of Religion* 50:4 (October 1970): 372-85.

145. "get in there": "Who Are We?" February 5, 1966, MLK, Atlanta.

145. "that's about it": "Remaining Awake During a Great Revolution," *Testament*, p. 273.

145. colloquialisms: "get it over": "What a Mother Should Tell Her Child," May 12, 1963; "lie on you"; "Judging Others," June 4, 1967, where he adds, "You know, haven't you ever gossiped on somebody?"; "Messed up the world" and "the sun got up": "Who Is My Neighbor?" February 18, 1968—all MLK, Atlanta.

145. "git his diploma on God": Pipes, p. 87; "couldn't trick the Son of the Living God": quoted in Drake and Cayton, 2:676.

146. "get out of my face!": "The Three Dimensions of a Complete Life," April 9, 1967, MLK, Atlanta. "That Was Not Enough!": "Remaining Awake . . ." March 31, 1968, National Cathedral Archives, Washington, D.C., and *Testament*, p. 275.

147. Whitefield: Quoted in Sobel, p. 103. On the black preacher's use of dramatic first-person accounts, see Mitchell, *The Recovery of Preaching*, pp. 35-38, and Gerald Davis, p. 110.

147. understatement: "What Are Your New Year's Resolutions?" January 7,

1968, MLK, Atlanta (preached in New York City). Another example of under-statement: "But every now and then people must hear the truth." He is speaking about the Vietnam War ("Nonconformist–J. Bond," January 16, 1966, MLK, Atlanta).

147. King's mail: Correspondence files, MLK, Boston. THANK YOU: Oates, p. 94.

147. "toil on his brow": "Remaining Awake Through a Great Revolution," an address to the General Conference of the AME Church, 1964, in Cincinnati, MLK, Duke, audiotape.

148. "another sound": A perceptive discussion of language is found in Asante, *Afrocentric*, p. 115.

140. deleted references to God: Note for example the distributed tape of "Interruptions," January 21, 1968, MLK, Atlanta. Also, Intr. Wyatt Tee Walker.

149. parallelisms: Taken from an informative study by Brygida Irena Rudzka-Ostyn, "The Oratory of Martin Luther King and Malcolm X: A Study in Linguistic Stylistics" (Ph.D. diss., University of Rochester, 1972), pp. 28-29. On the "wedge of difference" see Robert Alter, *The Art of Biblical Poetry* (New York: Basic Books, 1984), pp. 10-11.

150. "misguided men": "The God of the Lost," September 18, 1966, MLK, Atlanta.

150. "fit to live": A paraphrase of one of King's favorite *sententiae* found in a transcript of an address given in Selma on March 9, 1965, MLK, Atlanta. For an excellent example of his use of antithesis, see "Pride Versus Humility," which contains the following sentences: "We may not rob a bank, [but] how many times have we robbed our brothers and sisters of their good names through malicious gossip? We may not get drunk, but how many times have we staggered before our children and our friends intoxicated by the wind of a bad temper? We may not murder anyone physically, but how many wives have spiritually murdered their husbands, and how many [husbands] have spiritually murdered their wives through the bullet of mental cruelty?" (October 9, 1966, MLK, Atlanta).

150-51. running style: As opposed to the highly crafted periodic style, which is characterized by inversion of the natural English sentence order and the with-holding of meaning until the end of the sentence—see Lanham, *Style*, pp. 119-21. On King's "nominality," see Spillers, 17.

150. "We see": "Standing by the Best in an Evil Time," August 6, 1967, MLK, Atlanta. In "Guidelines for a Constructive Church," June 5, 1966, MLK, Atlanta, he begins nineteen consecutive sentences with the phrase "the acceptable year of the Lord."

150. KJV: See Rudzka-Ostyn, p. 33. On the additive style, see Ong, *Orality and Literacy*, p. 37.

151. civil religion: Indispensable to this study are the essays collected in Russell E. Richey and Donald G. Jones, eds., *American Civil Religion* (New York: Harper & Row, 1974).

151. "God Almighty is wrong": "Holt Street Address," December 5, 1955, MLK, Atlanta, audiotape and transcript.

151. Fatherhood of God: Prayer Pilgrimage program, MLK, Atlanta.

151. "true to our native land": "Give Us the Ballot," in *Testament*, p. 200. "Behind the dim unknown": James Russell Lowell quoted in the 1956 address "Facing the Challenge of a New Age," *Testament*, p. 141. "Brother Kennedy": "Nonconformist–J. Bond," January 16, 1966, MLK, Atlanta.

152. "universal religion of the nation": The phrase is Sidney Mead's. It is commented upon in the introduction to Richey and Jones. In the same volume see Will Herberg, "America's Civil Religion: What It Is and Whence It Comes," pp. 76–88. He defines civil religion as "the American Way of Life Religion" and "hold-out groups" as those who are "incompletely enculturated" and therefore resist the predominant civil religion. "These groups are very small, and are rapidly diminishing" (p. 85). He does not mention black people in America!

152. The Declaration of Independence: On Lincoln's preference for the Declaration above the Constitution, see Wills, p. 100.

152. "All men": "Great . . . But," July 2, 1967, MLK, Atlanta. Jon Michael Spencer made helpful comments on the glissando.

153. "always said the same": James Baldwin, "The Highroad to Destiny," in Lincoln, *Martin Luther King, Jr.*, p. 91.

153. Southern Baptists: His remarks are a part of the Gay Lectures that he gave at Southern Baptist Theological Seminary in Louisville. See "The Church on the Frontier of Racial Tension," April 19, 1961, MLK, Atlanta, transcript of audiotape.

153–55. the rhetoric of Malcolm and Martin: See Rudzka-Ostyn, pp. 42, 54–58, 68–75, 82, 95–102, 112–14. "we got a good house"; *Malcolm X Speaks*, ed. George Breitman (New York: Grove Press, 1965), p. 10. "deaf, dumb, and blind": Rudzka-Ostyn, p. 82. "friends and enemies": *Malcolm X Speaks*, p. 24.

154. few imperatives in King: See Warren, "A Rhetorical Study of the Preaching of Doctor Martin Luther King, Jr. . . . ," p. 200.

154. snake: *Malcolm X Speaks*, p. 135.

155. Augustine's advice: *On Christian Doctrine*, IV, 12.

155. shame and guilt: "Kenneth B. Clark Interview," *Testament* p. 336.

155. "guilt-ridden man": Stanlcy Levison quoted in Garrow, *Bearing*, p. 588.

155. "hypocrisy" of Bible-believing racists: See James H. Smylie, "On Jesus, Pharaohs, and the Chosen People: Martin Luther King as Biblical Interpreter and Humanist," *Interpretation* 24:1 (January 1970): 90. Walter Rauschenbusch said that effective social evangelism "must hold up a moral standard so high above their [unbelievers'] actual lives that it will smite them with conviction of sin" (quoted in Winthrop Hudson's introduction to *Walter Rauschenbusch, Selected Writings*, ed. Winthrop S. Hudson (New York: Paulist Press, 1984), p. 31.

155. "get rid of his guilt": See the transcript of a talk he gave in Grenada, Mississippi, June 16, 1966, MLK, Atlanta.

155. against black-on-white violence: Mass-meeting speech, March 19, 1968, given in Clarksdale, Mississippi, MLK, Atlanta.

155–56. the ridicule of critics: One of the most articulate of the early critics of his rhetorical strategy was August Meier. In "The Conservative Militant," in Lincoln, *Martin Luther King Jr.*, Meier argues, "He has faith that the white man will redeem himself. Negroes must not hate whites, but love them. In this manner, King first arouses the guilt feelings of whites, and then relieves them—though always leaving the lingering feeling in his white listeners that they should support his nonviolent crusade. Like a Greek tragedy, King's performance provides an extraordinary catharsis for the white listener" (p. 148; this analysis originally appeared in 1965).

156-57. the signifying monkey: See Henry Louis Gates, Jr., *The Signifying Monkey: A Theory of Afro-American Literary Criticism* (New York: Oxford University Press, 1988), pp. 4-56. Gates develops his account of signifying without explicit reference to Martin Luther King, Jr. On loud talking, see pp. 82-83. To mix the metaphor: Walter E. Fauntroy said in an interview that King knew how to deal with "Tarzan liberals" in white suburbia who want to rescue African Americans from the bad white people (Intr. Moorland-Spingarn).

156-57. "white liberalism . . . in his own black voice": Variations on this general approach to the interpretation of King's manipulation of his audiences have been voiced by Keith D. Miller and James Cone, whose theories are discussed in Chapter 4. One of the earliest expressions of this interpretation occurs in the terse assessment of August Meier, first published during King's lifetime:

> At the heart of King's continuing influence and popularity are two facts. First, better than anyone else, he articulates the aspirations of Negroes who respond to the cadence of his addresses, his religious phraseology and manner of speaking, and the vision of his dream for them and for America. King has intuitively adopted the style of the old-fashioned Negro Baptist preacher and transformed it into a new art form; he has, indeed, restored oratory to its place among the arts. Second, he communicates Negro aspirations to white America more effectively than anyone else. His religious terminology and manipulation of the Christian symbols of love and nonresistance are partly responsible for his appeal among whites. To talk in terms of Christianity, love, nonviolence, is reassuring to the mentality of white America. At the same time, the very superficialities of his philosophy—that rich and eclectic amalgam of Jesus, Hegel, Gandhi, and others as outlined in his *Stride Toward Freedom*—make him appear intellectually profound to the superficially educated middle-class white American. Actually, if he were a truly profound religious thinker, like Tillich or Niebuhr, his influence would of necessity be limited to a select audience. But by uttering moral clichés, the Christian pieties, in a magnificent display of oratory, King becomes enormously effective.
>
> If his success with Negroes is largely due to the style of his utterance, his success with whites is a much more complicated matter. For one thing, he unerringly knows how to exploit to maximum effectiveness their growing feeling of guilt. ("The Conservative Militant," in Lincoln, *Martin Luther King, Jr.*, p. 147)

157. "splattered with the blood": "Crisis and a Political Rally in Alabama," May 3, 1963, MLK, Atlanta.

157. "issues to the surface": "Remaining Awake . . . ," 1964, MLK, Duke, audiotape.

157. TV era: Reporter Pat Watters was an eyewitness observer of many of the mass meetings in churches and demonstrations. He soon discovered that the media was preoccupied with violence and reserved its most thorough coverage for it (Watters, pp. 109–12). He regrets all that America missed: "I sit and lament anew that the movement did not reach southern whites, lament the southern cultural proscriptions that made it impossible for whites to enter such churches, hear such eloquence, feel the southernness of those meetings, and lament as much the forces, the compulsions of American culture that prevented any serious attempt by the media (television being surely the most appropriate) to present what was said and felt by the Negro people in those meetings" (p. 191).

157. "manipulation of the media": Garrow, *Bearing*, p. 261.

157–58. irony of violence: See Garrow, *Protest at Selma: Martin Luther King, Jr., and the Voting Rights Act of 1965* (New Haven: Yale University Press, 1978), pp. 226, 235–36.

158. lancing the boil: See King's classic "Letter" in *Testament*, p. 295, in which he insists,

> Like a boil that can never be cured as long as it is covered up but must be opened with all its pus-flowing ugliness to the natural medicines of air and light, injustice must likewise be exposed, with all the tension its exposing creates, to the light of human conscience and the air of national opinion before it can be cured.

Jonathan Edwards:

> To blame a minister for thus declaring the truth to those who are under awakenings, and not immediately administering comfort to them, is like blaming a surgeon because, when he has begun to thrust in his lance, whereby he has already put his patient in great pain, and he shrieks and cries out with anguish, he is so cruel that he will not stay his hand, but goes on to thrust it in further, until he comes to the core of the wound. (Jonathan Edwards, *Thoughts on the Revival of Religion in New England, 1742* . . . [New York: Dunning & Spalding, 1832], pp. 237–38)

King:

> Now I wonder if anybody in here tonight would condemn a physician who through his skills . . . revealed to a person that they had cancer and found that cancer and set out to cure it. Are you gonna say that that doctor caused the cancer? Indeed, you would praise the doctor for having the wisdom to bring it out into the open and deal with it. Now that's all we are. We are the social physicians of Chicago revealing that there is a terrible cancer. ("Why I Must March," August 18, 1966, MLK, Atlanta)

158. secrets of their hearts: R. G. Collingwood quoted in James A. Sanders, *God Has a Story Too* (Philadelphia: Fortress Press, 1979), p. 70.

158. "national punishments": Benjamin Rush quoted in Robert N. Bellah, *The Broken Covenant: American Civil Religion in Time of Trial* (New York: Seabury Press, 1975), p. 43.

158. "born in genocide": *Why We Can't Wait* (New York: Harper & Row, 1963), pp. 130–31.

158-59. "America is a racist country": Mass-meeting address, March 22, 1968, Augusta, Georgia, MLK, Duke, audiotape. Cf. his assertion "Racism is very deep," mass-meeting address, February 16, 1968, Montgomery, Alabama, MLK, Atlanta.

159. America's Bicentennial: "Who Is My Neighbor?" February 18, 1968, MLK, Atlanta.

159. "never was America" . . . : Langston Hughes, "Let America Be America Again," in Langston Hughes and Anna Bontemps, eds., *The Poetry of the Negro, 1746-1949* (Garden City, N.Y.: Doubleday, 1949), p. 108. Cf. "A Time to Break Silence" (Clergy and Laity Concerned, Riverside Church, New York), April 4, 1967, *Testament*, p. 234.

159. Emancipation . . . "bootstraps": "The Meaning of Hope," December 10, 1967, MLK, Atlanta.

159. "concentration camp": Mass-meeting address, February 16, 1968, MLK, Atlanta.

160. "and hated him": King apparently came to an understanding of that hatred in the Chicago campaign. He told David Halberstam that Chicago convinced him that most whites didn't want integration. "Up to now he had thought he was reaching the best in white America" (Oates, p. 404).

160. "sick with racism still": "The Meaning of Hope," December 10, 1967, MLK, Atlanta, audiotape and transcript.

160. "which will you choose?" Address, September 21, 1966, MLK, Atlanta.

160. "*seen* a demonstration": Reported in Oates, pp. 40-41.

161. "guerrilla warfare": "Showdown for Nonviolence," *Testament*, p. 69.

161. on Vietnam: "Why I Am Opposed to the War in Vietnam," (sermon preached at Ebenezer) April 30, 1967, MLK, Atlanta. Cf. *Testament*, p. 236. America on nonviolence: See the speech entitled "A Time to Break Silence," in *Testament*, pp. 231-44. Garrow describes the reaction to the speech in *Bearing*, pp. 552-55. Three weeks later King complained to Ebenezer that the nation "applauded" nonviolence toward "Bull" Connor and Jim Clark, but the media "curse and damn you" when you say, "Be nonviolent toward little brown Vietnamese children." ("Why I Am Opposed to the War in Vietnam," April 30, 1967, MLK, Atlanta).

161. from reform to revolution: David J. Garrow, *The FBI and Martin Luther King, Jr.: From "Solo" to Memphis* (New York: Norton, 1981), traces King's deepening radicalism (pp. 214-15).

161. "tired of race": "Who Is My Neighbor?" February 18, 1968, MLK, Atlanta.

161. "two hostile societies": "Showdown for Nonviolence," *Testament*, p. 64. He is alluding to the report of the President's National Advisory Commission on Civil Disorders. Both the report and King's comments have been borne out by the research of Andrew Hacker, *Two Nations: Black and White, Separate, Hostile, Unequal* (New York: Macmillan, 1992).

161. redistribution of wealth: See "New Wine in Old Bottles," January 22,

1966, MLK, Atlanta. On the same theme, see "Is the Universe Friendly?" December 12, 1965, MLK, Atlanta. In a 1968 interview with David Lewis, Pius Barbour insisted that King "was economically a Marxist. . . . He thought the capitalist system was predicated on exploitation and prejudice, poverty, and that we wouldn't solve these problems until we got a new social order" (quoted in Garrow, *The FBI and Martin Luther King, Jr.*, p. 304, n).

161–62. King's view of the Watts riots: "A Christian Movement in a Revolutionary Age," September 28, 1965, given in Europe, MLK, Atlanta. For similar views on the meaning of the riots in America, see the sermons "The Meaning of Hope," December 10, 1967, and "Great . . . But," July 2, 1967, MLK, Atlanta. Cf. "Showdown for Nonviolence," *Testament*, p. 68.

162. "selling a little dope": "Judging Others," June 4, 1967, MLK, Atlanta.

162. sad stories: In "Judging Others" he tells a pointless story about a chef who didn't deserve criticism for his bad food because his wife had died (June 4, 1967, MLK, Atlanta). In "Interruptions" he tells of a boy who commits suicide by jumping off a bridge (January 21, 1968, MLK, Atlanta); in "The Meaning of Hope" he has another suicide story—about a woman who kills herself when her welfare check does not arrive, only for it to arrive the next day (March 1968, Archive audiotape, Central Methodist Church, Detroit).

Chapter 7: The Masks of Character

Page

163. Mehta: quoted in Oates, p. 282.

163. Halberstam: "When 'Civil Rights' and 'Peace' Join Forces," in Lincoln, *Martin Luther King, Jr.*, p. 194.

163. "instinct for symbolic action": Bennett, p. 187.

163. King as actor: From one of King's handwritten notes at the beginning of the Selma campaign: "Also please don't be too soft. We have the offensive. It was a mistake not to march today. In a crisis we must have a sense of drama" (note to an aide, MLK, Boston).

163–64. "front": Erving Goffman, *The Presentation of Self in Everyday Life* (Garden City, N.Y.: Doubleday Anchor, 1959), p. 22.

163–64. Aristotle on rhetoric: *Rhetoric*, II, 1. The basis of a speaker's "ethical appeal" is sound sense (*phronesis*), high moral character (*arete*), and benevolence (*eunoia*). Cf. Corbett, p. 93.

164. "*look* right": *Rhetoric*, I, 2; II, 1, italics added.

164. "not their *cause*": Goffman, pp. 252–53.

165. "that makes me *me*?": Cf. "The Dimensions of a Complete Life," *Measure*, pp. 51–53.

166. "the way of history": "Sermon on Gandhi," March 22, 1959, preached at Dexter, MLK, Atlanta.

166. on Nkrumah: "The Birth of a New Nation," April 1957, preached at Dexter, MLK, Atlanta.

166-67. "Unplumbed passion for justice": *Stride*, p. 59.

167. "majestic group demonstration of nonviolence": *Stride*, p. 138.

167. "rewrite": Garrow, *Bearing*, p. 105. For a chronicle of the contributions of others in Montgomery, including those of the Women's Political Council, see Jo Ann Robinson, *The Montgomery Bus Boycott and the Women Who Started It: The Memoir of Jo Ann Gibson Robinson*, ed. David J. Garrow (Knoxville: University of Tennessee Press, 1987).

167. "hard feelings" in Montgomery: Garrow, *Bearing*, p. 90. Intr. James Orange, Moorland-Spingarn.

167. "away from my children": "Standing by the Best in an Evil Time," August 6, 1967, MLK, Atlanta.

167. "first Negro here": "Ingratitude," June 18, 1967, MLK, Atlanta.

167-68. "you didn't realize was there": "Interruptions," January 21, 1968, MLK, Atlanta.

168. "Habits are easy to develop": "The God of the Lost," September 18, 1966, MLK, Atlanta.

168. "tragic habit . . . gripped you?": "Is the Universe Friendly?" December 12, 1965, MLK, Atlanta.

168. "old habit was still there": "Answer to a Perplexing Question," March 3, 1963, MLK, Atlanta.

169. sexual habits: Garrow, *Bearing*, pp. 374-76; 587; 638-39, n. Cf. Garrow, *The FBI and Martin Luther King, Jr.*, p. 116; Ralph David Abernathy, *And the Walls Came Tumbling Down* (New York: Harper & Row, 1989), pp. 434-36; 471-72. Intrs. Wyatt Tee Walker, Bernard Lee.

169. containment efforts: Garrow, *Bearing*, pp. 374-75.

169. "stymied his associates": Intr. William Rutherford, Moorland-Spingarn.

169. awe and shame: Goffman, p. 70.

169. "casual sex": Garrow, *Bearing*, p. 375; Intr. Rutherford, Moorland-Spingarn.

169. "lesser sins of the flesh": "Pride versus Humility," October 9, 1966, MLK, Atlanta.

169. "want to do it themselves": "Judging Others," June 4, 1967, MLK, Atlanta.

170. "you had the desire": "Unfulfilled Dreams," March 3, 1968, MLK, Atlanta.

170. "Grace has no receptacle": Luther, "Answer to . . . Emser," *Works of Martin Luther*, vol. 3 (Philadelphia: Muhlenberg Press, 1930), p. 360.

170. spiritual malaise: Intr. William Rutherford, Moorland-Spingarn; Oates, pp. 438-42. See Garrow's assessment, including a segment of his transcribed interview with Stanley Levison in *The FBI and Martin Luther King, Jr.*, pp. 216-19.

170. heckled in Watts: Garrow, *Bearing*, p. 439.

170-71. central and peripheral prophet: Robert R. Wilson, *Prophecy and Society in Ancient Israel* (Philadelphia: Fortress Press, 1980), pp. 42-48, 52, 72-74, 200, 215-220, 242-48, 293.

171. Against LBJ: One of King's many vitriolic attacks on Johnson is recorded in Oates, p. 429.

171. "out there alone": Intr. Bernard Lee.

171. "Martin was finished": Lewis, *King*, 363; cf. 383.

171. one Big Win: Downing, p. 272; Oates, p. 439; Garrow, *Bearing*, p. 582.

171. fifty plots: Oates, p. 439. See COINTELPRO files, U.S. Library of Congress.

171. "kill you some day": Correspondence, MLK, Boston.

171. "threat of death": "Nonconformist–J. Bond," January 16, 1966, MLK, Atlanta.

171. "it never deterred him": Intr. John Gibson, Moorland-Spingarn.

172. *And I know He watches me: Songs of Zion*, #33.

172. "fear of death, really": "Mastering Our Fears," September 10, 1967, MLK, Atlanta.

172. "wake up crying sometime": "Interruptions," January 21, 1968, MLK, Atlanta. See "How to Deal with Grief and Disappointment," May 23, 1965, MLK, Howard, audiotape, and MLK, Atlanta.

172. "not fearing any man": Documentary film, *King, a Filmed Record: Montgomery to Memphis*, Texture Films, 1970.

173. "truest self": See Goffman, p. 19.

173. "Moses of the twentieth century": See Donald H. Smith, "Martin Luther King, Jr.: Rhetorician of Revolt" (Ph.D. diss., University of Wisconsin, 1964), p. 251.

173. "qualifies" as Moses: Mays, "Eulogy," in *Disturbed About Man*, p. 13.

173. "even as Moses": Intr. Fred Shuttlesworth. Moorland-Spingarn.

173. "twentieth-century prophet": Martin Luther King., Sr., quoted in Halberstam, in Lincoln, *Martin Luther King*, p. 203.

173. "prophet of our time": Intr. C. T. Vivian.

173. "plumb line": Intr. T. M. Alexander, in Howell Raines, *My Soul Is Rested: Movement Days in the Deep South Remembered* (New York: G. P. Putnam's Sons, 1977), p. 63.

173. "Moses to the black race": Intr. Minnie Showers, May 19, 1987.

173–74. Moses and Jesus: Spencer writes, "Mary Magdalene need not murmur because Christ-Moses, even in his crucifixion, could not be subjugated by Pharaoh's army; for he victoriously passed through the womb of creation/Red Sea/gates of hell/eye of the needle, leaving Pharaoh swallowed up in the fiery abyss" (*Protest and Praise: Sacred Music of Black Religion* (Minneapolis: Fortress Press, 1990), p. 31. See Genovese, pp. 252–54; cf. p. 213. The earliest association of Jesus and Moses occurs in the Gospel of Matthew where Jesus is portrayed as Israel's new lawgiver and teacher.

174. another Moses: See Wilson, pp. 157–59.

174. "black Moses": Asante, *Afrocentric*, p. 110. Cf. Sarah Bradford, *Harriet Tubman: The Moses of Her People* (New York: Citadel Press, 1961); Edmund David Cronon, *Black Moses: The Story of Marcus Garvey and the Universal Negro Improvement Association* (Madison: University of Wisconsin Press, 1968). On black messianism, see Lewis V. Baldwin, pp. 252–68. Michael Walzer traces the political lineage of Moses through American history (without mentioning King) in *Exodus and Revolution* (New York: Basic Books, 1985).

174. "to set his chillun free": From *Joggin' Erlong* quoted in John Brown Childs, *The Political Black Minister: A Study in Afro-American Politics and Religion* (Boston: G. K. Hall, 1980), p. 23.

174. "courageously for righteousness": *Stride*, p. 210.

174. King on the prophets: "Letter" in *Testament*, p. 290.

174. importance of the "call": See Drake and Cayton, 2:630.

174-76. pattern of conversion: Sobel, pp. 108-16; cf. Johnson, *God Struck Me Dead*. King published an account of his call in *Stride*, pp. 134-35 and frequently alluded to it in his sermons.

175. "never again be confounded": Sobel, pp. 108-9.

175. "they seek my life": 1 Kings 19:10.

175. "like lightning": Isabella (Sojourner Truth) reported her conversion in the same terms. See Theophus H. Smith, p. 205.

175. "God struck me": Sobel, p. 111. Cf. Johnson, ed., pp. x-xi, 76, 112-13.

175-76. "Never Alone": *New National Baptist Hymnal* (Nashville: National Baptist Publishing Board, 1977), #127.

176. at Mount Pisgah: "Thou Fool," August 27, 1967, preached at Mount Pisgah Missionary Baptist Church, Chicago, MLK, Atlanta. The earliest account occurs in *Stride*, pp. 134-35. An abbreviated version appears in "Our God Is Able" in *Strength*, p. 113.

176. "free at last": Sobel, p. 116. "believing in him": "Thou Fool," August 27, 1967, MLK, Atlanta.

176-77. "imaginative picture": See Walter Brueggemann, *The Prophetic Imagination* (Philadelphia: Fortress Press, 1978) esp. chap. 1, "The Alternative Community of Moses."

177. "whole world . . . a slum": Abraham J. Heschel, *The Prophets,* 2 vols. (New York: Harper & Row, 1962), 1:3.

177. "maladjusted": Heschel, 2:188.

177. "your long prayers": "A Knock at Midnight," June 25, 1967, preached at All Saints Community Church, Los Angeles, MLK, Atlanta.

177. on prophets as "pathological": Gerhard von Rad, *Old Testament Theology*, vol. 2, *The Theology of Israel's Prophetic Traditions*, trans. D. M. G. Stalker (New York: Harper & Row, 1965), p. 8; Wilson, pp. 7-8; Heschel (quoting Gerhard Kittel) 2:175; cf. pp. 186-87.

177. "*must* tell the nation this": "Nonconformist—J. Bond," January 16, 1966, MLK, Atlanta.

177. "sons of former slaves": "I Have a Dream," *Testament*, p. 219.

178. "almost-chosen people": See Paul Johnson, "The Almost-Chosen People: Why America Is Different," in Richard John Neuhaus, general ed., *Unsecular America* (Grand Rapids: William B. Eerdmans, 1986), p. 1.

178. Jefferson's address: quoted in Bellah, *The Broken Covenant*, pp. 24-25.

178. conservative function of a prophet: See W. Sibley Towner, "On Calling People 'Prophets' in 1970," *Interpretation* 24:4 (October 1970): 492-509.

178. "the dream it used to be": Hughes, p. 108.

178-79. Bible and patriotic lyrics: "I Have a Dream" in *Testament*, p. 219. See the analysis in Bellah et al., *Habits of the Heart: Individualism and Commitment in American Life* (New York: Harper & Row, 1985), p. 249.

179. "die to make men free": E.g., "Our God Is Marching On" in *Testament*, p. 230. Cf. Bellah, *The Broken Covenant*, p. 53.

179. *"knowing"*: "I Have a Dream," *Testament*, p. 219, italics added.

179. God and moral law: Heschel, 1:216.

179. *"all* his ways are justice": Deuteronomy 32:4, Heschel, 1:198, italics added.

179. "guardian of the moral order": Heschel, 1:217.

179-80. *"her great ideas"*: "What a Mother Should Tell Her Child," May 12, 1963, MLK, Atlanta, italics added.

180. "fatherhood of God and the brotherhood of man": *Stride*, p. 210.

180. "ethical insights": Address to SCLC Convention, August 15, 1967, quoted in Keele, p. 175.

180. "prophetism into social ethics": See Hauerwas, pp. 31, 37.

180. "sufficient religious act": Rauschenbusch, *Christianity*, p. 91.

180. "transforming the nation": In a 1966 interview King said that he had both a priestly and a prophetic role. The prophet, he said, brings the Judeo-Christian faith to bear on society in a concrete and vigorous way. (Warren, "A Rhetorical Study of the Preaching of Doctor Martin Luther King, Jr.," p. 167). Rauschenbusch also understood prophecy as having a social function, but he was more concerned than King to reform the *church*'s message and practice of Christianity. King more directly attacked what he called "corporate structures."

181. "terrible ambivalence": Mass-meeting address, February 16, 1968, Maggie Street Baptist Church, Montgomery, MLK, Atlanta.

181. "judgment of God": Ibid.

181. "Repent": Quoted in Garrow, *Bearing*, p. 593.

181. "our authority": "Is the Universe Friendly?" December 12, 1965, MLK, Atlanta.

181. *"God told me to tell you"*: "Why I am Opposed to the War in Vietnam," April 30, 1967, MLK, Atlanta.

181. "who can but prophesy?": Ibid.

182. "hell": Mass-meeting speech, March 9, 1968, Greenwood, Mississippi, MLK, Atlanta, audiotape.

182. "treating his children right": "I See the Promised Land," *Testament*, p. 283.

182. "his calling": E.g., MLK, Massey Lectures, Reigner Recording Library, Union Theological Seminary, Richmond, Virginia, audiotape.

182. "preacher of the Gospel": "To Serve the Present Age," June 25, 1967, quoted in Keele, p. 180.

182. "my people": "Our God Is Marching On!" *Testament*, p. 229, Cf. Isaiah 10:24, 40:1.

182. prophet's grief: See "How to Deal with Grief and Disappointment," May 23, 1965, MLK, Howard, audiotape, and MLK, Atlanta.

182. "tears of love": "*Playboy* Interview," *Testament*, p. 345.

182. "Today": E.g., "Letter," *Testament*, p. 296. Cf. Exodus 32:29; Deuteronomy 15:15; Zechariah 9:12; Hebrews 3:7, 15.

182-83. "born again": "Where Do We Go from Here?" *Testament*, pp. 250-51.

183. parabolic actions: von Rad, 2:95-97. Cf. Isaiah 20:3; Ezekiel 24:15-24.

183. "wilderness of state troopers": Prayer, March 9, 1965, MLK, Atlanta.

184. "firebombed Negro church": Branch, p. 630. Intr. James Bevel.

184. convert the jailer: Acts 16:25-40.

184. "eyes on the prize": Quoted with slightly altered punctuation from Watters, p. 141. The song is a reworking of the spiritual *Paul and Silas*. Mark Miles Fisher interprets its words to signify the singer's desire to return to West Africa:

> Paul and Silas, *bound* in jail,
> Christians pray *both* night and day,
> *And* I hope dat trump *might* blow me home
> To my new Jerusalem.

(*Negro Slave Songs in the United States* [Ithaca: Cornell University Press, 1953, p. 116]).

184-85. "mortified": Intr. J. T. Porter. Cf. Acts 14. An observer of the Holt Street meeting remembers the effect King had on the audience: "When we had the first meeting, the church was so full, there were so many people. It was like a revival starting. That's what it was like. And Reverend King prayed so that night, I'm telling you the goddam truth, you had to hold people to keep them from gettin' to him" (quoted in Henry Hampton and Steve Fayer, *Voices of Freedom: An Oral History of the Civil Rights Movement from the 1950s through the 1980s* (New York: Bantam Books, 1990), p. 24.

185. "all things to all people": 1 Corinthians 9:22.

185. "highest authorities in the government": See Acts 24:22-27. In confronting the authorities King also assumed the role of his namesake, Martin Luther. In Chicago he tacked the Movements's demands on the door of City Hall (Oates, pp. 392-93).

185. "Macedonian call": "Letter," *Testament*, p. 290. Cf. the same explanation in a Chicago address of July 26, 1965, MLK, Atlanta.

185. *We Shall Overcome*: Watters, p. 54.

185-86. Paul's and King's sufferings: 2 Corinthians 11:21-29, and "The Three Dimensions of a Complete Life," April 9, 1967, MLK, Atlanta.

186. death in us but life in you: 2 Corinthians 4:12.

186. "My gospel": Romans 2:16; 2 Timothy 2:8.

186. on Hitler's Germany and Vietnam: See "But If Not . . . ," November 5, 1967, MLK, Atlanta, and "What Are Your New Year's Resolutions?" January 7, 1968, MLK, Atlanta.

186. divine rather than human perspective: 1 Corinthians 5:16.

187. "small thing" to be judged: 1 Corinthians 4:3.

187. "in him who strengthens me": Philippians 4:12-13.

187. "*suffer* courageously": *Stride*, p. 210, italics added.

187. "vocation of agony": "Why I Am Opposed to the War in Vietnam," April 30, 1967, MLK, Atlanta.

187. "nailed to the cross for you": Reddick, p. 145.

187. Come Thou Almighty King: Donald Smith, pp. 254-55.

187. "blood and body-of-Christ": quoted in Reddick, p. 151. Abernathy once introduced King to a crowd as "conceived by God" (Oates, p. 353).

187-88. "sainthood in man's church": Claude Thompson quoted in Reddick, pp. 165-69.

188. King's prayer: *Stride*, pp. 177-78.

188. Sermon on the Mount: Reddick, p. 15. For its influence on his opposition to the Vietnam War, see the address of April 4, 1967, at Riverside Church in New York, MLK, Atlanta, originally published in *Ramparts* and republished as "A Time to Break Silence," *Testament*, pp. 234, 242-43. "Love your enemy": Mass-meeting speech, June 21, 1963, Gadsden, Alabama, MLK, Atlanta.

188. King on seminary training: Intr. Ralph David Abernathy.

188. unmerited suffering: See Heschel, 1:149-51.

189. "If it means dying for them": "Good Samaritan," August 28, 1966, MLK, Atlanta.

189. "I'll die for them": "Guidelines for a Constructive Church," MLK, Atlanta.

189. "the cross . . . you die on": Speech to staff retreat, May 22, 1967, quoted in Garrow, *Bearing*, p. 564. For another expression of his radical new proposals for America, see his last SCLC presidential address, "Where Do We Go from Here?" in *Testament*, pp. 245-52.

189. Nat Turner's question: *The Confessions of Nat Turner* . . . , quoted in Sernett, *Afro-American*, p. 92.

189-90. "spreading false rumors on me": "Standing by the Best in an Evil Time," August 6, 1967, MLK, Atlanta.

180. "fatigue of despair": "Interruptions," January 21, 1968, MLK, Atlanta.

190. "tried to love and serve humanity": "The Drum Major Instinct," *Testament*, pp. 266-67, and audiotape, MLK, Atlanta. Cf. Wallis, p. 40.

190. the congregation cried: Intr. Shirley Showers Barnhart.

190. "three stories about suicide": "The Meaning of Hope," March 1968, Central United Methodist Church, Detroit, audiotape.

190. Levison on King: Garrow, *Bearing*, p. 588; cf. 601-20, passim.

190-91. "darkly brooding element": Intr. Gardner C. Taylor. Cf. *How Shall They Preach* (Elgin, IL: Progressive Baptist Publishing House, 1977), p. 71. On King's "interior torture": Intr. Gardner C. Taylor.

191. on preachers' depression: Taylor, pp. 69-71.

191. "Compulsion . . . to sacrifice": Allison Davis, *Leadership, Love, and Aggression*, quoted in Garrow, *Bearing*, p. 716, n.

191. "moral leader of our nation": documentary film, *Montgomery to Memphis*.

191. "many a white—would like to be": Quoted in Oates, p. 111.

191. he "*was* the Movement": Intr. C. T. Vivian.

191. "I shall *pray* for him": Intr. Bernard Lee.

191. "believe his myth": Halberstam, "The Second Coming of Martin Luther King," retitled "When 'Civil Rights' and 'Peace' Join Forces" in Lincoln, *Martin Luther King, Jr.*, p. 202.

193. "brutal schedule of appearances": For example, in 1963 he traveled 275,000 miles and gave more than 350 speeches. "I have lost freshness and creativity," he complained. "I cannot write speeches each time I talk, and it is a great frustration to have to rehash old stuff again and again" (Oates, pp. 271-72).

193-94. "*not the blood of our white brothers*": Transcript of testimony in *Williams v Wallace*, pp. 107-8, MLK, Boston, italics added. In 1964 speech to AME religious leaders he said, "But even if he [the white oppressor] beats you, you develop a quiet courage, accepting blows without retaliation" ("Remaining Awake Through a Great Revolution," 1964, MLK, Duke audiotape).

194. suffering as redemption: Asante recalls the uses of the myth in African-American history: "Like Jesus Christ, the African race was going through the Valley of the Shadow of Death to rise again at the new dawn, having saved the world through its substantive, creative experience of pain" (*Afrocentric*, p. 104). See Theophus H. Smith, pp. 239-41, on the interpretation of the Ethiopian eunuch reading from the account in Isaiah 53 of the Suffering Servant (Acts 8). For the views of Richard Allen and Daniel Coker on redemptive suffering, see Chapter 1.

194. "unbearable symbol": Lincoln, *Martin Luther King, Jr.*, "Introduction," p. vii. Lincoln is not specifically referring to the religious personae adopted by King.

Chapter 8: In the Mirror of the Bible

Page

197. Marx: "The Eighteenth Brumaire of Louis Bonaparte," *Selected Works* New York: International Publishers, 1951 [1869]), 2:315-16.

197. "movement made Martin": Quoted in Garrow, *Bearing*, p. 625.

197. performance depends on situation: Lloyd Bitzer writes, "Rhetorical works belong to the class of things which obtain their character from the circumstances of the historic context in which they occur" ("The Rhetorical Situation," in Richard L. Johannesen, ed., *Contemporary Theories of Rhetoric: Selected Readings* [New York: Harper & Row, 1971], p. 384).

198. "declare its meaning": Intr. C. T. Vivian, Moorland-Spingarn.

198. "night of captivity": Holt Street address, December 5, 1955, audiotape and transcript, MLK, Atlanta.

198-99. King at Crozer: See Smith and Zepp, pp. 16-18. His class notes from Enslin's course in the Synoptic Gospels indicate that he received the standard historical-critical introduction to the Gospels (MLK, Boston, n.d.).

198. Jesus the enigma: Albert Schweitzer, *The Quest of the Historical Jesus*, trans. W. Montgomery (New York: Macmillan, 1968 [1906]), p. 399. Jesus an apocalyptic figure: Schweitzer, pp. 223-69.

198. Moses a "code-word": See Branch, p. 72.

199. King's essays on the Bible: "What Experiences of Christians Living in the Early Christian Century Led to the Christian Doctrines of the Divine Sonship of Jesus, the Virgin Birth, and the Bodily Resurrection," *Papers*, 1:225-30. On "unscientific views in the Bible": "How to Use the Bible in Modern Theological Construction, ibid., pp. 251-52. "Objective and disinterested": Ibid., p. 253.

199. on progressive revelation: "How to Use the Bible in Modern Theological Construction," ibid., pp. 254-56. Cf. Introd., pp. 49-50.

200. lessons on moral behavior: For a good summary, see Hans W. Frei, *The Eclipse of Biblical Narrative: A Study in Eighteenth and Nineteenth Century Hermeneutics* (New Haven: Yale University Press, 1974), pp. 1-16.

200-201. on the African-American world of the Bible: Vincent Wimbush, "The Bible and African Americans: An Outline of an Interpretive History," in Cain Hope Felder, ed., *Stony the Road We Trod: African American Biblical Interpretation* (Minneapolis: Fortress Press, 1991), p. 83, and William C. Turner, "The Musicality of Black Preaching," unpublished manuscript version, pp. 23-24.

201. "figural" interpretation: See Frei, especially the chapter, "Precritical Interpretation of Biblical Narrative," pp. 17-50. In his discussion of figuralism Frei draws heavily upon the groundbreaking work of Erich Auerbach. Auerbach's essay, "Figura," in *Scenes from the Drama of European Literature*, trans. Ralph Manheim, vol. 9 of *Theory and History of European Literature* (Minneapolis: University of Minnesota Press, 1955 [1944]), is the most thorough discussion available (pp. 11-76). Cf. Theophus H. Smith, pp. 3-5, 173-208, passim.

201. "language of comparison": Trulear and Richey, 7.

201. "they interpreted literally": Quoted in Raboteau, pp. 208-9; also cited in Michael G. Cartwright, *The Practice and Performance of Scripture: Toward an Ecclesial Hermeneutic for Christian Ethics* (forthcoming), p. 351.

201. "enrolls all the instruments of culture": Paul Ricoeur, *The Conflict of Interpretations* (Evanston: Northwestern University Press, 1974), p. 385.

201. "everyday problems of a black man": Mitchell, *Black Preaching*, p. 113, with minor modification.

202. "in the wilderness of America": Gardner Taylor, sermon, 1978, Duke Divinity School Media Center, audiotape.

202. the difference between typology and allegory: See Frei, pp. 29-30. Also Henri de Lubac, *The Sources of Revelation*, trans. Luke O'Neill (New York: Herder & Herder, 1968): "Everything told in Scripture actually occurred in visible reality, but the account which stems from that reality does not have its end in itself: it must all yet be accomplished and it is in fact accomplished daily in *us*, through the mystery of spiritual understanding" (p. 88).

202-4. King's use of figuralism: "The Birth of a New Nation," April 1957, preached at Dexter Avenue, MLK, Atlanta; the following references are taken from that sermon. On the black reading of Exodus: Theophus H. Smith, pp. 1, 231-38. The "law of history": Cf. "Sermon on Gandhi," March 22, 1959, preached at Dexter Avenue, MLK, Atlanta.

204. "in the contemporary world": "A Knock at Midnight," August 27, 1967, preached at All Saints Community Church, Los Angeles, Keele, Appendix 2, p. 305. Compare a version of the same sermon preached at Mount Zion Church in Rocky Mount, N.C., n.d. (MLK, Duke, audiotape).

204. Good Samaritan: "Good Samaritan," August 28, 1966, MLK, Atlanta. sanitation workers: "I See the Promised Land," *Testament*, pp. 284-85.

204. Shadrach et al.: "But If Not . . . ," November 5, 1967, MLK, Atlanta. In a March 6, 1988, sermon, King's successor at Dexter Avenue Church, G. Murray Branch, preached a sermon on 2 Kings 6:16: "Fear not, for they that be with us are more than they that be with them." The sermon exemplifies the figural tradition in African-American preaching. Taking Elisha's assurance that the angelic chariots far outnumber the Syrian king's, Branch applied that divine assurance to a contemporary situation, that of Jesse Jackson's presidential candidacy. Though our candidate appears to be outnumbered, he said, an even greater array of voters and moral forces are with us. We have only to open our eyes to the multitudes around us.

204-5. Lazarus and Dives: "Lazarus and Dives," March 10, 1963, MLK, Atlanta.

205. against black materialism: "Discerning the Signs of History," November 15, 1964, MLK, Atlanta. Benjamin Mays reproduces a sermon containing the same sentiments in *The Negro's Church*, p. 68.

205-6. King and allegory: For an example of a strictly allegorical and symbolic treatment of a text, see "The Death of Evil upon the Seashore," *Strength*, pp. 76-85. "Egypt" symbolizes evil, the "Israelites" symbolize goodness, and so on. The sermon employs allegory but does not allude to the church's traditional *typological* reading of the Exodus with reference to Christ's and the Christian's resurrection victory.

206. Bible as almanac: See Genovese, p. 242.

206. "good preaching material": Mitchell, *Black Preaching*, p. 113.

206. medieval interpretation: See de Lubac, pp. 12-13.

206. on the laws of history: "Discerning the Signs of History," November 15, 1964, MLK, Atlanta.

206-7. Good Samaritan interpretation: "The Good Samaritan," August 28, 1966, MLK, Atlanta. The third-century interpreter Origen, designated the following values: the man who fell among thieves = Adam; Jerusalem = heaven; Jericho = the world; robbers = the devil; priest = the law; the Levite = the prophets; the Samaritan = Christ; the beast = Christ's body; the inn = the church; two pence = the Father and the Son; the Samaritan's promise to return = Christ's Second Coming. One hundred fifty years later, Augustine added that the innkeeper = St. Paul (see Archibald M. Hunter, *Interpreting the Parables* (Philadel-phia: Westminster Press, 1960), pp. 25–26.

207–8. "keep the bread fresh": "A Knock at Midnight," 1967 or 1968, Mount Zion Church, Rocky Mount, N.C.: MLK, Duke, audiotape.

208. "religious profundity": Frei, p. 106.

208. "*prior* to God's self-revelation": See "The Death of Evil upon the Sea-shore," *Strength*, p. 77.

208–9. King's psychologism: "failure complex": "Interruptions," January 21, 1968; on Freud and Adler: "The Drum Major Instinct," *Testament*, p. 260; "phobia-phobia": "Mastering Our Fears," September 10, 1967; "id and superego": "Who Are We?" February 5, 1966; importance of self-esteem: "Training Your Child in Love," May 8, 1966; on wife abuse: "Interruptions"; against repression: "Guide-lines for a Constructive Church," June 5, 1966; cf. "Judging Others," June 4, 1967; dangers of daydreaming: "Interruptions" (all sermon references, MLK, Atlanta). On Rabbi Liebman, See chapter 5.

209. "about fifty years out of date": A paraphrase of Northrop Frye's remark "modern usually means about a hundred years out of date" in *The Great Code: The Bible and Literature* (New York: Harcourt Brace Jovanovich, 1982), p. 41.

210. on the interpreter and the church: David H. Kelsey reminds us that the biblical interpreter always works with a "logically prior imaginative judgment about how best to construe the mode of God's presence" (*The Uses of Scripture in Recent Theology* [Philadelphia: Fortress Press, 1975], p. 166).

210. "Christ in *this* world": "Revolution and Redemption," August 16, 1964, delivered at the Baptist Convocation in Amsterdam, MLK, Atlanta.

211. the Puritans as antitypes: Stout, p. 173, italics added. In his sermon preached on board ship in 1630, John Winthrop, the first leader of the Massa-chusetts Bay Colony, compared the ocean crossing to the crossing of the Red Sea and the Jordan River. See Bellah, *The Broken Covenant*, pp. 13–15.

211. how to "take land what was already took?": quoted in Moses, p. 76, spelling altered.

211. becoming a biblical antitype: Werner Sollors writes in *Beyond Ethnicity: Consent and Descent in American Culture* (1986), "In post-Puritan America, white and black, business and labor, Jew and Gentile, could follow typological pat-terns and *become biblical antitypes*" (quoted in Theophus H. Smith, p. 244).

211. on the black use of Exodus symbolism in America: See Wilmore, who also alludes to other biblical motifs employed in the service of liberation, includ-ing Gabriel Prosser's use of the story of Samson, Denmark Vesey's allusions to Joshua, David Walker's reliance on the prophets as well as the Exodus, and Nat Turner's identification with the apocalyptic Son of Man (pp. 37–38, 54–58, 65). Cf. Cone, *God of the Oppressed*, pp. 55–60. The Birmingham Police report of an April 5, 1963, mass meeting at Thurgood Baptist Church on 7th Avenue opens with a quotation from the Reverend Ed. Gardner: "The Lord said to Moses, 'Go down to Egypt and set my people free.' Everyone in the United States is free, Chinese, Japanese, everyone but the Negro. The Negro has been in this country

for 300 years and is still not free, but a great day is coming" (Eugene "Bull" Connor Papers, Birmingham Public Library Archives, Birmingham, Alabama, hereafter cited as Connor Collection).

211. "crimes of the blackest hue": Sobel, p. 159. On Vesey's interpretation of the Exodus, see Harding, *There Is a River*, p. 69.

211. King's infrequent allusions to Exodus: Smylie, p. 81.

211-12. King's frequent use of Exodus: See "A Christian . . . ," December 12, 1965, MLK, Atlanta: "Answer to . . . ," March 3, 1963, MLK, Atlanta; "Facing . . . ," March 1962, Rocky Mount N.C., MLK, Duke, audiotape; "Desirability . . . ," January 13, 1958, presented in a synagogue in Chicago, MLK, Atlanta; "I See the Promised Land," *Testament*, p. 286; and too many other allusions to count. For Smylie's comments on "The Death . . . ," see pp. 81-85. King's allusions to the Exodus in *Where Do We Go from Here?* and *Why We Can't Wait* indicate that even King's ghostwriters understood the significance of the Exodus for the Civil Rights Movement!

212. "I may not get there": Compare the conclusions of "The Birth of a New Nation," April 1957, MLK, Atlanta, and "I See . . . ," *Testament*, p. 286.

212. "enrich and illumine one another": Raboteau writes, "Categorizing sacred and secular elements is of limited usefulness in discussing the spirituals because the slaves, following African and biblical traditions, believed that the supernatural continually impinged on the natural, that divine action constantly took place within the lives of men, in the past, present and future" (p. 250).

212. not to *illustrate* God but to *create* God: See Kelsey for a discussion of Karl Barth's use of story in such a way that it does not merely illustrate a proclamation about God but renders an agent, God, who is known only through the development of a narrative (pp. 34-41). Nicholas Lash has argued that the two poles of interpretation are not the text and a written interpretation of it but the text and the community's enactment or performance of it in the common life of the community. See his essay "Performing the Scriptures" in *Theology on the Way to Emmaus* (London: SCM Press, 1986), pp. 40-42. King's sermon, "The Meaning of Hope," December 10, 1967, MLK, Atlanta, contains a good example of the process of *rendering* of the agent of deliverance.

213. signifying: Cartwright makes much of King's signifying reading of the Exodus in his unpublished manuscript, *The Practice and Performance of Scripture*.

213. "hands are clean": "The Birth of a New Nation," April 1957, preached at Dexter, MLK, Atlanta.

213. "it is the word 'love'": "Levels of Love," May 21, 1967, MLK, Atlanta. See Mays, *Disturbed about Man*, p. 54.

214. make it plain to all: Even passages that do not mention love King reinterpreted to yield the key principle of his theology. A Mother's Day sermon on Proverbs 22:6, "Train up a child in the way that he should go and when he is old he will not depart from it," is entitled "Training Your Child in Love" (May 9, 1966, MLK, Atlanta).

214. "God our Father . . . ": See Wimbush in Felder, pp. 90-92. In one of his student outlines King wrote, "The Fatherhood of God and the Brotherhood of man is the starting point of the Christian ethic" (*Papers* 1:281).

214. love and Gandhi: See Bennett, p. 73.

215. role of love: Mitchell, *The Recovery of Preaching*, p. 146.

215-16. on black apocalypticism: For example, see references to the Apocalypse in Walker, "Our Wretchedness . . . ," in Sernett, *Afro-American*, p. 192, and "The Confessions of Nat Turner" in Sernett, *Afro-American*, p. 92.

216. Ethiopia: Wilmore, pp. 116-22, especially p. 121. See also Moses, p. 196, Theophus H. Smith, p. 241, and M. Cartwright. King might have known the sermon on the Ethiopian theme reproduced by Benjamin Mays in *The Negro's Church*, p. 69.

216. "Curse of Ham": King's one fleeting reference is in "Paul's Letter to American Christians," November 4, 1956, Dexter, reprinted in Evans and Alexander, pp. 251-52.

216. other popular Bible stories: See Theophus H. Smith, pp. 173-86.

218. 1 Corinthians 13 and the Movement: See King, *Stride*, p. 161, and Robert S. Graetz, *Montgomery: A White Preacher's Memoir* (Minneapolis: Fortress Press, 1991), p. 106.

218. "interpretive community": Stanley Fish, *Is There a Text in This Class? The Authority of Interpretive Communities* (Cambridge: Harvard University Press, 1980), pp. 14-16.

218. antiphonal effect: See Spencer, *Sacred Symphony*, p. 8. "humming a tune": Rosenberg, p. 61.

219. "match to gasoline": Pipes, p. 42.

219. "word . . . prevailed": Acts 19:20.

Chapter 9: The Ebenezer Gospel

Page

222. "'love your enemies'": "Loving Your Enemies," *Strength*, p. 47.

222. "desire for attention": "The Drum Major Instinct," February 4, 1968, MLK, Atlanta. Unless otherwise noted, all sermons in the remainder of the chapter were preached at Ebenezer Baptist Church in Atlanta and are housed in the Archives of the Martin Luther King, Jr. Center for Nonviolent Social Change, Atlanta.

222. "lives are interrupted": "Interruptions," January 21, 1968.

222. "unfinishable": "Unfulfilled Dreams," March 3, 1968.

222. "hopes fulfilled": "How to Deal with Grief and Disappointment," May 23, 1965. Cf. MLK, Howard, audiotape.

222. "destroying ourselves": "Discerning the Signs of History," November 15, 1964.

222. "'thank you'": "Ingratitude," June 18, 1967.

222. white Southerner's problem: "Pride versus Humility," October 6, 1966.

223. "instruments of robbery": "Good Samaritan," August 28, 1966.

223. "no greater robbery": "Good Samaritan," August 28, 1966.

223. "this brother's problem": "How to Deal with Grief and Disappointment," May 23, 1965.

223. "stench of back waters": "Ingratitude," June 18, 1967.

224. "habit structure": "New Wine in Old Bottles," January 2, 1966. Cf. "Is the Universe Friendly?" December 12, 1965.

224. Image of God: "Why I Am Opposed to the War in Vietnam," April 30, 1967. Cf. "Desirability of Being Maladjusted," an address to a Jewish congregation in Chicago, January 13, 1958, MLK, Atlanta.

224. "child of mine has worth": "The God of the Lost," September 18, 1966.

224. "not made for that": "Who Are We?" February 5, 1966. Cf. "The Man Who Was a Fool, " January 29, 1961, Chicago Sunday Evening Club, MLK, Atlanta.

224. "earth is the Lord's": "A Knock at Midnight," June 25, 1967, All Saints Community Church, Los Angeles, Keele, Appendix 2. Cf. "Three Dimensions of a Complete Life," April 9, 1967. This and countless expressions like it do not appear in the published sermons.

224. idolatry: "Why I Am Opposed to the War in Vietnam," April 30, 1967.

225. "God the original segregationist": Carey Daniel (sermon), in DeWitte Holland, ed., *Sermons in American History* (Nashville: Abingdon Press, 1971), pp. 513-22.

225. "'Go forward!'": "Answer to a Perplexing Question," March 3, 1963.

225. "searching for the lost": "The God of the Lost," September 18, 1966.

225. God is love: "Is the Universe Friendly?" December 12, 1965.

225. "because Jesus Christ is merciful": "We Would See Jesus," May 5, 1967.

226. Lily of the Valley: See Chapter 5.

226. "How I got over": "Ingratitude," June 18, 1967.

226. on eternal life: "Shattered Dreams," *Strength*, p. 95.

226. "He is able": King, Sr., audiotape, n.d., Ebenezer Baptist Church. Cf. "What a Mother Should Tell Her Child," May 12, 1963.

227. "rescue in the Savior": "A Knock at Midnight," March 14, 1965, Chicago, MLK, Atlanta.

227. "Can't He make you *pretty*": Pipes, p. 49, italics added.

227. "into a woman of honor": "We Would See Jesus," May 5, 1967.

228. "put his hands on me": "Answer to a Perplexing Question," March 3, 1963. Cf. "Who Are We?" February 5, 1966: "Great . . . But," July 2, 1967.

228. "many nameless prodigals": "Answer to a Perplexing Question," March 3, 1963.

228. "way out of no way": "What Are Your New Year's Resolutions?" January 7, 1968, New York. Cf. "Interruptions," January 21, 1968. This is perhaps the most ubiquitous phrase in all black preaching.

228. "entertainment centers": "Guidelines for a Constructive Church," June 5, 1966.

228. "born again": See "Where Do We Go From Here?" *Testament*, p. 249, but also "New Wine in Old Bottles," January 2, 1966.

229. cooperation with God: "Answer to a Perplexing Question," March 3, 1963.

229. "Honey, I'm sorry": "But If Not," November 5, 1967.

229. "America a better nation": "Answer to a Perplexing Question," March 3, 1963. Cf. "Nonconformist–J. Bond," January 16, 1966.

229. "That's all it takes": "We Would See Jesus," May 5, 1967. The transcription mistakenly reads "No matter *how* you are."

230. Nebuchadnezzar and Duke student: "But If Not," November 5, 1967.

230-31. critique of love: James E. Sellers, "Love, Justice, and the Non-Violent Movement," *Theology Today*, 18:4 (January 1962): 426-33.

231. "God's photograph": "Levels of Love," May 21, 1967.

231. "taking up the cross": "Why I Am Opposed to the War in Vietnam," April 30, 1967.

231. "well-adjusted personality": "Mastering Our Fears," September 10, 1967.

231. Good Friday/Easter: "Why I Am Opposed to the War in Vietnam," April 30, 1967. Cf. "Guidelines for a Constructive Church," June 5, 1966.

231. meaning of the cross: "Mastering Our Fears," September 10, 1967.

231. Niebuhr's remark: Fox, p. 103.

231. "lynching me": E.g., "Who Is My Neighbor?" February 18, 1968.

232. integration as "true spiritual affinity": "Who Is My Neighbor?" February 18, 1968. See "Sermon on Gandhi," March 22, 1959, Dexter, MLK, Atlanta, where he describes integration as "a new friendship and reconciliation."

232. "it's right to be just": "But If Not," November 5, 1967.

232. "somebody . . . gonna get hit": King, Sr., conclusion of "What a Mother Should Tell Her Child," May 12, 1963.

233. "crackers" and "niggers": "Nonconformist–J. Bond," January 16, 1966. Cf. "Sermon on Gandhi," March 22, 1959.

233. Blackstone Rangers: "Judging Others," June 4, 1967.

233. "ain't goin' to study war": "Standing by the Best in an Evil Time," August 6, 1967.

233. "I would rather die": "Levels of Love," May 21, 1967.

233. "on the right road": "The God of the Lost," September 18, 1966. Cf. "Unfulfilled Dreams," March 3, 1968. The quotation is attributed to Luther. See the *Weimar Ausgabe*, Vol. 4:364, for an approximation.

234. "New Atlanta": "Who Are We?" February 5, 1966. See Chapter 2. For examples of traditional Negro eschatology, see Drake and Cayton, 2:618-19.

234. "life eternal": "Shattered Dreams," *Strength*, p. 95.

234. "home today with God": "After the Death of Three Children," September 22, 1963, Sixth Avenue Baptist Church, Birmingham.

234. "Beloved Community": See "The Birth of a New Nation," April 1957, Dexter, MLK, Atlanta, and Ansbro's discussion, pp. 190-92, 319 n.

234. qualities in persons: John H. Cartwright, "The Social Eschatology of Martin Luther King, Jr.," in Cartwright, ed., *Essays in Honor of Martin Luther King, Jr.*, (Evanston, Ill.: Garrett Evangelical Theological Seminary, 1971); p. 10.

234. Community to Kingdom: See Ansbro, p. 190, and Smith and Zepp, pp. 128-31.

235. "in the midst of you": "The Death of Evil upon the Seashore," *Strength*, p. 83.

235. "old order": "Training Your Child in Love," May 8, 1966.

235. "Lurleen Wallace": "Three Dimensions of a Complete Life," April 9, 1967.

235. God, the Supreme Court: "After the Death of Three Children," September 22, 1963.

235. "kingdom right here": "Good Samaritan," August 28, 1966. Negroes took the promise of the Kingdom literally, he told the Baptist Assembly in Amsterdam in 1964, "Now they are demanding that the promise of the kingdom be fulfilled here and now" ("Revolution and Redemption," August 16, 1964, MLK, Atlanta). In Chicago he prayed that the Movement "hast taught us anew that thy kingdom shall come; indeed, thou hast taught us to work and pray for its coming" (prayer and speech, July 26, 1965, Chicago, MLK, Atlanta).

235-36. The Acceptable Year: "Guidelines for a Constructive Church," June 5, 1966.

236. "guided by their example": "The Meaning of Hope," December 10, 1967, Dexter anniversary service, MLK, Atlanta. In the sermon King acknowledges the influence of Tillich's sermon "The Right to Hope."

237. ecumenical awakening: King, Jr., *Where?*, p. 9. In 1967 he referred to the Selma campaign as a "shining moment" ("Why I Am Opposed to the War in Vietnam," April 30, 1967). Cf. "The Trumpet of Conscience" in *Testament*, p. 635.

237. "hopes dashed": "Letter," *Testament*, p. 299.

237. black man in the choir: "A Knock at Midnight," June 25, 1967, Keele, Appendix 2.

238. Wallace a Sunday School teacher: "Pride versus Humility," October 9, 1966.

238. on white preachers: "A Knock at Midnight," 1967 or 1968, North Carolina, MLK, Duke, audiotape.

238. "worship of the same God": *Where?*, King, Jr., p. 151.

238. King and the NBC: Garrow, *Bearing*, pp. 165-66; Branch, pp. 500-503.

238. "ashamed they are black": "A Knock at Midnight," 1967 or 1968, North Carolina, MLK, Duke, audiotape. Cf. "Lazarus and Dives," March 10, 1963; "What a Mother Should Tell Her Child," May 12, 1963; "Remember Who You Are," July 7, 1963; "Who Are We?" February 5, 1966; "Guidelines for a Constructive Church," June 5, 1966.

238. Messianic assignment: see Wilmore, "Blackness as Sign and Assignment," Appendix B, in Kelly Miller Smith, *Social Crisis Preaching* (Macon: Mercer

University Press, 1984). "your great opportunity": "New Wine in Old Bottles," January 2, 1966.

239. "colony of heaven": "Letter," *Testament*, p. 300, Cf. "Paul's Letter to American Christians," in Evans and Alexander, p. 250.

239. church's true task: "Guidelines for a Constructive Church," June 5, 1966.

239. *ecclesia*: "Letter," *Testament*, p. 300.

239. church must suffer: "A Knock at Midnight," 1967 or 1968, North Carolina, MLK, Duke, audiotape.

239. church and nonviolence: "Letter," *Testament*, p. 297.

240. "calling for you and for me": "The God of the Lost," September 18, 1966.

240. "Wherever you are": "Answer to a Perplexing Question," March 3, 1963. Cf. "What a Mother Should Tell Her Child," May 12, 1963; "Who Are We?" February 5, 1966.

241. open the doors of the church: See Pipes, p. 30; Mays, *The Negro's Church*, p. 152; and Spencer, *Sacred Symphony*, p. 78, for the same formula.

241. "Is there another?": "What a Mother Should Tell Her Child," May 12, 1963. The transcript erroneously reads "thankful."

241. "Where He Leads Me": "Remember Who You Are," July 7, 1963.

241. "care of things next Sunday": "How to Deal with Grief and Disappointment," May 23, 1965.

241. "Right Quick!": "What a Mother Should Tell Her Child," postservice announcement, May 12, 1963.

Chapter 10: Bearing "The Gospel of Freedom": The Mass Meeting

Page

243. Billy Graham: Branch, pp. 594-95; Garrow, *Bearing*, p. 97.

244. "by 4": Intr. R. D. Nesbett.

244. "victory celebration": Intr. Graetz.

244. "we could change it": Coretta Scott King in Hampton and Fayer, p. 30.

245. "*made* them sermons": Intr. Zelia Evans.

245. "out of your body": quoted in Oates, p. 76.

245. "better than church": Watters, p. 24; cf. p. 50. One of the pastors confessed, "I saw more God in the movement than in twenty years in the church" (p. 51).

245. "dignified life here": quoted in Watters, pp. 165-66.

246. Jackie Robinson: Connor Collection, May 14, 1963.

246. "all over": Watters, p. 191.

246. meeting locations: The meetings of the Alabama Christian Movement

for Human Rights were held at the following churches in and around Birmingham: Saint Luke's, Saint James, Tabernacle Baptist, The New Pilgrim Church, New Hope Baptist, First Baptist of Ensley, Abyssinia Baptist, Greater Seventeenth Street Church of God, Sixth Avenue Baptist, Thurgood Baptist, Sixteenth Street Baptist.

246. "with their flags": Connor Collection, May 22, 1963.

246. displays of patriotism: See Spencer, *Protest and Praise*, p. 91.

247. Negroes "sitting in the windows": Connor Collection, April 10, 1963.

247. "sleet and ice": Ibid., February 13, 1963.

247. "between 1,800 and 2,000": Ibid., May 3, 1963.

247. "gave us the treatment": Ibid., May 9, 1963.

247. "I had a dream tonight": Ibid., April 11, 1963.

247. "his dry jokes": Ibid., May 15, 1963.

247. "someone must have loved him": Ibid., May 17, 1963.

247. "you might be crucified": Ibid., May 14, 1963.

248. "dog bitten for *nothing*" and the tank: "Crisis and a Political Rally in Alabama," May 3, 1963, MLK, Atlanta.

248. "doors of the church are open": Selma meeting, n.d., Birmingham Public Library Archives (hereafter cited as Birmingham).

249. "Beautiful, beautiful Zion": Watters, pp. 208-9.

249-50. freedom songs: See Spencer, *Protest and Praise*, pp. 83-87; Cone, *Speaking the Truth*, p. 140; King, *Why We Can't Wait*, pp. 61-62; Wyatt Tee Walker, *"Somebody's Calling My Name": Black Sacred Music and Social Change* (Valley Forge: Judson Press, 1979), p. 154.

250. "gonna let it shine": Watters, p. 169.

250. "Sunday morning altar call": King, *Why We Can't Wait*, p. 59.

250. Selma's opportunity: Birmingham, January 14, 1965.

250. "too busy swinging": *Malcolm X Speaks*, p. 9.

250. "They were mad": quoted in Spencer, *Protest and Praise*, p. 99.

250. mockery of worship: Watters, p. 391.

251. "Martin was a Christian": Intr. C. T. Vivian.

251. marching: "Our God Is Marching On," *Testament*, p. 230, and audiotape, MLK, Atlanta.

251. "We love ole' Bull": Connor Collection, April 12, 1963. Cf. "As we prepare to leave this morning let us be sure that our hearts are right," "King's Plan for Selma," January 19, 1965, MLK, Atlanta.

251. so many meetings: Connor Collection, May 15, 1963.

251. Young's speech: Ibid., May 7, 1963.

252. "blood of our white brother": Ibid., May 14, 1963.

252. "heart changing business": "The Future of Integration," January 19, 1958, Keele, Appendix 3.

252. "general appeals for sacrifice": "I See the Promised Land," *Testament*, p. 281.

252. "they can't stop it": rally speech, Yazoo, Mississippi, June 21, 1966, MLK, Atlanta.

253. get down and preach: See Oates on the reaction to King's barnstorming, p. 385.

253. "they were ready": Intr. C. T. Vivian.

253. civic address: King's most famous civic address was, of course, "I Have a Dream." Its concept came from the sermons of the Hebrew prophets, and most of its famous passages originated in mass meetings held in Albany, Birmingham, Rocky Mount, N.C., Detroit, and scores of other cities. The Washington version of "I Have a Dream" was a mass-meeting speech writ large.

253. "the same, the same": Intr. C. T. Vivian.

253. "It's a church": Adam Fairclough, *To Redeem the Soul of America: The Southern Christian Leadership Conference and Martin Luther King, Jr.* (Athens: University of Georgia Press, 1987), p. 1. See "The Preachers and the People: The Origins of the SCLC," pp. 11–35.

254. "foremost preacher": Branch, p. 559.

254. "into the streets": Intr. C. T. Vivian.

254. "contain the fires they had set": Wilmore, p. 50.

254. "preachers did us in": Intr. C. T. Vivian.

255. Bevel: Connor Collection, April 18, 1963. Intr. James Bevel.

255. "Ph.D. from Boston": Selma speech, n.d., Birmingham.

255-57. Abernathy's "doohicky": Selma speech, n.d., Birmingham.

257. "meetin's not over yet!": Selma speech, January 1, 1965, Birmingham.

257. King's arrival: Watters, p. 12.

258-60. King's speech at Brown's Chapel, Selma: Surveillance recordings made by Dallas County, Alabama, Sheriff's Department, Birmingham, January 14, 1965, audiotape. All references in the next several pages are to that speech. Author's transcription and poetic arrangement.

261. Bull to a steer: Connor Collection, April 5, 1963.

261. Clark's "posse": Birmingham, February 10, 1965.

261. "get with this drive": Police report of meeting, February 2, 1966, Birmingham, MLK, Atlanta.

261. "now": Connor Collection, police summary, April 3, 1963.

261-62. *eros* and *agape*; "keep moving": Birmingham meeting, audiotape, May 5, 1963, MLK, Atlanta.

262. "Christ died for the segregationist": Selma meeting, n.d., Birmingham.

262. tape of Malcolm's press conference: Birmingham, February 10, 1965.

263-64. King's Chicago speech: "Why I Must March," August 18, 1966, MLK, Atlanta. Author's poetic arrangement.

264-65. Vivian's reflections on the Movement: Intr. C. T. Vivian, Moorland-Spingarn. See also C. T. Vivian, *Black Power and the American Myth* (Philadelphia: Fortress Press, 1970), "The Old Assumptions," pp. 55–126.

265. "guardian of the community": See Duncan, pp. 292; cf. pp. 295–97.

Epilogue

Page

267. "a man from Georgia": "He Had a Dream" (editorial), *New York Times*, April 7, 1968, p. 12 (E).

268. "touched my heart and soul": Reported and tape-recorded by Sarah Kenyon Lischer, 1993.

269. "God could be trusted": Quoted in Spencer, *Protest and Praise*, p. 92.

269. "Can it?": Sermon, "Rich Young Ruler," Barbour Collection, n.d.

Bibliography

Books

Abernathy, Ralph David. *And the Walls Came Tumbling Down.* New York: Harper & Row, 1989.

Allen, Richard. *The Life Experience and Gospel Labors of the Rt. Rev. Richard Allen.* New York: Abingdon Press, 1960 [1833].

Alter, Robert. *The Art of Biblical Narrative.* New York: Basic Books, 1981.

Alter, Robert. *The Art of Biblical Poetry.* New York: Basic Books, 1984.

Andrews, William L., ed. *Sisters of the Spirit: Three Black Women's Autobiographies of the Nineteenth Century.* Bloomington: Indiana University Press, 1986.

Ansbro, John J. *Martin Luther King, Jr.: The Making of a Mind.* Maryknoll, N.Y.: Orbis Books, 1982.

Aristotle. *Rhetoric and Poetics,* trans. W. Rhys. Oxford: Modern Library, 1924.

Asante, Molefi Kete. *The Afrocentric Idea.* Philadelphia: Temple University Press, 1987.

Auerbach, Erich. *Scenes from the Drama of European Literature.* Translated by Ralph Manheim. Vol. 9 of *Theory and History of European Literature.* Minneapolis: University of Minnesota Press, 1955 [1944].

Augustine. *On Christian Doctrine.* Translated by D. W. Robertson, Jr. Indianapolis: Bobbs-Merrill, 1958.

Baker-Fletcher, Garth. *Somebodyness: Martin Luther King, Jr. and the Theory of Dignity.* Minneapolis: Fortress Press, 1993.

Baldwin, Lewis V. *There Is a Balm in Gilead: The Cultural Roots of Martin Luther King, Jr.* Minneapolis: Fortress Press, 1991.

Barth, Karl. *Church Dogmatics.* vol. 1, pt. 1 of *The Doctrine of the Word of God.* Translated by G. W. Bromiley. Edinburgh: T. & T. Clark, 1975 [1936].

Baxter, Batsell Barrett. *The Heart of the Yale Lectures.* New York: Macmillan, 1947.

Bellah, Robert N. *The Broken Covenant: American Civil Religion in Time of Trial.* New York: Seabury Press, 1975.

Bellah, Robert N., et al. *Habits of the Heart: Individualism and Commitment in American Life.* New York: Harper & Row, 1985.

Bennett, Lerone, Jr. *What Manner of Man: A Biography of Martin Luther King, Jr.* Chicago: Johnson, 1968.

Berger, Peter L. *The Sacred Canopy: Elements of a Sociological Theory of Religion.* Garden City, N.Y.: Doubleday, 1967.

Best Sermons. Edited by Joseph Fort Newton. New York: Harcourt Brace, 1926 [and succeeding years to 1964].

Birmingham, Stephen. *Certain People: America's Black Elite.* Boston: Little, Brown, 1977.

Bitzer, Lloyd F., and Edwin Black, eds. *The Prospect of Rhetoric.* Englewood Cliffs, N.J.: Prentice-Hall, 1971.

Blackwood, Andrew W. *The Preparation of Sermons.* New York: Abingdon-Cokesbury Press, 1948.

Boddie, Charles Emerson. *God's "Bad Boys".* Valley Forge: Judson, 1972.

Bormann, Ernest G., ed. *Forerunners of Black Power: The Rhetoric of Abolition.* Englewood Cliffs, N.J.: Prentice-Hall, 1971.

Boulware, Marcus H. *The Oratory of Negro Leaders, 1900–1968.* Westport, Conn.: Negro Universities Press, 1969.

Bradford, Sarah. *Harriet Tubman: The Moses of Her People.* New York: Citadel Press, 1961.

Branch, Taylor. *Parting the Waters: America in the King Years, 1954-63.* New York: Simon & Schuster, 1988.

Brightman, Edgar Sheffield. *Moral Laws.* New York: Abingdon Press, 1933.

Brilioth, Yngve. *A Brief History of Preaching.* Translated by Karl E. Mattson. Philadelphia: Fortress Press, 1965.

Brink, William, and Louis Harris. *The Negro Revolution in America.* New York: Simon & Schuster, 1964.

Brooks, Phillips. *Lectures on Preaching.* Grand Rapids: Baker Book House, 1969 [1877].

Brooks, Phillips. *Selected Sermons.* Edited by W. Scarlett. New York: Dutton, 1949 [n.d.].

Brooks, Phillips. *Sermons Preached in English Churches.* New York: Dutton, 1883.

Brueggemann, Walter. *The Prophetic Imagination.* Philadelphia: Fortress Press, 1978.

Burke, Kenneth. *Counter-Statement.* Berkeley: University of California Press, 1968 [1937].

Burke, Kenneth. *A Grammar of Motives.* Berkeley: University of California Press, 1969 [1945].

Burke, Kenneth. *A Rhetoric of Motives.* Berkeley: University of California Press, 1969 [1950].

Buttrick, George A. *Jesus Came Preaching.* New York: Scribner's, 1931.

Buttrick, George A. *The Parables of Jesus.* Garden City, N.Y.: Doubleday, Doran, 1929.

Buttrick, George A. *Sermons Preached in a University Church.* New York: Abingdon Press, 1959.

Caird, G. B. *The Language and Imagery of the Bible.* Philadelphia: Westminster Press, 1980.

Cartwright, John H., ed. *Essays in Honor of Martin Luther King, Jr.* Evanston, Ill.: Garrett Evangelical Theological Seminary, 1971.

Cauthen, Kenneth. *The Impact of American Religious Liberalism.* New York: Harper & Row, 1962.

Childs, John Brown. *The Political Black Minister: A Study in Afro-American Politics and Religion.* Boston: G. K. Hall, 1980.

Coker, Daniel. *A Dialogue Between a Virginian and an African Minister.* In *Negro Protest Pamphlets: A Compendium.* New York: Arno Press and New York Times, 1969 [1810].

Cone, James H. *Martin and Malcolm and America: A Dream or a Nightmare?.* Maryknoll, N.Y.: Orbis Books, 1991.

Cone, James H. *For My People: Black Theology and the Black Church.* Maryknoll, N.Y.: Orbis Books, 1984.

Cone, James H. *God of the Oppressed.* New York: Seabury Press, 1975.

Cone, James H. *Speaking the Truth: Ecumenism, Liberation and Black Theology.* Grand Rapids: William B. Eerdmans, 1986.

Corbett, Edward P. J. *Classical Rhetoric for the Modern Student,* 2d ed. New York: Oxford University Press, 1971.

Cronon, Edmund David. *Black Moses: The Story of Marcus Garvey and the Universal Negro Improvement Association.* Madison: University of Wisconsin Press, 1968.

Davis, Allison. *Leadership, Love, and Aggression.* New York: Harcourt Brace Jovanovich, 1983.

Davis, Gerald L. *I got the Word in me and I can sing it, you know: A Study of the Performed African-American Sermon.* Philadelphia: University of Pennsylvania Press, 1985.

DeWolf, L. Harold. Rev. ed. *A Theology of the Living Church.* New York: Harper & Row, 1960.

Downing, Frederick L. *To See the Promised Land: The Faith Pilgrimage of Martin Luther King, Jr.* Macon: Mercer University Press, 1986.

Drake, St. Clair, and Horace R. Cayton. *Black Metropolis: A Study of Negro Life in a Northern City.* 2 vols. Rev. ed. New York: Harcourt, Brace & World, 1962 [1945].

Du Bois, W. E. B., ed. *The Negro Church.* Atlanta: Atlanta University Press, 1903.

Du Bois, W. E. B. *The Souls of Black Folk.* In *Three Negro Classics.* New York: Avon Books, 1965 [1900].

Duncan, Hugh Dalziel. *Communication and Social Order.* New York: Bedminster Press, 1962.

Dvorak, Katharine L. *An African-American Exodus: The Segregation of the Southern Churches*. Brooklyn: Carlson, 1991.

Ebenezer Baptist Church: The Centennial Celebration. Atlanta, 1986.

Edwards, Jonathan. *Thoughts on the Revival of Religion in New England, 1742 . . . .* New York: Dunning & Spalding, 1832.

English, James W. *The Prophet of Wheat Street: The Story of William Holmes Borders*. Elgin, Ill.: David C. Cook, 1967.

Epstein, E. L. *Language and Style*. London: Methuen, 1978.

Evans, Zelia S., with J. T. Alexander, eds. *The Dexter Avenue Baptist Church, 1877–1977*. 1978.

Fairclough, Adam. *To Redeem the Soul of America: The Southern Christian Leadership Conference and Martin Luther King, Jr*. Athens: University of Georgia Press, 1987.

Farmer, James. *Lay Bare the Heart: An Autobiography of the Civil Rights Movement*. New York: Arbor House, 1985.

Felder, Cain Hope, ed. *Stony the Road We Trod: African American Biblical Interpretation*. Minneapolis: Fortress Press, 1991.

Fish, Stanley. *Is There a Text in This Class? The Authority of Interpretive Communities*. Cambridge: Harvard University Press, 1980.

Fisher, Miles Mark. *Negro Slave Songs in the United States*. Ithaca: Cornell University Press, 1953.

Fosdick, Harry Emerson. *The Hope of the World*. New York: Harper, 1933.

Fosdick, Harry Emerson. *On Being Fit to Live With. Sermons on Post-War Christianity*. New York: Harper, 1946.

Fosdick, Harry Emerson. *Riverside Sermons*. New York: Harper, 1958.

Fosdick, Harry Emerson. *The Secret of Victorious Living*. New York: Harper, 1934.

Fox, Richard Wightman. *Reinhold Niebuhr: A Biography*. San Francisco: Harper & Row, 1985.

Franklin, John Hope. *From Slavery to Freedom: A History of Negro Americans*. 7th ed. New York: Knopf, 1994 [1947].

Frazier, E. Franklin. *The Negro Church in America*. New York: Schocken Books, 1963.

Frei, Hans W. *The Eclipse of Biblical Narrative: A Study in Eighteenth and Nineteenth Century Hermeneutics*. New Haven: Yale University Press, 1974.

Frye, Northrop. *The Great Code: The Bible and Literature*. New York: Harcourt Brace Jovanovich, 1982.

Garrow, David J. *Bearing the Cross: Martin Luther King, Jr., and the Southern Christian Leadership Conference*. New York: Morrow, 1986.

Garrow, David J. *The FBI and Martin Luther King, Jr.: From "Solo" to Memphis*. New York: Norton, 1981.

Garrow, David, ed. *The Martin Luther King Jr., FBI File. Black Studies Research Sources. Microfilms from Major Archival and Manuscript Collections*. Frederick, Md.: University Publications of America, 1984.

Garrow, David J. *Protest at Selma: Martin Luther King, Jr. and the Voting Rights Act of 1965*. New Haven: Yale University Press, 1978.

Gates, Henry Louis, Jr. *The Signifying Monkey: A Theory of Afro-American Literary Criticism.* New York: Oxford University Press, 1988.

Geertz, Clifford. *The Interpretation of Cultures.* New York: Basic Books, 1973.

Genovese, Eugene D. *Roll, Jordan, Roll: The World the Slaves Made.* New York: Vintage Books, 1972.

Goffman, Erving. *The Presentation of Self in Everyday Life.* Garden City, N.Y.: Doubleday Anchor Books, 1959.

Graetz, Robert S. *Montgomery: A White Preacher's Memoir.* Minneapolis: Fortress Press, 1991.

Hacker, Andrew. *Two Nations: Black and White, Separate, Hostile, Unequal.* New York: Macmillan, 1992.

Hamilton, Charles V. *The Black Preacher in America.* New York: Morrow, 1972.

Hamilton, J. Wallace. *Horns and Halos in Human Nature.* Westwood, N.J.: Fleming H. Revell, 1954.

Hampton, Henry, and Steve Fayer. *Voices of Freedom: An Oral History of the Civil Right's Movement from the 1950s Through the 1980s.* New York: Bantam Books, 1990.

Harding, Vincent. *There Is a River: The Black Struggle for Freedom in America.* New York: Harcourt Brace Jovanovich, 1981.

Hatcher, William E. *John Jasper: The Unmatched Negro Philosopher and Preacher.* New York: Fleming H. Revell, 1908.

Havelock, Eric A. *Preface to Plato.* Cambridge: Harvard University Press, Belknap Press, 1963.

Haygood, Wil. *King of the Cats: The Life and Times of Adam Clayton Powell, Jr.* Boston: Houghton Mifflin, 1993.

Herskovits, Melville J. *The Myth of the Negro Past.* New York: Harper, 1941.

Heschel, Abraham J. *The Prophets.* 2 vols. New York: Harper & Row, 1962.

Hicks, H. Beecher, Jr. *Images of the Black Preacher: The Man Nobody Knows.* Valley Forge: Judson, 1977.

Holland, Dewitte, ed. *Sermons in American History.* Nashville: Abingdon Press, 1971.

Hughes, Langston, and Anna Bontemps, eds. *The Poetry of the Negro, 1746–1949.* Garden City, N.Y.: Doubleday, 1949.

Hunter, Archibald M. *Interpreting the Parables.* Philadelphia: Westminster Press, 1960.

Jahn, Janheinz. *Muntu: An Outline of Neo-African Culture.* Translated by Marjorie Grene. London; Faber & Faber, 1958, 1961.

Johannesen, Richard L., ed. *Contemporary Theories of Rhetoric: Selected Readings.* New York: Harper & Row, 1971.

Johns, Vernon. *Human Possibilities: A Vernon Johns Reader.* Edited by Samuel Lucius Gandy. Washington, D.C.: Hoffman Press, 1977.

Johnson, Clifton H., ed. *God Struck Me Dead: Religious Conversion Experiences and Autobiographies of Ex-slaves.* Philadelphia: Pilgrim Press, 1969.

Johnson, James Weldon. *God's Trombones: Seven Negro Sermons in Verse.* New York: Viking, 1927.

Kelsey, David H. *The Uses of Scripture in Recent Theology*. Philadelphia: Fortress Press, 1975.

King, Coretta Scott. *My Life with Martin Luther King, Jr.* New York: Holt, Rinehart & Winston, 1969.

King, Martin Luther, Jr. *The Measure of a Man*. Philadelphia: Fortress Press, 1988 [1959].

King, Martin Luther, Jr. *Called to Serve, January 1929-June 1951*. Vol. 1 of *The Papers of Martin Luther King, Jr.*, edited by Clayborne Carson. Berkeley: University of California Press, 1992.

King, Martin Luther, Jr. *Stride Toward Freedom: The Montgomery Story*. New York: Harper, 1958.

King, Martin Luther, Jr. *Strength to Love*. Philadelphia: Fortress Press, 1981 [1963].

King, Martin Luther, Jr. *A Testament of Hope: The Essential Writings of Martin Luther King, Jr.* Edited by James M. Washington. San Francisco: Harper & Row, 1986.

King, Martin Luther, Jr. *The Trumpet of Conscience*. New York: Harper & Row, 1967.

King, Martin Luther, Jr. *Where Do We Go from Here: Chaos or Community*. New York: Harper & Row, 1967.

King, Martin Luther, Jr. *Why We Can't Wait*. New York: New American Library, 1963.

King, Martin Luther, Sr. (with Clayton Riley). *Daddy King*. New York: Morrow, 1980.

Kittel, Gerhard, ed. *Theological Dictionary of the New Testament*. Vol. 6. Translated by Geoffrey W. Bromiley. Grand Rapids: William B. Eerdmans, 1968.

Lanham, Richard A. *The Motives of Eloquence: Literary Rhetoric in the Renaissance*. New Haven: Yale University Press, 1976.

Lanham, Richard A. *Style: An Anti-Textbook*. New Haven: Yale University Press, 1974.

Lash, Nicholas. *Theology on the Way to Emmaus*. London: SCM Press, 1986.

Lewis, David L. *King: A Critical Biography*. New York: Praeger, 1970.

Liebman, Joshua. *Peace of Mind*. New York: Simon & Schuster, 1946.

Lincoln, C. Eric. *The Black Church since Frazier*. New York: Schocken Books, 1974.

Lincoln, C. Eric, ed. *The Black Experience in Religion*. Garden City, N.Y.: Anchor Press, 1974.

Lincoln, C. Eric. *Martin Luther King, Jr.: A Profile*. Rev. ed. New York: Hill & Wang, 1984.

Lincoln, C. Eric. *Sounds of the Struggle: Persons and Perspectives in Civil Rights*. New York: Morrow, 1967.

Lincoln, C. Eric, and Lawrence H. Mamiya. *The Black Church in the African American Experience*. Durham: Duke University Press, 1990.

Lischer, Richard. *A Theology of Preaching*. Rev. ed. Durham, N.C.: Labyrinth, 1992 [1981].

Lischer, Richard. *Theories of Preaching: Selected Readings in the Homiletical Tradition*. Durham, N.C.: Labyrinth, 1987.

Lischer, Richard, and William Willimon, eds. *Concise Encyclopedia of Preaching.* Louisville: Westminster/John Knox Press, 1995.

Lord, Albert B. *The Singer of Tales.* Harvard Studies in Comparative Literature, vol. 24. Cambridge: Harvard University Press, 1960.

de Lubac, Henri. *The Sources of Revelation.* Translated by Luke O'Neill. New York: Herder & Herder, 1968.

Luccock, Halford E. *Communicating the Gospel.* New York: Harper, 1954.

Luccock, Halford E. *In the Minister's Workshop.* New York: Abingdon-Cokesbury, 1944.

Luker, Ralph E. *The Social Gospel in Black and White: American Radical Reform, 1885-1912.* Chapel Hill: University of North Carolina Press, 1991.

Lynch, Hollis R. *The Black Urban Condition: A Documentary History, 1866-1971.* New York: Crowell, 1973.

Luther, Martin. *Works of Martin Luther.* Vol. 3. Philadelphia: Muhlenberg, 1930.

The Autobiography of Malcolm X. With the assistance of Alex Haley. New York: Ballantine Books, 1964.

Malcolm X Speaks. Edited by George Breitman. New York: Grove Press, 1965.

Marx, Karl. *Selected Works.* Vol. 2. New York: International, 1951.

Maser, Frederick E. *Richard Allen.* Lake Junaluska, N.C.: Commission on Archives and History, the United Methodist Church, 1976.

Mays, Benjamin E. *Disturbed About Man.* Richmond: John Knox Press, 1969.

Mays, Benjamin E., and Joseph W. Nicholson. *The Negro's Church.* New York: Negro Universities Press, 1933.

Mays, Benjamin E. *The Negro's God.* Boston: Chapman & Grimes, 1938.

Mbiti, John S. *An Introduction to African Religion.* London: Heinemann, 1975.

Meier, August, and Elliott Rudwick, eds. *The Making of Black America.* Vol.2. New York: Atheneum, 1969.

Miller, Keith D. *Voice of Deliverance: The Language of Martin Luther King, Jr. and Its Sources.* New York: Free Press, 1992.

Miller, Robert Moats. *Harry Emerson Fosdick: Preacher, Pastor, Prophet.* New York: Oxford University Press, 1985.

Mitchell, Henry H. *Black Belief: Folk Beliefs of Blacks in America and West Africa.* New York: Harper & Row, 1975.

Mitchell, Henry H. *Black Preaching.* Philadelphia: Lippincott, 1970.

Mitchell, Henry. *The Recovery of Preaching.* San Francisco: Harper & Row, 1977.

Moses, Wilson Jeremiah. *Black Messiahs and Uncle Toms: Social and Literary Manipulations of a Religious Myth.* University Park: Pennsylvania State University Press, 1982.

Motter, Alan M., ed. *Great Preaching Today.* New York: Harper, 1955.

Narrative of Sojourner Truth. Battle Creek, Mich., 1878.

Neuhaus, Richard J., ed. *Unsecular America.* Grand Rapids: William B. Eerdmans, 1986.

The New National Baptist Hymnal. Nashville: National Baptist Publishing Board, 1977.

Niebuhr, Reinhold. *Moral Man and Immoral Society*. New York: Scribner's, 1952 [1932].

Oates, Stephen B. *Let the Trumpet Sound: The Life of Martin Luther King, Jr.* New York: New American Library, 1982.

Ong, Walter J. *Orality and Literacy*. London: Methuen, 1982.

Ong, Walter J. *The Presence of the Word*. New Haven: Yale University Press, 1967.

Perelman, C. H., and L. Olbrechts-Tyteca. *The New Rhetoric: A Treatise on Argumentation*. Translated by John Wilkinson and Purcell Weaver. Notre Dame: University of Notre Dame Press, 1969.

Pipes, William H. *Say Amen, Brother! Old-Time Negro Preaching: A Study in Frustration*. New York: William Frederick Press, 1951.

Postman, Neil. *Amusing Ourselves to Death: Public Discourse in the Age of Show Business*. New York: Viking, 1985.

Proctor, Samuel D. *Preaching about Crises in the Community*. Philadelphia: Westminster Press, 1988.

Putnam, Carleton. *Race and Reason: A Yankee View*. Washington, D.C.: Public Affairs Press, 1961.

The Pulpit. Chicago: Christian Century Foundation, 1948–1955.

Raboteau, Albert J. *Slave Religion: The "Invisible Institution" in the Antebellum South*. New York: Oxford University Press, 1978.

von Rad, Gerhard. *Old Testament Theology*. Vol. 2, *The Theology of Israel's Prophetic Traditions*. Translated by D. M. G. Stalker. New York: Harper & Row, 1965.

Raines, Howell. *My Soul Is Rested: Movement Days in the Deep South Remembered*. New York: Putnam's, 1977.

Rauschenbusch, Walter. *Christianity and the Social Crisis*. Edited by Robert D. Cross. New York: Harper & Row, 1964 [1907].

Walter Rauschenbusch: Selected Writings. Edited by Winthrop S. Hudson. New York: Paulist Press, 1984.

Ray, Sandy F. *Journeying through a Jungle*. Nashville: Broadman, 1979.

Reddick, L. D. *Crusader Without Violence: A Biography of Martin Luther King, Jr.* New York: Harper, 1959.

Richey, Russell E., and Donald G. Jones, eds. *American Civil Religion*. New York: Harper & Row, 1974.

Ricoeur, Paul. *The Conflict of Interpretations*. Evanston: Northwestern University Press, 1974.

Robinson, Jo Ann. *The Montgomery Bus Boycott and the Women Who Started It: The Memoir of Jo Ann Gibson Robinson*. Edited by David J. Garrow. Knoxville: University of Tennessee Press, 1987.

Rosenberg, Bruce A. *Can These Bones Live? The Art of the American Folk Preacher*. Rev. ed. Urbana: University of Illinois Press, 1988.

Rosenstock-Huessy, Eugen. *Speech and Reality*. Norwich, Vt.: Argo Books, 1970.

Sandeen, Ernest, ed. *The Bible and Social Reform*. Philadelphia: Fortress Press, 1982.

Sangster, W. E. *The Craft of Sermon Construction*. Philadelphia: Westminster Press, 1951.

Schor, Joel. *Henry Highland Garnet: A Voice of Black Radicalism in the Nineteenth Century*. Westport, Conn.: Greenwood Press, 1977.

Scott, Bernard B. *Hear Then the Parable*. Minneapolis: Fortress Press, 1989.

Scott, Manuel L. *From a Black Brother*. Nashville: Broadman, 1971.

Schwartz, Regina, ed. *The Book and the Text: The Bible and Literary Theory*. Oxford: Basil Blackwell, 1990.

Schweitzer, Albert. *The Quest of the Historical Jesus*. Translated by W. Montgomery. New York: Macmillan, 1968 [1906].

Sernett, Milton C., ed. *Afro-American Religious History: A Documentary Witness*. Durham: Duke University Press, 1985.

Shelp, Earl E., and Ronald H. Sunderland, eds. *The Pastor as Prophet*. New York: Pilgrim Press, 1985.

Smith, Arthur L. (Molefi Kete Asante). *The Rhetoric of Black Revolution*. Boston: Allyn & Bacon, 1969.

Smith, Kelly Miller. *Social Crisis Preaching*. Macon: Mercer University Press, 1984.

Smith, Kenneth L., and Ira G. Zepp, Jr. *Search for the Beloved Community: The Thinking of Martin Luther King, Jr*. Valley Forge: Judson, 1974.

Smitherman, Geneva. *Talkin and Testifyin: The Language of Black America*. Boston: Houghton Mifflin, 1977.

Sobel, Mechal. *Trabelin' On: The Slave Journey to an Afro-Baptist Faith*. Westport, Conn.: Greenwood Press, 1979.

Songs of Zion. Supplemental Worship Resources 12. Nashville: Abingdon, 1981.

Spencer, Jon Michael. *Protest and Praise: Sacred Music of Black Religion*. Minneapolis: Fortress Press, 1990.

Spencer, Jon Michael. *Sacred Symphony: The Chanted Sermon of the Black Preacher*. Contributions in Afro-American and African Studies, no. 111. New York: Greenwood Press, 1987.

Stout, Harry S. *The New England Soul: Preaching and Religious Culture in Colonial New England*. New York: Oxford University Press, 1986.

Taylor, Gardner C. *How Shall They Preach*. Elgin, Ill.: Progressive Baptist Publishing House, 1977.

Thurman, Howard. *Deep River*. New York: Harper, 1955 [1945].

Vivian, C. T. *Black Power and the American Myth*. Philadelphia: Fortress Press, 1970.

Wagner, Clarence M. *Profiles of Black Georgia Baptists*. Atlanta: Bennett Brothers, 1980.

Walker, Wyatt Tee. *"Somebody's Calling My Name": Black Sacred Music and Social Change*. Valley Forge: Judson Press, 1979.

Wallis, Charles L. *A Treasury of Sermon Illustrations*. New York: Abingdon-Cokesbury Press, 1950.

Walzer, Michael. *Exodus and Revolution*. New York: Basic Books, 1985.

Warren, Mervyn A. *Black Preaching: Truth and Soul*. Washington, D.C.: University Press of America, 1977.

Washington, James Melvin. *Frustrated Fellowship: The Black Baptist Quest for Social Power*. Macon: Mercer University Press, 1986.

Washington, Joseph R., Jr. *Black Religion: The Negro and Christianity in the United States*. Boston: Beacon Press, 1964.

Watters, Pat. *Down to Now: Reflections on the Southern Civil Rights Movement*. New York: Pantheon, 1971.

Wheat Street Baptist Church Forty-fifth Pastoral Anniversary, 1937–1982. Atlanta, 1982.

Williams, Melvin D. *Community in a Black Pentecostal Church: An Anthropological Study*. Pittsburgh: University of Pittsburgh Press, 1974.

Wilmore, Gayraud S. *Black Religion and Black Radicalism: An Interpretation of the Religious History of Afro-American People*. 2d ed. rev. and enlarged. Maryknoll, N.Y.: Orbis Books 1983.

Wills, Garry. *Lincoln at Gettysburg: The Words That Remade America*. New York: Simon & Shuster, 1992.

Wilson, Robert R. *Prophecy and Society in Ancient Israel*. Philadelphia: Fortress Press, 1980.

Woodson, Carter G. *The History of the Negro Church*. Washington, D.C.: Associated Publishers, 1972 [1921].

Woodson, Carter G. *Negro Orators and Their Orations*. Washington, D.C.: Associated Publishers, 1925.

Woodward, C. Vann. *The Strange Career of Jim Crow*. Rev. ed. New York: Oxford University Press, 1974.

Young, Henry J. *Major Black Religious Leaders, 1755–1940*. Nashville: Abingdon Press, 1977.

Periodicals and Chapters in Books

Baldwin, James. "The Highroad to Destiny." In *Martin Luther King Jr.: A Profile*. edited by C. Eric Lincoln, pp. 90–112. Rev. ed. New York: Hill & Wang, 1984.

Bitzer, Lloyd. "The Rhetorical Situation." In *Contemporary Theories of Rhetoric: Selected Readings*, edited by Richard L. Johannesen, pp. 381–94. New York: Harper & Row, 1971.

Brockriede, Wayne E., and Douglas Ehninger. "Toulmin on Argument: An Interpretation and Application." In *Contemporary Theories of Rhetoric: Selected Readings*, edited by Richard L. Johannesen, pp. 241–55. New York: Harper & Row, 1971.

Brueggemann, Walter. "The Prophet as a Destabilizing Presence." In *The Pastor as Prophet*, edited by Earl E. Shelp and Ronald H. Sunderland, pp. 49–77. New York: Pilgrim Press, 1985.

Carson, Clayborne, et al. "Martin Luther King, Jr. as Scholar: A Re-examination of His Theological Writings." *Journal of American History* 78:1 (June 1991): 93–105.

Cartwright, John H. "The Social Eschatology of Martin Luther King, Jr." In *Essays in Honor of Martin Luther King, Jr.*, pp. 1-13. Evanston, Ill.: Garrett Evangelical Theological Seminary, 1971.

Cleague, Albert B., Jr. "A New Time Religion." In *The Black Experience in Religion*, edited by C. Eric Lincoln, pp. 167-80. Garden City, N.Y.: Anchor Press, 1974.

Cleghorn, Reese. "Crowned with Crises." In *Martin Luther King Jr.: A Profile*, edited by C. Eric Lincoln, pp. 113-127. Rev. ed. New York: Hill & Wang, 1984.

Cone, James H. "Martin Luther King, Jr., Black Theology—Black Church," *Theology Today* 40:4 (January 1984): 409-20.

Daniel, Carey. "God the Original Segregationist" (sermon). In *Sermons in American History*, edited by DeWitte Holland, pp. 513-22. Nashville: Abingdon Press, 1971.

Downing, Frederick L. "Martin Luther King, Jr. as Public Theologian," *Theology Today* 55:1 (April 1987): 15-31.

Garrow, David. "King's Plagiarism: Imitation, Insecurity, and Transformation," *Journal of American History* 78:1 (June 1991): 86-92.

Halberstam, David. "When 'Civil Rights' and 'Peace' Join Forces." In *Martin Luther King, Jr.: A Profile*, edited by C. Eric Lincoln, pp. 187-211. Rev. ed. New York: Hill & Wang, 1984.

Hauerwas, Stanley M. "The Pastor as Prophet: Ethical Reflections on an Improbable Mission." In *The Pastor as Prophet*, edited by Earl E. Shelp and Ronald H. Sunderland, pp. 27-48. New York: Pilgrim Press, 1985.

Henderson, Alexa, and Eugene Walker. "Sweet Auburn: The Thriving Hub of Black Atlanta, 1900-1960." Denver: U.S. Department of the Interior, National Park Service, n.d.

Holloran, Peter. "Recovering Lost Values: Transcribing an African-American Sermon." *Documentary Editing*, September 1991, pp. 49-53.

Jones, Kirk Byron. "The Activism of Interpretation: Black Pastors and Public Life." *Christian Century*, September 13-20, 1989, 817-18.

King, Martin Luther, Jr. "The UnChristian Christian." *Ebony* 20:10 (August 1965): 76-80.

King, William McGuire. "The Biblical Base of the Social Gospel." In *The Bible and Social Reform*, edited by Ernest Sandeen, 59-84. Philadelphia: Fortress Press, 1982.

Lewis, David L. "Failing to Know Martin Luther King, Jr." *Journal of American History* 78:1 (June 1991): 81-85.

Lischer, Richard. "'What Language Shall I Borrow?': The Role of Metaphor in Proclamation." *Dialog* 26:4 (Fall 1987): 281-86.

Lischer, Richard. "The Word That Moves: The Preaching of Martin Luther King, Jr." *Theology Today* 45:2 (July 1989): 169-82.

Lomax, Louis. "When 'Nonviolence' Meets 'Black Power.'" In *Martin Luther King Jr.: A Profile*, edited by C. Eric Lincoln, pp. 157-180. Rev. ed. New York: Hill & Wang, 1984.

Marbury, Carl H. "An Excursus on the Biblical and Theological Rhetoric of Martin Luther King." In *Essays in Honor of Martin Luther, Jr.*, pp. 14-28. Evanston, Ill.: Garrett Evangelical Theological Seminary, 1971.

Marks, Herbert. "On Prophetic Stammering." In *The Book and the Text: The Bible and Literary Theory*, edited by Regina M. Schwartz, pp. 60-80. Oxford: Basil Blackwell, 1990.

Marx, Gary T. "Religion: Opiate or Inspiration of Civil Rights Militancy Among Negroes." In *The Making of Black America*, vol. 2, edited by August Meier and Elliott Rudwick. New York: Atheneum, 1969.

McClendon, James Wm., Jr. "M. L. King: Politican or American Church Father? A Review-Article," *Journal of Ecumenical Studies* 8:1 (Winter 1971): 115-21.

Meier, August. "The Conservative Militant." In *Martin Luther King: A Profile*, edited by C. Eric Lincoln, pp. 144-56. Rev. ed. New York: Hill & Wang, 1984.

Miller, Keith D. "Composing Martin Luther King, Jr." *PMLA* 105:1 (January 1990): 70-82.

Miller, Keith D. "Martin Luther King, Jr. and the Black Folk Pulpit." *Journal of American History* 78:1 (June 1991): 120-23.

Miller, Keith D. "Martin Luther King, Jr. Borrows a Revolution: Argument, Audience, and Implications of a Secondhand Universe." *College English* 48:2 (February 1986): 249-65.

Morris, Calvin. "Martin Luther King, Jr., Exemplary Preacher." *Journal of the Interdenominational Theological Center* 4:1 (Spring 1977): 61-66.

Paris, Peter. "The Bible and the Black Churches." In *The Bible and Social Reform*, edited by Ernest Sandeen, pp. 133-54. Philadelphia: Fortress Press, 1982.

Perelman, Chaim, "The New Rhetoric." In *The Prospect of Rhetoric*, edited by Lloyd F. Bitzer and Edwin Black, pp. 115-22. Englewood Cliffs, N.J.: Prentice-Hall, 1971.

Pitts, Walter. "West African Poetics in the Black Preaching Style." *American Speech* 64:2 (Summer 1989): 137-49.

Reynolds, David S. "From Doctrine to Narrative: The Rise of Pulpit Story-telling in America." *American Quarterly* 32:1 (Spring 1980): 479-98.

Sellers, James E. "Love, Justice, and the Non-violent Movement." *Theology Today* 18:4 (January 1962): 422-34.

Southern, Eileen. "The Religious Occasion." In *The Black Experience in Religion*, edited by C. Eric Lincoln, pp. 51-63. Garden City, N.Y.: Anchor Press, 1974.

Smylie, James H. "On Jesus, Pharaohs, and the Chosen People: Martin Luther King as Biblical Interpreter and Humanist." *Interpretation* 24 (January 1970): 74-91.

Spillers, Hortense J. "Martin Luther King and the Style of the Black Sermon." *Black Scholar* 3:1 (September 1971): 14-27.

"The Student Papers of Martin Luther King, Jr.: A Summary Statement on Research," *Journal of American History* 78:1 (June 1991): 23-31.

Tannehill, Robert. "The 'Focal Instance' as a Form of New Testament Speech: A Study of Matthew 5: 39b-42." *The Journal of Religion* 50:4 (October 1970), 372-85.

Towner, W. Sibley. "On Calling People 'Prophets' in 1970." *Interpretation* 24:4 (October 1970), 492-509.

Trulear, H. Dean, and Russell E. Richey. "Two Sermons by Brother Carper: 'The Eloquent Negro Preacher.'" *American Baptist Quarterly* 6 (March 1987): 3-16.

Turner, William C. "The Musicality of Black Preaching: A Phenomenology," *Journal of Black Sacred Music* 2:1 (Spring 1988): 21-29.

Williams, Preston. "The Problem of a Black Ethic." In *The Black Experience in Religion*, edited by C. Eric Lincoln, pp. 180-86. Garden City, N.Y.: Anchor Press, 1974.

Wimbush, Vincent L. "The Bible and African Americans: An Outline of an Interpretative History." In *Stony the Road We Trod: African American Biblical Interpretation*, edited by Cain Hope Felder, pp. 81-97. Minneapolis: Fortress Press, 1991.

Theses and Dissertations

Franklin, Marion J., Jr. "The Relationship of Black Preaching to Black Gospel Music." D.Min. thesis, Drew University, 1982.

Garber, Paul Russell. "Martin Luther King, Jr.: Theologian and Precursor of Black Theology." Ph.D. diss., Florida State University, 1973.

Kane, Brian M. "The Influence of Boston Personalism on the Thought of Dr. Martin Luther King, Jr." Th.M. thesis, School of Theology, Boston University, 1985.

Keele, Lucy Anne McCandlish. "A Burkeian Analysis of the Rhetorical Strategies of Dr. Martin Luther King, Jr., 1955-1968." Ph.D. diss., University of Oregon, 1972.

King, Martin Luther, Jr. "A Comparison of the Conceptions of God in the Thinking of Paul Tillich and Henry Nelson Weiman." Ph.D. diss., Boston University, 1955.

Miller, Keith D. "The Influence of a Liberal Homiletic Tradition on *Strength to Love* by Martin Luther King, Jr." Ph.D. diss., Texas Christian University, 1984.

Mikelson, Thomas, J. S. "The Negro's God in the Theology of Martin Luther King, Jr.: Social Commentary and Theological Discourse." Th.D. diss., Harvard University, 1988.

Montgomery, William Edwards. "Negro Churches in the South, 1865-1915." Ph.D. diss., University of Texas, 1975.

Rudzka-Ostyn, Brygida Irena. "The Oratory of Martin Luther King and Malcolm X: A Study in Linguistic Stylistics." Ph.D. diss., University of Rochester, 1972.

Sernett, Milton. "Black Religion and American Evangelicalism: White Protestants, Plantation Missions, and the Independent Negro Church, 1787–1865." Ph.D. diss., University of Delaware, 1972.

Sloan, Rose Mary. "'Then My Living Will Not Be in Vain': A Rhetorical Study of Dr. Martin Luther King, Jr., and the Southern Christian Leadership Conference in the Mobilization for Collective Action Toward Nonviolent Means to Integration, 1954–1964." Ph.D. diss., Ohio State University, 1977.

Smith, Donald H. "Martin Luther King, Jr.: Rhetorician of Revolt." Ph.D. diss., University of Wisconsin, 1964.

Smith, Theophus H. "The Biblical Shape of Black Experience; An Essay in Philosophical Theology." Ph.D. diss., Graduate Theological Union, 1987.

Warren, Mervyn A. "A Rhetorical Study of the Preaching of Doctor Martin Luther King, Jr., Pastor and Pulpit Orator." Ph.D. diss., Michigan State University, 1966.

Film Documentary

King, a Filmed Record: Montgomery to Memphis, Texture Films, 1970.

Repositories of Unpublished Materials, Interviews, Transcripts, and Audio Recordings

J. Pius Barbour Collection. Typed and holograph sermons, sermon notes, outlines, and audio recordings of the Reverend J. Pius Barbour. Philadelphia, Pennsylvania.

Eugene "Bull" Connor Papers. Interoffice communications of the Birmingham Police Department. Houses tape recordings of Birmingham mass meetings. Birmingham Public Library Archives. Birmingham, Alabama.

Ralph J. Bunche Oral History Collection, Moorland-Spingarn Research Center. Houses tape-recorded interviews concerning the Civil Rights Movement and tapes and documents of African-American history. Howard University, Washington, D.C.

Birmingham Public Library Archives. Houses documents of African-American history, and records and audiotapes of Selma mass meetings. Birmingham, Alabama.

Duke Divinity School Media Center. Audiotapes of sermons by Martin Luther King, Jr., Martin Luther King, Sr., Howard Thurman, and Gardner C. Taylor. Duke University, Durham, North Carolina.

Howard Divinity School Tape Recording Collection. Houses audiotapes of sermons by Martin Luther King, Jr. and other African-American preachers. Howard University, Washington, D.C.

Martin Luther King, Jr. Center for Nonviolent Social Change, Library and Archives. Houses documents, audio and video recordings of sermons and speeches of Martin Luther King, Jr., and series of tape-recorded interviews pertaining to African-American history, the Civil Rights Movement, and the life of Martin Luther King, Jr. Atlanta, Georgia.

Special Collections of the Mugar Memorial Library. Houses documents and papers of Martin Luther King, Jr. to 1964. Boston University, Boston, Massachusetts.

Reigner Recording Library. Contains audiotapes of sermons by Martin Luther King, Jr. and other African-American preachers. Union Theological Seminary, Richmond, Virginia.

Additional Sources of Tape Recordings and Transcripts

Abbey Memorial Chapel, Mount Holyoke College, South Hadley, Massachusetts.
Central United Methodist Church, Detroit, Michigan.
Concord Baptist Church of Christ, Brooklyn, New York.
Cornerstone Baptist Church, Brooklyn, New York.
Duke University Chapel, Durham, North Carolina.
Ebenezer Baptist Church, Atlanta, Georgia.
Grace Cathedral, San Francisco, California.
Martin Luther King, Jr. *Papers* Project, Stanford University, Stanford, California.
National Cathedral, Washington, D.C.
Unitarian Church of Germantown, Philadelphia, Pennsylvania.
Wheat Street Baptist Church, Atlanta, Georgia.

Interview Records

In addition to the tape-recorded interviews housed in the repositories listed above, this study drew on the following interviews conducted by Richard Lischer:

Ralph David Abernathy, May 19, 1987
Almanina Barbour, May 22, 1989; November 27–28, 1989
Shirley Showers Barnhart, May 19, 1987
James Bevel, April 14, 1993
Juel Pate Borders, November 2, 1989
G. Murray Branch, March 7, 1988
Ralph Bryson, January 12, 1989
Evans Crawford, December 12, 1988
Zelia Evans, January 12, 1989
Robert Graetz, September 3, 1988

Jesse Jackson, August 29, 1988
Bernard Lee, July 23, 1987
Lillian Lewis, May 19, 1987
Gordon Midgette, November 28, 1988; August 3, 1989
R. D. Nesbitt, March 7, 1988
J. T. Porter, March 7, 1988
Samuel Proctor, January 18, 1992
Sarah Reed, February 25, 1987
Thelma Rice, January 14, 1989
Joseph Roberts, February 25, 1987
Minnie Showers, May 19, 1987
Kenneth L. Smith, July 6, 1991
William E. Smith, December 12, 1988
Gardner C. Taylor, February 5, 1988
C. T. Vivian, May 20-21, 1987
Wyatt Tee Walker, April 8, 1987
Mary Lucy Williams, January 12, 1989
Prathia Hall Wynn, March 8, 1991
Franklin Young, July 19, 1989

Index